CMOS Cascade Sigma-Delta Modulators for Sensors and Telecom

ANALOG CIRCUITS AND SIGNAL PROCESSING SERIES
Consulting Editor*: Mohammed Ismail. *Ohio State University

Titles in former series International Series in Engineering and Computer Science:

SIGMA DELTA A/D CONVERSION FOR SIGNAL CONDITIONING
Philips, K., van Roermund, A.H.M.
Vol. 874, ISBN 1-4020-4679-0

CALIBRATION TECHNIQUES IN NYQUIST A/D CONVERTERS
van der Ploeg, H., Nauta, B.
Vol. 873, ISBN 1-4020-4634-0

ADAPTIVE TECHNIQUES FOR MIXED SIGNAL SYSTEM ON CHIP
Fayed, A., Ismail, M.
Vol. 872, ISBN 0-387-32154-3

WIDE-BANDWIDTH HIGH-DYNAMIC RANGE D/A CONVERTERS
Doris, Konstantinos, van Roermund, Arthur, Leenaerts, Domine
Vol. 871 ISBN: 0-387-30415-0

METHODOLOGY FOR THE DIGITAL CALIBRATION OF ANALOG CIRCUITS AND SYSTEMS: WITH CASE STUDIES
Pastre, Marc, Kayal, Maher
Vol. 870, ISBN: 1-4020-4252-3

HIGH-SPEED PHOTODIODES IN STANDARD CMOS TECHNOLOGY
Radovanovic, Sasa, Annema, Anne-Johan, Nauta, Bram
Vol. 869, ISBN: 0-387-28591-1

LOW-POWER LOW-VOLTAGE SIGMA-DELTA MODULATORS IN NANOMETER CMOS
Yao, Libin, Steyaert, Michiel, Sansen, Willy
Vol. 868, ISBN: 1-4020-4139-X

DESIGN OF VERY HIGH-FREQUENCY MULTIRATE SWITCHED-CAPACITOR CIRCUITS
U, Seng Pan, Martins, Rui Paulo, Epifânio da Franca, José
Vol. 867, ISBN: 0-387-26121-4

DYNAMIC CHARACTERISATION OF ANALOGUE-TO-DIGITAL CONVERTERS
Dallet, Dominique; Machado da Silva, José (Eds.)
Vol. 860, ISBN: 0-387-25902-3

ANALOG DESIGN ESSENTIALS
Sansen, Willy
Vol. 859, ISBN: 0-387-25746-2

DESIGN OF WIRELESS AUTONOMOUS DATALOGGER IC'S
Claes and Sansen
Vol. 854, ISBN: 1-4020-3208-0

MATCHING PROPERTIES OF DEEP SUB-MICRON MOS TRANSISTORS
Croon, Sansen, Maes
Vol. 851, ISBN: 0-387-24314-3

LNA-ESD CO-DESIGN FOR FULLY INTEGRATED CMOS WIRELESS RECEIVERS
Leroux and Steyaert
Vol. 843, ISBN: 1-4020-3190-4

SYSTEMATIC MODELING AND ANALYSIS OF TELECOM FRONTENDS AND THEIR BUILDING BLOCKS
Vanassche, Gielen, Sansen
Vol. 842, ISBN: 1-4020-3173-4

LOW-POWER DEEP SUB-MICRON CMOS LOGIC SUB-THRESHOLD CURRENT REDUCTION
van der Meer, van Staveren, van Roermund
Vol. 841, ISBN: 1-4020-2848-2

WIDEBAND LOW NOISE AMPLIFIERS EXPLOITING THERMAL NOISE CANCELLATION
Bruccoleri, Klumperink, Nauta
Vol. 840, ISBN: 1-4020-3187-4

CMOS PLL SYNTHESIZERS: ANALYSIS AND DESIGN
Shu, Keliu, Sánchez-Sinencio, Edgar
Vol. 783, ISBN: 0-387-23668-6

SYSTEMATIC DESIGN OF SIGMA-DELTA ANALOG-TO-DIGITAL CONVERTERS
Bajdechi and Huijsing
Vol. 768, ISBN: 1-4020-7945-1

OPERATIONAL AMPLIFIER SPEED AND ACCURACY IMPROVEMENT
Ivanov and Filanovsky
Vol. 763, ISBN: 1-4020-7772-6

STATIC AND DYNAMIC PERFORMANCE LIMITATIONS FOR HIGH SPEED D/A CONVERTERS
van den Bosch, Steyaert and Sansen
Vol. 761, ISBN: 1-4020-7761-0

DESIGN AND ANALYSIS OF HIGH EFFICIENCY LINE DRIVERS FOR Xdsl
Piessens and Steyaert
Vol. 759, ISBN: 1-4020-7727-0

CMOS Cascade Sigma-Delta Modulators for Sensors and Telecom

Error Analysis and Practical Design

By

R. del Río

F. Medeiro

B. Pérez-Verdú

J.M. de la Rosa

and

Á. Rodríguez-Vázquez

Spanish Microelectronics Center IMSE-CNM
and University of Seville, Spain

Springer

A C.I.P. Catalogue record for this book is available from the Library of Congress.

ISBN-10 94-007-8726-X (HB)
ISBN-13 978-94-007-8726-1 (HB)
ISBN-10 1-4020-4776-2 (eBook)
ISBN-13 978-1-4020-4776-2 (eBook)

Published by Springer,
P.O. Box 17, 3300 AA Dordrecht, The Netherlands.

www.springer.com

Printed on acid-free paper

Printed in the Netherlands

CONTENTS

CHAPTER 3

CHAPTER 4

A ΣΔ Modulator in 2.5-V 0.25-μm CMOS for ADSL/ADSL+ 193

LIST OF ABBREVIATIONS

ΠΣΔ	Parallel Sigma-Delta
ΣΔ	Sigma-Delta
ΣΔM	Sigma-Delta Modulator
A/D, A-to-D	Analog-to-Digital
ADC	Analog-to-Digital Converter
ADSL	Asymmetric Digital Subscriber Line
AFE	Analog Front-End
AGC	Automatic Gain Control
BiCMOS	Bipolar and Complementary Metal-Oxide-Semiconductor
BPΣΔM	Band-Pass Sigma-Delta Modulator
CAD	Computer Aided Design
CDMA	Code Division Multiple Access
CLA	Clocked Averaging
CMFB	Common-Mode Feedback
CMOS	Complementary Metal-Oxide-Semiconductor
CPE	Customer Premises Equipment
CPU	Central Processing Unit
CT	Continuous-Time
D/A, D-to-A	Digital-to-Analog
DAC	Digital-to-Analog Converter
DC	Direct Current
DDS	Data Directed Scrambling
DEM	Dynamic Element Matching
DMT	Discrete Multi-Tone
DNL	Differential Non-Linearity
DOR	Digital Output Rate
DR	Dynamic Range
DSL	Digital Subscriber Line
DSP	Digital Signal Processing
DT	Discrete-Time
DWA	Data Weighted Averaging
EDGE	Enhanced Data-rates for Global Evolution
ENOB	Effective Number Of Bits
ESD	Electrostatic Discharge

FFT	Fast Fourier Transform
FIR	Finite Impulse Response
FOM	Figure Of Merit
FS	Full Scale
GB	Gain-Bandwidth Product
GPS	Global Positioning System
GSM	Global System for Mobile-Communications
HD	Harmonic Distortion
HDSL	High-data-rate Digital Subscriber Line
IBE	In-Band Error
IC	Integrated Circuit
IIR	Infinite Impulse Response
ILA	Individual Level Averaging
INL	Integral Non-Linearity
IO	Input/Output
ISDN	Integrated Services Digital Network
LNA	Low-Noise Amplifier
LPΣΔM	Low-Pass Sigma-Delta Modulator
LSB	Least Significant Bit
MASH	Multi-Stage Noise Shaping
MEMS	Micro-Electro-Mechanical System
MiM	Metal-insulator-Metal
MOS	Metal-Oxide-Semiconductor
MOSFET, MOST	Metal-Oxide-Semiconductor Field-Effect Transistor
MSB	Most Significant Bit
MTPR	Multi-Tone Power Ratio
nMOS	N-channel MOS
NTF	Noise Transfer Function
Opamp	Operational Amplifier
OS	Output Swing
OSR	Oversampling Ratio
OTA	Operational Transconductance Amplifier
PCB	Printed Circuit Board
PDF	Probability Density Function
PDM	Pulse-Density Modulated
PLC	Power Line Communications

PLL	Phase-Locked Loop
pMOS	P-channel MOS
PROM	Programmable Read-Only Memory
PSD	Power Spectral Density
RAM	Random Access Memory
RF	Radio Frequency
ROM	Read-Only Memory
S/H	Sample-and-Hold
SC	Switched-Capacitor
SDLS	Symmetrical Digital Subscriber Line
SFDR	Spurious-Free Dynamic Range
SI	Switched-Current
SNR	Signal-to-Noise Ratio
SNDR	Signal-to-(Noise+Distortion) Ratio
SoC	System-on-Chip
SR	Slew Rate
STF	Signal Transfer Function
THD	Total Harmonic Distortion
TIΣΔ	Time-Interleaved Sigma-Delta
UMTS	Universal Mobile Telecommunications System
UWB	Ultra-Wide Band
VDSL	Very-high-data-rate Digital Subscriber Line
VLSI	Very Large Scale of Integration
WLAN	Wireless Local Area Network
WMAN	Wireless Metropolitan Area Network
xDSL	All/any Digital Subscriber Line

PREFACE

The rapid evolution experimented by microelectronics during the last decades has propitiated the birth and spread of lots of electronic systems with increasing presence in different aspects of our everyday life: consumer electronics, information technology, communications, automotion, medicine, leisure, etc. Probably, communications has been one of the areas with largest expansion; many applications have been developed, both for wireline systems—DSL technologies for broadband access to the Internet, PLC technology, etc. [Gagn97]—and for wireless systems—mobile telephony, GPS, WLAN, WMAN, UWB, etc. [Abidi95].

No doubt, the continuous scaling of VLSI technologies has been a determinant factor for this rapid evolution. Technology scaling has allowed miniaturization, portability, increased functionality, and cost reduction of these systems. Nowadays, it is possible to integrate millions of transistors in a single chip using submicron CMOS processes and, simultaneously, the speed of digital circuits has increased up to the gigahertz range. This technological advances have enabled monolithic integration of complete electronic systems on a single chip (SoC), in which digital signal processing (DSP) techniques are extensively used for robust implementation of complex algorithms within reduced computational times.

In these systems, the present trend is to move the border between the analog and digital parts, usually called interface, as close as possible to the point where information is received or emitted. In this way, most of the SoC functionalities are implemented in the digital domain, where the system benefits from the reduction of silicon area, supply voltage, and power consumption, and from the increased operation speed that are peculiar to the progressive technology scaling. The application of analog circuits is then restricted, in most cases, to interface tasks: signal conditioning, filtering, and analog-to-digital (A-to-D) and digital-to-analog (D-to-A) conversion. In addition, the trend to massive digital processing and to an earlier digitalization of signals leads to an increase of the dynamic range and bandwidth requirements in the interface circuits.

On the other hand, the design of high resolution, high bandwidth converters is greatly involved when they are integrated together with the DSP circuits, mainly because the designers must use mainstream digital CMOS processes, in which analog primitives are not fully optimized [Bult00] [Malo01]. Thus, these converters have to operate with low voltage supply and transistors whose threshold voltages are comparatively high, with no use of extra process steps to improve the linearity or the matching of the devices, and, above all, in an hostile environment full of noisy digital circuits.

Among the existing techniques to perform the A-to-D conversion, those based on ΣΔ modulation [Inose62] offer key advantages for their implementation in SoCs. Unlike traditional converters, which require high accuracy in their building blocks in order to achieve overall high accuracy, the oversampling and noise-shaping techniques employed in ΣΔ converters allow to trade speed for accuracy. In this way, an operation that is relatively insensitive to imperfections on the analog circuit can be obtained at the cost of increased complexity and speed in the associated digital circuitry (needed for post-processing) [Nors97].

These demanding requirements on the digital part, which were a handicap for the integration of ΣΔ converters before the development of VLSI technologies, now relax the implementation of the analog section, whose requirements are more difficult to achieve in processes with a clear digital orientation. This has motivated that, although being originally conceived for low-frequency, high-resolution applications like audio [Candy85] [Adams86] [Boser88] [Bran91a] and precision measurement [Sign90] [Yama94], the usage of ΣΔ converters has progressively spreaded across medium- and high-frequency applications [Bran91b] [OptE91] [Yin94] [Broo97].

Fig. 1 illustrates the state of the art in A-to-D converters in CMOS technologies reported up to year 2000 and places them in the resolution—bandwidth plane. The ranges of the specifications for the main applications are depicted on this plane in an approximated way. It can be observed that ΣΔ modulation-based A-to-D converters cover a wide frequency

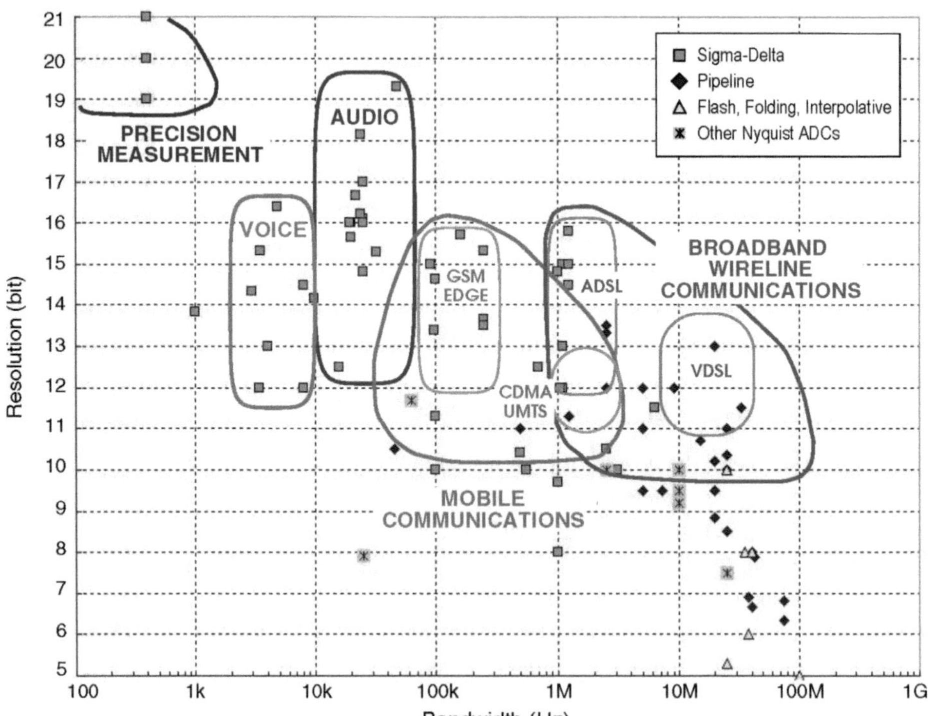

FIGURE 1 State of the art in A-to-D converters in CMOS technologies reported up to year 2000. (The ranges of the applications shown are approximated).

interval, ranging from 100Hz to 10MHz. Higher speed applications are still dominated by Nyquist-type converters (especially, pipeline, folding-interpolative, and flash) [Plas94] [Raza95]. Oversampling techniques are little efficient in these applications, because of the excessive operation speed that is required in the analog blocks. However, ΣΔ converters are clearly dominant in measurement, voice, and audio systems, and coexist with algorithmic, subranging, and pipeline converters in systems for mobile communications and broadband wireline applications like ADSL. Furthermore, it is commonly accepted that whenever an industrial application can be covered by using a ΣΔ converter, the simpler the better, this solution should be considered as the optimum one, for feasibility, yield, robustness, and time-to-market reasons [Rivo03].

In these applications, the implementation of Nyquist converters gets involved as the technology scales down: calibration techniques that consume considerable area and power are required in order to achieve resolutions larger than 13 bits [Mayes96] [Opris00] [Guil01]. As an alternative, the use of ΣΔ converters has gained ground and architectures that are oriented to high-frequency applications have been successfully implemented. In these architectures, the weakened benefits of oversampling (necessarily moderate) are compensated by resorting to high-order topologies—either in a single loop [Geer00] or in a multi-stage cascade [Yin94] [Feld98] [Marq98] [Geer99] [Mori00]—, which often incorporate multi-bit quantization—either pure [Geer00] [Fuji00] or by means of dual-quantization techniques [Bran91b] [Broo97] [Mede99].

Nevertheless, the prototypes implemented so far demonstrate the viability of ΣΔ converters for high-frequency applications (>1MHz), but not that their incorporation to SoCs is still robust in deep-submicron processes. Indeed, only a few prototypes [Geer99] [Fuji00] [Mori00] have been integrated in modern deep-submicron CMOS technologies, but they are mixed-signal oriented—they offer better device matching and supply voltages of 3.3V, or even 5V, together with low-Vt transistors [Fuji00].

In this scenario, the work presented in this book tries to demonstrate the viability of robust high-frequency, high-resolution ΣΔ converters using deep-submicron CMOS technologies oriented to the development of SoCs. This encompasses an adequate selection of architectures, techniques, and building blocks that allow, not only to obtain high-performance ΣΔ modulators, but also to solve the problems associated to their practical implementation in digital-oriented VLSI technologies (low supply voltage, poor linearity and matching of devices, etc.).

The results of this work are demonstrated through two prototypes for broadband applications that are integrated in deep-submicron CMOS technologies. They have been developed in the frame of the EU ESPRIT Project 29261 (MIXMODEST) and the Spanish CICYT Projects TIC97-05080 and TIC2001-0929 (ADAVERE), devoted to the investigation of architectures and techniques for the implementation of A-to-D converters in last generation CMOS technologies.

The first prototype is a wideband 2-1-1 ΣΔ cascade with dual quantization of 1 and 4 bits, which has been implemented in a 0.35-μm standard digital CMOS process with epitaxial (low-ohmic, conductive) substrate. The modulator operates with an oversampling ratio of 16 and exhibits a differential full-scale range of 4V using the 3.3-V nominal supply voltage. It achieves an effective resolution of 13bit at 4MS/s and consumes 78mW, while operated at 64-MHz internal clock frequency.

The second design is conceived to be incorporated to a CPE modem for ADSL and ADSL+ in a 2.5-V 0.25-μm CMOS process. The selected topology is a 2-1-1 cascade with quantization of 1 and 3 bits, which operates with an oversampling ratio of 32 or 16, and exhibits a differential full-scale range of 3V. The prototype achieves an effective resolution of 13.8bit at 2.2MS/s and 12.7bit at 4.4MS/s, with a power consumption of 66mW, while operating with a sampling frequency of 70.4MHz.

The book also presents the design of a third prototype to be included in an automotive sensor interface in a 3.3-V 0.35-μm CMOS process. The modulator topology is a 2-1 single-bit ΣΔ cascade that can be digitally programmed to yield four gain values— $\times 0.5$, $\times 1$, $\times 2$, and $\times 4$—in order to obtain a better fitting to the different sensor outputs. This prototype has been developed in the frame of the EU ESPRIT Project 34283 (TAMES-2), whose objective is to improve the industrial testability of high-resolution A-to-D interfaces embedded in SoCs. The modulator achieves an effective resolution of 18bit at 40kS/s and consumes 14.7mW, while operated at 5.12-MHz internal clock frequency.

The three prototypes presented in the book avoid the use of calibration techniques, non-standard transistors, or on-chip voltages larger than the nominal supply, and their performances are competitive to the current state of the art.

The contents of the book are organized in five chapters.

Chapter 1 presents an introduction to ΣΔ A-to-D converters, showing the principles of operation, the basic architectures, and the ideal performance of ΣΔ modulators. Topologies for their practical implementation are introduced and their pros and cons are discussed. The state of the art in low-pass ΣΔ modulators in CMOS technologies is then revised, showing existing trends in present designs.

Chapter 2 is dedicated to the exhaustive analysis of the main non-idealities that affect the performance of ΣΔ modulators. System considerations, behavioral models, and closed expressions are obtained for the impact of the different non-idealities, which can be used as estimable guidelines for practical implementation of ΣΔ modulators.

Chapters 3 and 4 describe the design of the two ΣΔ modulators intended for broadband applications, whereas Chapter 5 describes the design of the ΣΔ modulator with programmable gain for automotive sensor interfaces. The topology selection, the requirements of the building blocks, and their design at the transistor level are deeply discussed. The measured performance for the prototypes is presented and compared with the state-of-the-art ΣΔ modulators.

The considerations presented through the book for the design of cascade ΣΔ modulators in deep-submicron CMOS are extended in Appendixes A and B. Appendix A proposes a family of cascade ΣΔ modulators that is easily expandible to high order, while preserving a low systematic loss of resolution and a high overload level. An analytical method to estimate its power consumption is presented in Appendix B.

References to the Preface

[Abidi95] A.A. Abidi, "Low-Power Radio-Frequency ICs for Portable Communications". *Proceedings of the IEEE*, vol. 83, pp. 544-569, April 1995.

[Adams86] R.W. Adams, "Design and Implementation of an Audio 18-bit Analog-to-Digital Converter Using Oversampling Techniques". Journal of Audio *Engineering Society*, vol. 34, pp. 153-166, March 1986.

[Boser88] B.E. Boser and B.A. Wooley, "The Design of Sigma-Delta Modulation Analog-to-Digital Converters". *IEEE Journal of Solid-State Circuits*, vol. 23. pp. 1298-1308, December 1988.

[Bran91a] B. Brandt, D.W. Wingard, and B.A. Wooley, "Second-Order Sigma-Delta Modulation for Digital-Audio Signal Acquisition". *IEEE Journal of Solid-State Circuits*, vol. 23. pp. 618-627, April 1991.

[Bran91b] B.P. Brandt and B.A. Wooley, "A 50-MHz Multibit Sigma-Delta Modulator for 12-b 2-MHz A/D Conversion". *IEEE Journal of Solid-State Circuits*, vol. 26, pp. 1746-1756, December 1991.

[Broo97] T.L. Brooks, D.H. Robertson, D.F. Kelly, A. Del Muro, and S. W. Harston, "A Cascaded Sigma-Delta Pipeline A/D Converter with 1.25 MHz Signal Bandwidth and 89 dB SNR". *IEEE Journal of Solid-State Circuits*, vol. 32, pp. 1896-1906, December 1997.

[Bult00] K. Bult, "Analog Design in Deep Sub-Micron CMOS". *Proc. of the IEEE European Solid-State Circuits Conference*, pp. 11-17, 2000.

[Candy85] J.C. Candy, "A Use of Double Integration in Sigma-Delta Modulation". *IEEE Transactions on Communications*. vol. 33, pp. 249-258, March 1985.

[Feld98] A.R. Feldman, B.E. Boser, and P.R. Gray, "A 13-Bit, 1.4-MS/s Sigma-Delta Modulator for RF Baseband Channel Applications". *IEEE Journal of Solid-State Circuits*, vol. 33, pp. 1462-1469, October 1998.

[Fuji00] I. Fujimori, L. Longo, A. Hairapetian, K. Seiyama, S. Kosic, J. Cao, and S.-L. Chan, "A 90-dB SNR 2.5-MHz Output-Rate ADC Using Cascaded Multibit Delta-Sigma modulation at 8x Oversampling Ratio". *IEEE Journal of Solid-State Circuits*, vol. 35, pp. 1820-1828, December 2000.

[Gagn97] M. Gagnaire, "An Overview of Broad-band Access Technologies". *Proceedings of the IEEE*, vol. 85, pp. 1958-1972, December 1997.

[Geer99] Y. Geerts, A. Marques, M. Steyaert, and W. Sansen, "A 3.3-V, 15-bit, Delta-Sigma ADC with a Signal Bandwidth of 1.1 MHz for ADSL Applications". *IEEE Journal of Solid-State Circuits*, vol. 34, pp. 927-936, July 1999.

[Geer00] Y. Geerts, M. Steyaert, and W. Sansen, "A High-Performance Multibit ΔΣ CMOS ADC". *IEEE Journal of Solid-State Circuits*, vol. 35, pp. 1829-1840, December 2000.

[Guil01] J. Guilherme, P. Figueredo, P. Azevedo, G. Minderico, A. Leal, J. Vital, and J. F ranca, "A Pipeline 15-b 10Msample/s Analog-to-Digital Converter for ADSL Applications". *Proc. of the IEEE International Symposium on Circuits and Systems*, vol. 1, pp. 396-399, May 2001.

[Inose62] H. Inose, Y. Yasuda, and J. Murakami, "A Telemetering System by Code Modulation – Δ-Σ Modulation". *IRE Transactions on Space Electronics and Telemetry*, vol. 8, pp. 204-209, September 1962.

[Malo01] F. Maloberti, *Analog Design for CMOS VLSI Systems*. Kluwer Academic Publishers, 2001.

[Marq98] A.M. Marques, V. Peluso, M.S.J. Steyaert, and W. Sansen, "A 15-b Resolution 2-MHz Nyquist Rate ΔΣ ADC in a 1-μm CMOS Technology". *IEEE Journal of Solid-State Circuits*, vol. 33, pp. 1065-1075, July 1998.

[Mayes96] M.K. Mayes and S.W. Chin, "A 200 mW, 1 Msample/s 16-b Pipelined A/D Converter with On-Chip 32 b Microcontroller". *IEEE Journal of Solid-State Circuits*, vol. 31, pp. 1862-1872, December 1996.

[Mede99] F. Medeiro, B. Perez-Verdú, and A. Rodríguez-Vázquez, "A 13-bit, 2.2-MS/s, 55-mW Multibit Cascade ΣΔ Modulator in CMOS 0.7-μm Single-Poly Technology". *IEEE Journal of Solid-State Circuits*, vol. 34, pp. 748-760, June 1999.

[Mori00] J.C. Morizio, M. Hoke, T. Kocak, C. Geddie, C. Hughes, J. Perry, S. Madhavapeddi, M.H. Hood, G. Lynch, H. Kondoh, T. Kumamoto, T. Okuda, H. Noda, M. Ishiwaki, T. Miki, and M. Nakaya, "14-bit 2.2-MS/s Sigma-Delta ADC's". *IEEE Journal of Solid-State Circuits*, vol. 35, pp. 968-976, July 2000.

[Nors97] S.R. Norsworthy, R. Schreier, and G.C. Temes (Editors), *Delta-Sigma Data Converters: Theory, Design and Simulation*. IEEE Press, 1997.

[Opris00] I.E. Opris, B.C. Wong, and S.W. Chin, "A Pipeline A/D Converter Architecture with Low DNL". *IEEE Journal of Solid-State Circuits*, vol. 35, pp. 281-285, February 2000.

[OptE91] F. Op't Eynde, G.M. Yin, and W. Sansen, "A CMOS Fourth-order 14b 500k-sample/s Sigma-delta ADC Converter". *Proc. of IEEE International Solid-State Circuit Conference*, pp. 62-63, 1991.

[Plas94] R. van de Plassche, *Integrated Analog-to-Digital and Digital-to-Analog Converters*. Kluwer Academic Publishers, 1994.

[Raza95] B. Razavi, *Principles of Data Converter System Design*. IEEE Press, 1995.

[Rivo03] R. Rivoir, "Nyquist-rate Converters: An Overview", Chapter 1 in *CMOS Telecom Data Converters (A. Rodríguez-Vázquez, F. Medeiro, E. Janssens, Editors)*. Kluwer Academic Publishers, 2003.

[Sign90] B.P. del Signore, D.A. Kerth, N.S. Sooch, and E.J. Swanson, "A Monolithic 20-b Delta-Sigma A/D Converter". *IEEE Journal of Solid-State Circuits*, vol. 25, pp. 1311-1317, December 1990.

[Yama94] K. Yamamura, A. Nogi, and A. Barlow, "A low power 20 bit instrumentation delta-sigma ADC". *Proc. of the IEEE Custom Integrated Circuits Conference*, pp. 519-522, 1994.

[Yin94] G. Yin and W. Sansen, "A High-Frequency and High-Resolution Fourth-Order ΣΔ A/D Converter in BiCMOS Technology". *IEEE Journal of Solid-State Circuits*, vol. 29, pp. 857-865, August 1994.

CHAPTER 1

ΣΔ ADCs: Principles, Architectures, and State of the Art

THE BASIC IDEA UNDERLYING SIGMA-DELTA CONVERTERS is the use of oversampling, error processing, and feedback for improving the effective resolution of a coarse quantizer. An early description of some of these ideas was given in a patent by Cutler in 1960 [Cutl60] and, obviously, a long way has been walked in more than 40 years, regarding the architectures of sigma-delta (ΣΔ) converters, their modeling, the circuit techniques to implement them, their field of application, etc.

This chapter is conceived as an introduction to ΣΔ analog-to-digital converters (ADCs). The operation principles of ΣΔ modulation—oversampling and noise-shaping—, as well as the basic architecture and ideal performance of ΣΔ ADCs, are presented and compared with Nyquist-rate ADCs at the beginning of the chapter (Sections 1.1 and 1.2).

Practical topologies for the implementation of ΣΔ modulation techniques are then discussed. Section 1.3 is dedicated to single-loop ΣΔ architectures. Low- and high-order single-loops are considered, taking into account issues related to their practical implementation and problems not addressed by linear models, like pattern noise or instabilities. Cascade ΣΔ topologies are covered in Section 1.4.

In Section 1.5 the study is extended to ΣΔ converters using multi-bit internal quantizers, analyzing their pros and cons. Techniques to circumvent the disadvantages, such as dual-quantization or dynamic element matching, are revised.

Finally, the state of the art in ΣΔ ADCs is reviewed in detail, considering integrated implementations reported in open literature, in order to highlight existing tendencies for achieving common specifications imposed to ADCs.

1.1 Analog-to-Digital Conversion: Fundamentals

Analog-to-digital converters (ADCs) are systems that transform signals that are continuous in time and in amplitude (analog) into signals that are discrete in time and in amplitude (digital). Fig. 1.1*a* shows the generic scheme of an ADC intended for the conversion of low-pass signals. It basically includes an anti-aliasing filter, a sampling circuit, and a quantizer.

The operation of these blocks is illustrated in Fig. 1.1*b*, showing the signal processing involved in the time and the frequency domain. First, the analog input signal $x_a(t)$ of the ADC passes through the anti-aliasing filter, in order to remove possible out-of-band components. This way, fold-over (*aliasing*) is avoided during the subsequent sampling process. The resulting filtered signal $x_f(t)$ is sampled at a frequency f_s by the sampler, thus yielding a discrete-time signal $x_s(n) = x_f(nT_s)$, where $T_s = 1/f_s$ [†1]. Finally, the values of $x_s(n)$ are

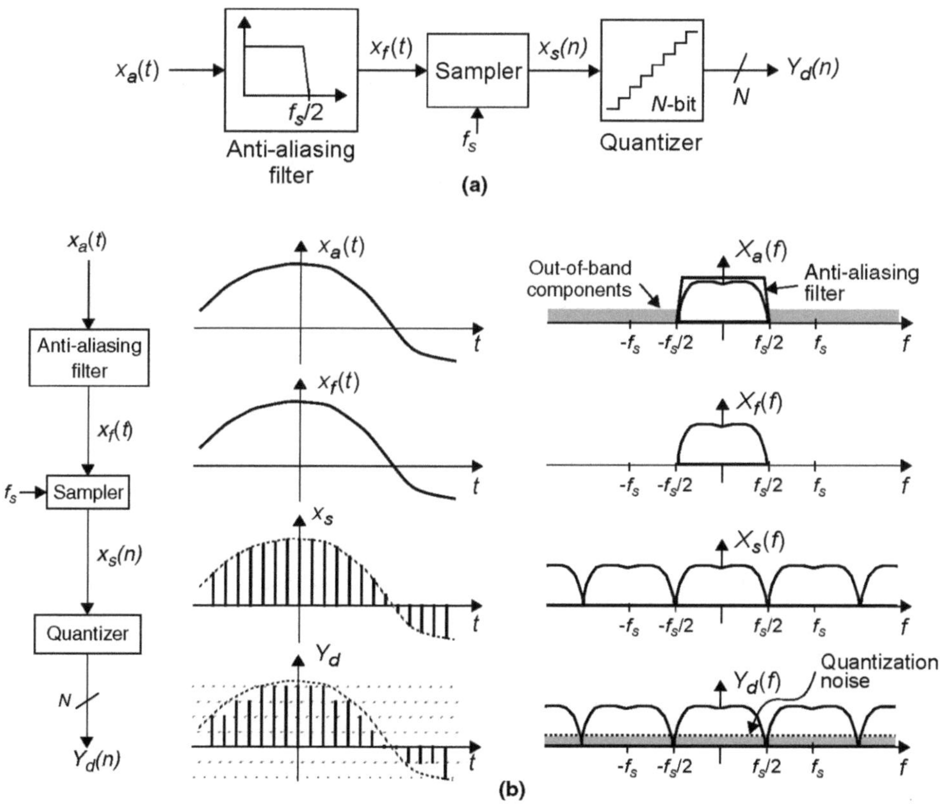

FIGURE 1.1 A-to-D conversion: (a) Generic scheme, (b) Illustration of the signal processing. (A Nyquist-rate ADC is assumed).

quantized using N bits; i.e., each continuous-valued input sample is mapped onto the closer discrete-valued level out of the 2^N covering the input signal variation interval. This process yields the converter digital output $Y_d(n)$.

As shown in Fig. 1.1, two fundamental operations are involved in the analog-to-digital conversion: *sampling* and *quantization*. The sampling process performs the continuous-to-discrete conversion of the input signal in time, whereas the quantization process performs the continuous-to-discrete conversion of the input signal in amplitude. These two transformations pose limitations to the performance of an ADC, even if they are realized ideally—i.e., using ideal circuit components.

1.1.1 Sampling

Sampling imposes a limit on the bandwidth of the analog input signal. According to the Nyquist theorem, the minimum frequency f_s required for sampling a signal with no loss of information is twice the signal bandwidth f_b; i.e., $f_N = 2f_b$, also called the Nyquist frequency. Based on this criterion, Those ADCs in that analog signals are sampled at minimum rate ($f_s = f_N$) are called *Nyquist-rate converters*.

Fig. 1.1*b* illustrated the operation of an ADC assuming a Nyquist-rate sampling. In the sampler, the filtered input signal $x_f(t)$ is multiplied by a train of Dirac impulses spaced out $T_s = 1/f_s$, which, in the frequency domain, corresponds to the convolution with a train of impulses located at multiples of f_s. Since in Nyquist-rate ADCs the input signal bandwidth f_b coincides with $f_s/2$, aliasing will occur if $x_a(t)$ contains frequency components above $f_s/2$. Therefore, high-order analog filters are required in order to remove the out-of-band components with no significant attenuation of the signal band.

1.1.2 Quantization

Quantization itself also introduces a limitation on the performance of an ideal ADC. It degrades the quality of the input signal whose continuous-valued levels are mapped onto a finite number of discrete levels. In this process an error is generated, called quantization error.

The quantizer operation is shown in Fig. 1.2. As a matter of example, Fig. 1.2*c* depicts the input-output curve of a quantizer with $N = 3$, although results apply to a generic N-bit quantizer. As the input signal changes from $-X_{FS}/2$ to $+X_{FS}/2$, the output is quantized ('rounded') to one out of the eight

---

1. In some cases, the sampled values are held during T_s using a sample-and-hold circuit.

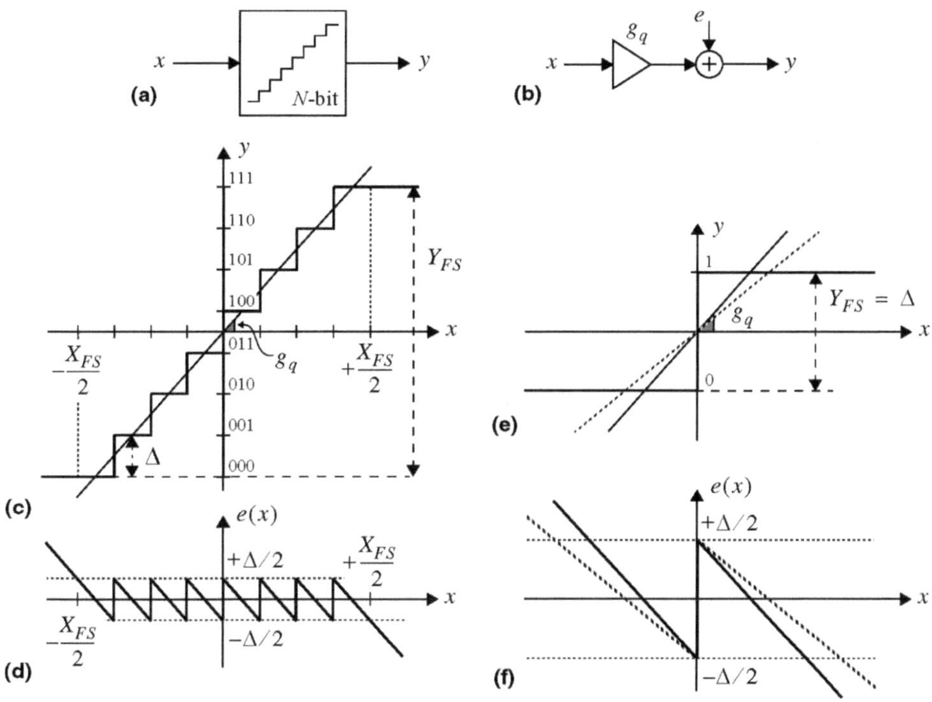

FIGURE 1.2 Ideal quantization process: (a) Symbolic representation, (b) Linear model of an ideal quantizer, (c) Ideal input-output curve and (d) quantization error of a 3-bit quantizer, (e) Ideal input-output curve and (f) quantization error of a 1-bit quantizer (comparator).

(2^N) different levels, each represented by a digital output code going from 000 to 111. The separation between adjacent output levels is defined by the quantization step, Δ. For an N-bit quantizer, $\Delta = Y_{FS}/(2^N - 1)$, with Y_{FS} being the full-scale output range of the quantizer, whereas X_{FS} is its full-scale input range. Since X_{FS} and Y_{FS} are not necessarily equal, the quantizer can exhibit a gain g_q, given by the slope of the line intersecting the code transitions. Thus, the quantizer operation can be described by a linear model (see Fig. 1.2*b*)

$$y = g_q x + e(x) \tag{1.1}$$

where g_q stands for the quantizer gain and $e(x)$ for the *quantization* (rounding) *error*. This error is a non-linear function of the input x, as shown in Fig. 1.2*d*. Note that, as long as x is confined in the range $\pm X_{FS}/2$, the quantization error is bounded by $\pm\Delta/2$. The maxima of $e(x)$ occur at the code transitions. For inputs outside the range $\pm X_{FS}/2$, the absolute value of the quantizer error grows monotonically. This situation is known as quantizer overload, whereas the input range $[-X_{FS}/2, +X_{FS}/2]$ is referred to as the non-overload region.

Fig. 1.2 also shows the operation of a 1-bit quantizer (comparator). Note from Fig. 1.2*e* that the main difference in comparison with a multi-bit quantizer is that its output only depends on the sign of the input, its absolute value being unimportant. Therefore, the gain g_q is undefined and can be arbitrarily chosen.

In order to evaluate the performance of an ideal quantizer, some assumptions are made on the properties of the quantization error [Benn48] [Widr60] [Srip77] [Gray90]. As shown in Fig. 1.2*d*, the quantization error is systematically defined by the input signal. Nevertheless, if the input is assumed to change randomly from sample to sample in the interval $\pm X_{FS}/2$, the error will also be uncorrelated from sample to sample. Under these requirements, quantization can be viewed as a random process, the quantization error being independent of the input, with an uniform distribution in the range $\pm\Delta/2$. Fig. 1.3*a* shows the probability density function (PDF) of such an error. Thus, the power associated to the quantization error is

$$\overline{e^2} = \sigma^2(e) = \int_{-\infty}^{+\infty} e^2 PDF(e) de = \frac{1}{\Delta}\int_{-\Delta/2}^{+\Delta/2} e^2 de = \frac{\Delta^2}{12} \tag{1.2}$$

Since the quantized signal is sampled at rate f_s, the power of the quantization error will be distributed in the band $[-f_s/2, +f_s/2]$. Moreover, the assumption of the quantization error being a random process with uniform PDF also implies that its power spectral density (PSD) is uniform, as shown in Fig. 1.3*b*. Since the error power can be also calculated as

$$\overline{e^2} = \int_{-\infty}^{+\infty} S_E(f) df = S_E \int_{-f_s/2}^{+f_s/2} df = \frac{\Delta^2}{12} \tag{1.3}$$

the power spectral density of the quantization error yields

$$S_E(f) = \frac{\overline{e^2}}{f_s} = \frac{\Delta^2}{12 f_s} \tag{1.4}$$

Since the input signal bandwidth of a Nyquist-rate converter spreads over the band $[-f_s/2, +f_s/2]$ (see Fig. 1.1*b*), all the quantization error power falls inside the signal band and passes to the ADC output as a part of the signal itself.

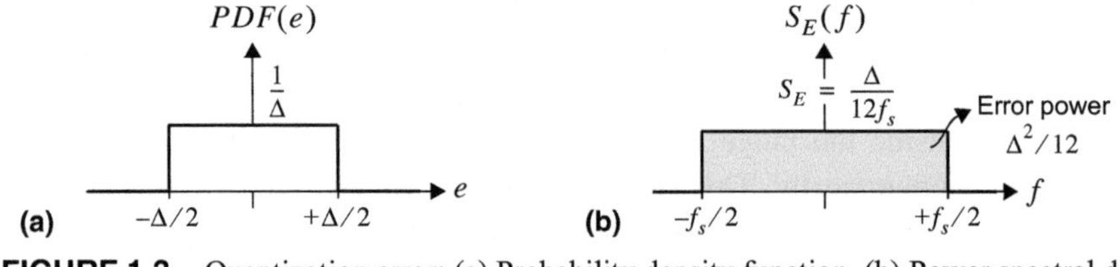

FIGURE 1.3 Quantization error: (a) Probability density function, (b) Power spectral density.

The assumptions made to this end are collectively referred to as the *additive white noise approximation* of the quantization error. Because of them, the quantization error is usually modeled as an additive white noise source and often called *quantization noise.* Although all these conditions are hardly met in practice and are not strictly valid, the additive noise assumption is commonly used in ADCs design and yields good results †2 —the larger the number of bits in the quantizer, the better the assumption. Rigorously speaking, the white noise approximation is not valid for single-bit quantizers. Nevertheless, experience shows that results obtained with this model are generally applicable to comparators [Nors97a]. This point will be further discussed in Section 1.3.

The degradation introduced by the quantizer in the performance of an ADC can be expressed through the *in-band quantization error power* P_Q; i.e. the power of the error caused by the quantization process within the signal band

$$P_Q = \int_{-f_b}^{+f_b} S_E(f)df = \int_{-f_s/2}^{+f_s/2} S_E(f)df = \frac{\Delta^2}{12} \tag{1.5}$$

The *dynamic range* DR of the ideal ADC can be determined as the ratio of the output power at the frequency of an input sinusoid with maximum amplitude to the in-band quantization error power. From Fig. 1.2*c*, the maximum input amplitude in the non-overload region of the quantizer is $X_{FS}/2$ and its corresponding power at the ADC output can be approximated to [Plas94] †3

$$P^{out}_{X_{FS}/2} \cong \frac{(Y_{FS}/2)^2}{2} \cong \frac{(2^N\Delta/2)^2}{2} = 2^{2N-3}\Delta^2 \tag{1.6}$$

The DR of the ideal ADC is then given by

$$DR = \frac{P^{out}_{X_{FS}/2}}{P_Q} = \frac{3}{2}2^{2N} \tag{1.7}$$

that, when expressed in decibels, leads to the well-known formula

$$DR|_{\mathrm{dB}} = 6.02N + 1.76 \tag{1.8}$$

2. There are two situations where these assumptions may not apply: DC inputs and inputs changing by multiples or submultiples of the step size from sample to sample [Candy97]. Nevertheless, a *dither* signal can be added, if needed, to modify the statistical properties of the quantizer input signal in order to satisfy the additive noise model [Lips92] [Gray97].

3. This approximation is only valid for moderate-resolution quantizers ($N \geq 5$), since the output signal is approximated to be also a sinusoid at the same frequency; i.e., the distortion introduced by the quantization is disregarded, neglecting the output power at multiples of the input frequency. For low-resolution quantizers, corrections to this approximation can be established based on the series expansion of the quantized output waveform [Plas94].

Therefore, besides the limitation sampling imposes on the signal band of an ideal ADC ($f_b \leq f_s/2$), its dynamic range is also limited by quantization noise. Note from eq(1.8) that each additional bit in the ideal quantizer results in an increase of approximately 6dB in the ADC DR.

1.2 Oversampling ΣΔ ADCs: Fundamentals

In contrast to Nyquist-rate ADCs, oversampling ΣΔ converters make use of two basic ideas to decrease the quantization error power within the signal band and increase the accuracy of the A-to-D conversion. These ideas are *oversampling* and *noise-shaping*.

1.2.1 Oversampling

Oversampling consists of sampling a signal faster than Nyquist rate—the minimum sampling frequency required to avoid aliasing. How much faster than required the signal is sampled is expressed through the *oversampling ratio*, defined as the ratio between the sampling frequency and the Nyquist frequency

$$OSR = \frac{f_s}{f_N} = \frac{f_s}{2f_b} \tag{1.9}$$

Oversampling has two noticeable effects:

- Since the signal bandwidth f_b is smaller than $f_s/2$, the images of the input created by the sampling process are more separated than in a Nyquist-rate converter. As shown in Fig. 1.4, frequency components of the input signal in the range $[f_b, f_s - f_b]$ do not alias within the signal band. Consequently, the transition from the pass- to the stop-band of the anti-aliasing filter can be smoother, what greatly simplifies its design.
- When an oversampled signal is quantized, the power of the quantization error is still distributed in the range $[-f_s/2, +f_s/2]$, but only part of the total error is placed within the signal band, since it extends to a frequency

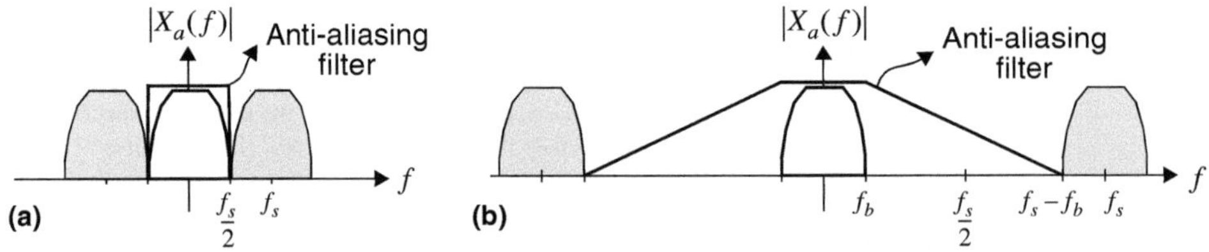

FIGURE 1.4 Anti-aliasing filter for: (a) Nyquist-rate converters, (b) Oversampling converters. Shadowed images correspond to those created by the sampling process.

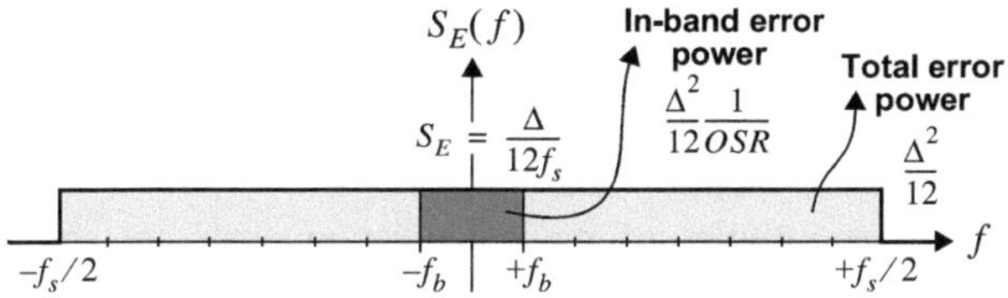

FIGURE 1.5 *PSD* of the quantization error in an oversampled converter ($OSR = 8$). Total power of the quantization error and power within the signal band are highlighted.

$f_b < f_s/2$. As illustrated in Fig. 1.5, the in-band power of quantization error now yields

$$P_Q = \int_{-f_b}^{+f_b} S_E(f) df = \int_{-f_b}^{+f_b} \frac{\Delta^2}{12 f_s} df = \frac{\Delta^2}{12 OSR} \tag{1.10}$$

so that the larger OSR, the smaller the portion of the total error power within the signal band.

As for Nyquist ADCs, the performance of an ideal *oversampling converter* will be limited by the quantization error of its N-bit quantizer. Following the same procedure as in Section 1.1.2, the dynamic range of an ideal oversampling converter can be calculated as

$$DR = \frac{P^{out}_{X_{FS}/2}}{P_Q} = \frac{3}{2} 2^{2N} OSR \tag{1.11}$$

that, when expressed in decibels, leads to

$$DR|_{\text{dB}} = 6.02N + 1.76 + 10\log_{10}(OSR) \tag{1.12}$$

Note that, just by oversampling, the DR of the ADC increases in approximately 3dB/octave, or equivalently 0.5bit/octave, with the oversampling ratio—i.e., the effect of multiplying OSR by a factor 4 is similar to having 1bit extra in the N-bit quantizer.

Therefore, an increase in OSR augments the effective resolution of the converter, but reduces the maximum signal frequency that can be processed (for given sampling frequency). So, in an oversampling converter, signal bandwidth and accuracy are exchanged. As we will see immediately on, this trade-off can be further exploited using oversampling in combination with noise-shaping.

1.2.2 Noise-shaping

An approach to further increase the accuracy of the A-to-D conversion consists of reducing the error power within the signal band through the *processing* of the quantization error. Let us consider an N-bit quantizer with an oversampled input signal. If the oversampling ratio is large enough, with the signal only changing slightly from sample to sample, most of the changes in the quantization error happen at high frequencies—i.e., low-frequency components of consecutive samples of the quantization error are similar. Hence, low-frequency in-band components of the quantization error can be attenuated by substracting the previous sample from the current one

$$e_{HP}(n) = e(n) - e(n-1) \tag{1.13}$$

The efficiency of this strategy is confirmed in Fig.1.6, showing a reduction of the in-band error power by a factor close to 20 in the particular case considered. Further reduction can be achieved by involving more previous error samples

$$\begin{aligned} e_{HP,1}(n) &= e(n) - e(n-1), && \text{1st-or error proc.} \\ e_{HP,2}(n) &= e(n) - 2e(n-1) + e(n-2), && \text{2nd-or error proc.} \\ e_{HP,3}(n) &= e(n) - 3e(n-1) + 3e(n-2) - e(n-3), && \text{3rd-or error proc.} \\ \ldots & && , \ldots \end{aligned} \tag{1.14}$$

The procedure can be formulated in an unified manner in the z-domain as

$$E_{HP,\,L}(z) = (1 - z^{-1})^{L} \cdot E(z) \tag{1.15}$$

showing that the processed error is a filtered version of the original. The filtering transfer function—often called *noise transfer function*—is therefore

$$NTF(z) = (1 - z^{-1})^{L} \tag{1.16}$$

where L denotes the *order* of the filtering realized on the quantization error.

For this transfer function

$$\left|NTF(e^{j\Theta})\right|^2 = \left|1 - e^{-j\Theta}\right|^{2L} = 2^{2L}\sin^{2L}\left(\frac{\Theta}{2}\right), \text{ with } \Theta = \frac{2\pi f}{f_s} = \frac{\pi f}{f_b OSR} \tag{1.17}$$

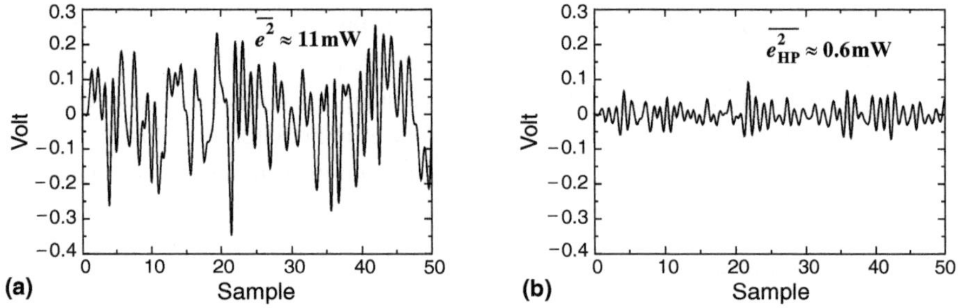

FIGURE 1.6 Error waveform: (a) Without processing, (b) With the processing in eq(1.13). The input is a full-scale (1V) random signal low-pass filtered, the quantizer is single-bit, and $OSR = 8$.

Within the signal band ($f \le f_b$), $\Theta \ll 1$ if OSR is large enough. Hence, the transfer function takes very small values in this range and the in-band power of the filtered quantization error results in

$$P_Q = \int_{-f_b}^{+f_b} \frac{\Delta^2}{12 f_s} |NTF(f)|^2 df \approx \frac{\Delta^2}{12} \cdot \frac{\pi^{2L}}{(2L+1) OSR^{(2L+1)}} \quad (1.18)$$

that is much smaller than only applying oversampling—see eq(1.10).

Fig. 1.7 depicts the amplitude of NTF for $L = 1$. Note that the error reduction happens due to:

- the high-pass shape of the transfer function, that yields large in-band attenuation, pushing noise to high frequencies.
- the inverse dependence of the integration interval with OSR.

Only the dark area at the bottom of Fig. 1.7 is integrated, so that the in-band error power is a small portion of the total thanks to the *noise-shaping* action—the smaller the larger OSR and/or L. Note from eq(1.18) that the high non-linearity of the shaping results into a non-linear dependency of P_Q with OSR.

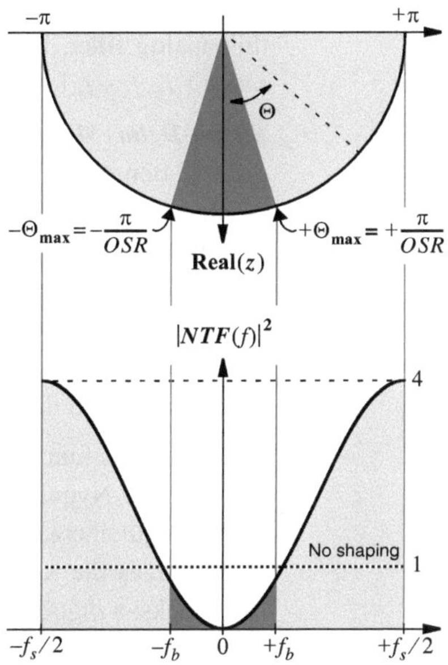

FIGURE 1.7 Amplitude of NTF for 1st-order error processing.

Using equations (1.6) and (1.18), the dynamic range of an ideal oversampling noise-shaping converter yields

$$DR = \frac{P^{out}_{X_{FS}/2}}{P_Q} \approx \frac{3}{2} 2^{2N} \cdot \frac{(2L+1) OSR^{(2L+1)}}{\pi^{2L}} \quad (1.19)$$

that, when expressed in decibels, leads to

$$DR|_{dB} \approx 6.02N + 1.76 + 10\log_{10}\left(\frac{2L+1}{\pi^{2L}}\right) + (2L+1)10\log_{10}(OSR) \quad (1.20)$$

Note that, when an Lth-order error filtering is used in combination with oversampling, the DR of the converter increases with OSR in approximately $(L + 1/2)$bit/octave—e.g., only by using a 1st-order error processing, the rate increases from 0.5bit/octave to 1.5bit/octave.

1.2.3 Basic architecture of oversampling ΣΔ ADCs

The basic scheme of a ΣΔ ADC intended for the conversion of low-pass signals is illustrated in Fig. 1.8. This diagram comprises three main blocks, whose operation in time and frequency domain is illustrated in Fig. 1.9:

- ***Anti-Aliasing Filter.*** The function of this block, identical to that in Nyquist-rate ADCs, is attenuating the out-of-band components of the input signal in order to avoid aliasing when it is sampled. However, as already explained, the use of oversampling relaxes the attenuation requirements for this analog filter, because the frequency components of the input in the range $[f_b, f_s - f_b]$ do not fold within the baseband.
- ***Sigma-Delta Modulator.*** It simultaneously performs the sampling and quantization of the filtered input signal. In addition, the quantization error is high-pass filtered by means of a noise-shaping technique. This is accomplished by placing an appropriate loop filter $H(z)$ before the quantizer and negative feedback around them. This, combined with oversampling, greatly enlarges the accuracy of the A-to-D conversion over that of the embedded quantizer †4. The output of the sigma-delta modulator (ΣΔM) is a B-bit digital stream at f_s sampling rate.
- ***Decimator.*** Its function is to reduce the rate of the ΣΔM output stream (f_s) down to the Nyquist frequency ($f_N = f_s/OSR$). At the same time, the word length increases from B to N ($N > B$), in order to preserve the resolution as the word rate decreases. The decimator basically consists of two blocks: a *digital filter* and a *downsampler.*
  The digital filter removes the frequency components of the digital input stream above f_b—basically shaped quantization error—, in order to avoid aliasing during downsampling. This must be done without degrading the baseband, so that usually high-selectivity filters are required. However, these tougher requirements are imposed on a digital filter, which is less

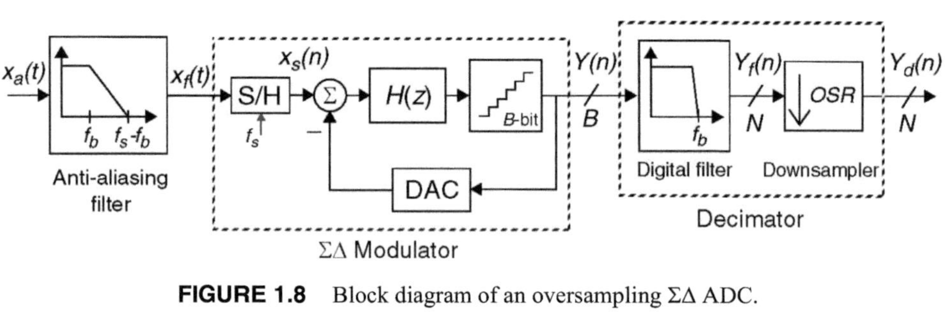

FIGURE 1.8 Block diagram of an oversampling ΣΔ ADC.

4. ΣΔ conversion can be used with no oversampling. However, we will implicitly assume $OSR > 1$. Although strictly speaking one should refer to these converters as *oversampling sigma-delta converters*, for simplicity they are often called just sigma-delta converters.

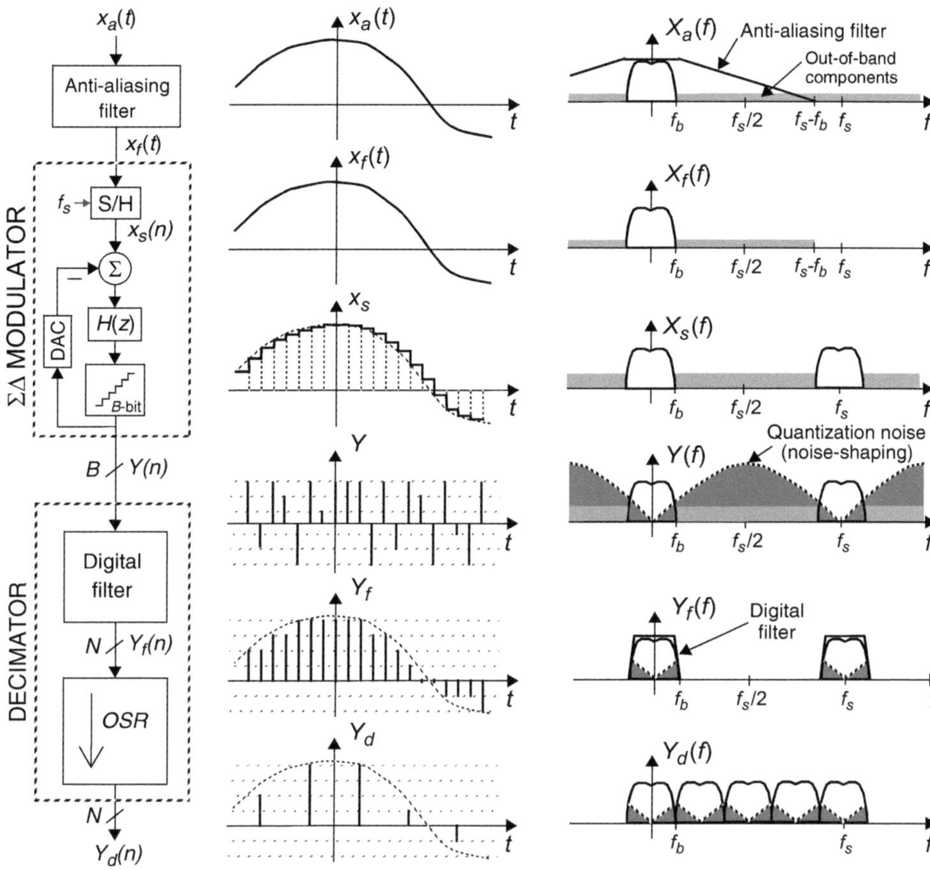

FIGURE 1.9 Illustration of the signal processing in a ΣΔ ADC.

difficult to implement than a high-selectivity analog one—as that required for anti-aliasing in Nyquist ADCs. In short, in ΣΔ ADCs the most difficult filtering is passed on to the digital domain, which is another reason for their well-known robustness.

The downsampler divides the rate of the digital stream by OSR. This can be done in a simple way by keeping only one sample of the stream and removing the next $OSR - 1$ samples.

In practice, the implementation of the decimator may differ from its conceptual representation in Fig. 1.8: the position of the digital filter and the downsampler can be swapped; both are often implemented in a single block; stages with moderate downsampling ratios can be cascaded if OSR is large, etc. [Croc83] [Vaid90] [Nors97b].

Out of these blocks, the one influencing most the ADC performance is the ΣΔ modulator, basically because it is the ultimate responsible for the accuracy

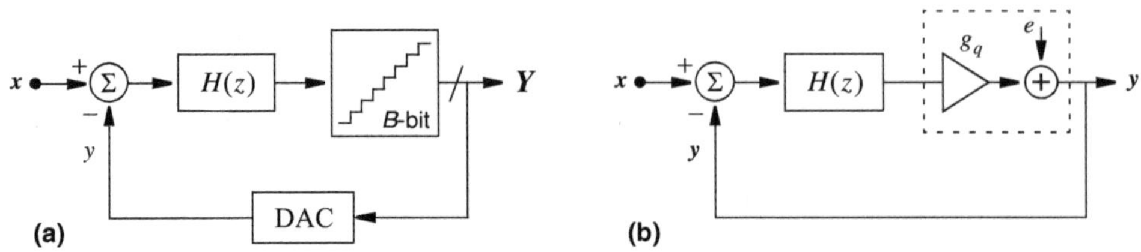

FIGURE 1.10 ΣΔM architecture: (a) Basic scheme, (b) Corresponding linear model.

in the A-to-D conversion. From now on, we will hence focus on this block—although keeping in mind that a ΣΔ ADC is more than just its ΣΔM.

Fig. 1.10*a* shows the basic scheme of a ΣΔ modulator. It consists of a feedforward path, formed by a loop filter $H(z)$ and a B-bit quantizer, and a negative feedback path around them, using a B-bit D-to-A converter [Inose62].

The operation of the ΣΔM can be explained as follows. Assume that the gain of $H(z)$ is large inside the signal band and small outside it. Due to the action of feedback, the error signal $x-y$ will become practically null in the signal band; i.e., the input signal x and the analog version of the modulator output y will practically coincide within this band. Consequently, most of the discrepancies between x and y will be placed at higher frequencies; i.e., the quantization error is shaped and pushed outside the signal band.

Fig. 1.10*b* shows the linear model corresponding to the ΣΔM in Fig. 1.10*a*. Note that the DAC is assumed to be ideal and the quantizer is replaced by the model in Fig. 1.2*b*, according to the additive white noise approximation of the quantization error. This way, the modulator can be viewed as a two-input (x, e) one-output (y) system that can be represented in the z-domain by

$$Y(z) = STF(z)X(z) + NTF(z)E(z) \tag{1.21}$$

where (Xz) and (Ez) are the z-transform of the input signal and the quantization noise, respectively, whereas $STF(z)$ and $NTF(z)$ are the respective transfer functions, given by

$$STF(z) = \frac{g_q H(z)}{1 + g_q H(z)} \qquad NTF(z) = \frac{1}{1 + g_q H(z)} \tag{1.22}$$

Since the signal and the noise are affected by different transfer functions, $H(z)$ can be chosen so that the noise-shaping is implemented without degrading the signal. Using a loop filter with large gain within the signal band, the signal and noise transfer functions can be approximated in that range to †5

5. Certain ΣΔMs are designed to provide a gain G in the converted signal, $STF(z) \approx G$—e.g., in applications with sensors, in which the input signal can be very small.

$$STF(z) \approx 1 \qquad NTF(z) \approx \frac{1}{g_q H(z)} \ll 1 \tag{1.23}$$

By properly selecting $H(z)$, the noise-shaping functions in eq(1.16) are built †6. The easiest loop filter that exhibits the desired frequency performance is an integrator, whose transfer function in the z-domain is

$$H(z) = \frac{z^{-1}}{1 - z^{-1}} \tag{1.24}$$

Assuming that the quantizer gain g_q equals unity, the modulator output yields

$$Y(z) = z^{-1}X(z) + (1 - z^{-1})E(z) \tag{1.25}$$

This modulator is called *1st-order modulator*, referring to the order of the noise-shaping function. Its filter, signal, and noise transfer functions are depicted in Fig. 1.11, whereas its time-domain operation is illustrated in Fig. 1.12.

Fig. 1.12*a* shows the output of a 1st-order ΣΔ modulator that uses a 1-bit internal quantizer when a ramp signal is applied at the input. Note that, due to the combined action of oversampling and feedback, the modulator output is a *pulse-density modulated* (PDM) digital signal whose local average tracks the input: when the input signal is low, the modulator output contains more -1 's than $+1$'s; when it is high, the $+1$'s are dominant; and when the input signal is close to zero, the density of $+1$'s and -1 's practically coincides.

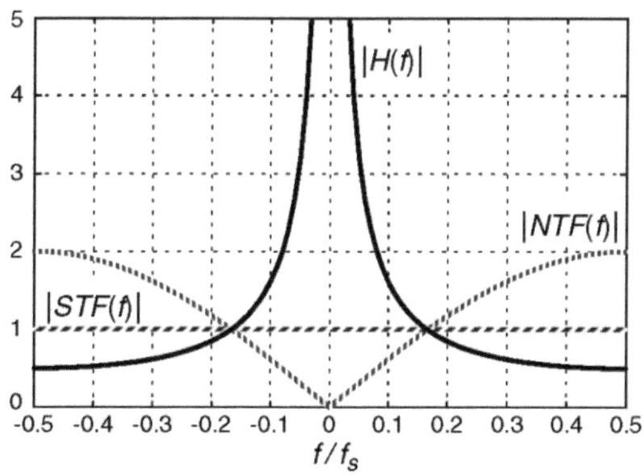

FIGURE 1.11 Transfer functions of $H(f)$, $STF(f)$, and $NTF(f)$ for a 1st-order ΣΔM.

6. Another way to obtain noise-shaping is placing the loop filter in the feedback path. Assuming $g_q = 1$, $Y(z) = X(z) + [1 - H(z)]E(z)$ and $H(z)$ can be arbitrarily chosen. This topology, called *error feedback*, is often used in ΣΔ DACs. However, it is never used in ADCs because: 1) it significantly compromises the stability of high-order shapings, 2) it is very sensitive to errors in the analog substractor used to extract e.

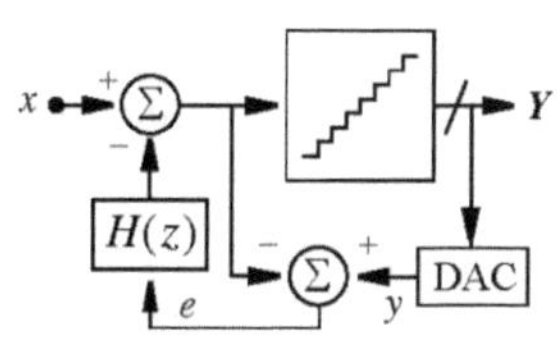

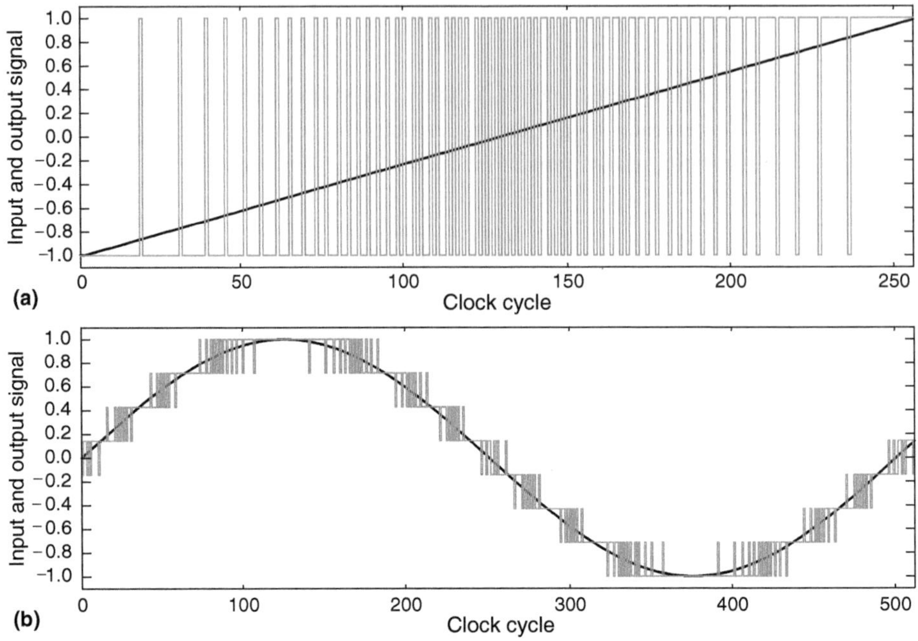

FIGURE 1.12 Illustration of the PDM output streams of a 1st-order ΣΔM: (a) Output for a ramp input using 1-bit internal quantizer, (b) Output for a sinusoidal input using 3-bit internal quantizer.

If the resolution of the internal quantizer is increased, the output will track the input much closer, since the separation between the output code levels decreases. This is illustrated in Fig. 1.12*b*, that shows the output of the 1st-order ΣΔM for an input sinewave when a 3-bit quantizer is used. Note that the behavior of the PDM output signal in an input range between adjacent code transitions is similar to that formerly described.

1.2.4 Performance metrics

At this point, it is convenient to introduce the most important specifications commonly used to characterize the performance of oversampling ΣΔ converters.

- ***Signal-to-noise ratio, SNR*.** It is the ratio of the output power at the frequency of an input sinusoide to the uncorrelated in-band error power. As shown in Chapter 2, due to non-idealities of the circuitry used to implement the converter, there are other error contributions—linear and non-linear—to the in-band error power, apart from quantization noise.
  The *SNR* commonly accounts for the linear performance of the converter, so that the in-band power associated to harmonics is not included. It is usually given in decibels.

For an ideal converter, accounting only for the quantization error, the SNR can be approximated to

$$SNR|_{\mathrm{dB}} = 10\log_{10}\left(\frac{A_y^2}{2P_Q}\right) \tag{1.26}$$

where A_y is the amplitude of the output sinusoide.

- ***Signal-to-(noise + distortion) ratio, SNDR.*** It is defined as the ratio of the output power at the frequency of an input sinusoid to the total in-band error power, also accounting for possible harmonics at the converter output.
- ***Dynamic range, DR.*** It can be defined as the ratio of the output power at the frequency of an input sinusoid with maximum amplitude to the output power for a small input for which $SNR = 0\mathrm{dB}$; i.e., so it cannot be distinguished from the error.
  Ideally, a sinusoid with maximum amplitude at the converter input $X_{FS}/2$ will provide an output sinusoid sweeping the full-scale range Y_{FS} of the modulator quantizer, and hence

$$DR|_{\mathrm{dB}} = 10\log_{10}\left[\frac{(Y_{FS}/2)^2}{2P_Q}\right] \tag{1.27}$$

- ***Effective number of bits, ENOB.*** Since the DR of an ideal N-bit Nyquist-rate converter is given by eq(1.8), a similar expression can be established for ΣΔMs

$$ENOB = \frac{DR|_{\mathrm{dB}} - 1.76}{6.02} \tag{1.28}$$

  where the $ENOB$ can be defined as the number of bits needed for an ideal Nyquist-rate converter to achieve the same DR as the ΣΔ converter. Thus, the performance of oversampled ΣΔ converters and Nyquist-rate ADCs can be compared in simple way [Boser88].
- ***Overload level, X_{OL}.*** As we show later on, for a B-bit quantizer embedded in a ΣΔM, overloading does not start when the amplitude of the modulator input signal A_x equals half the full-scale input range of the quantizer ($X_{FS}/2$—see Fig. 1.2*c* and Fig. 1.2*d*), but before. Thus, the SNR of a ΣΔ modulator does not increase monotonously for input amplitudes in the range $[0, X_{FS}/2]$. For large amplitudes close to $X_{FS}/2$, overloading occurs, causing an increase of the in-band noise and a sharp drop in the SNR. The maximum value of the SNR before that drop is labelled as SNR_{peak} and its corresponding input signal level will be referred to as the overload level of the ΣΔM, X_{OL}.

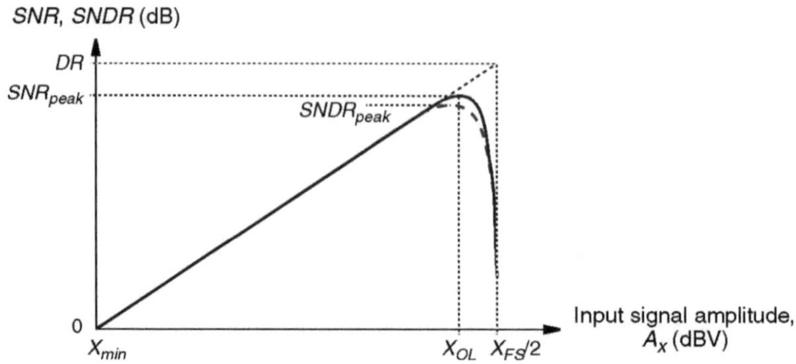

FIGURE 1.13 Illustration of the performance of a ΣΔ modulator on a typical *SNR* curve.

The previous parameters are illustrated in Fig. 1.13, that shows typical curves for the *SNR* and the *SNDR* of a ΣΔ modulator as a function of the amplitude of a sinusoid signal applied at the modulator input. Usually, both curves coincide for small and medium input levels, since the distortion due to non-linear effects is submerged into the modulator noise floor. For large input levels, harmonic distortion becomes more evident, causing performance degradation and the deviation of the *SNDR* curve.

1.2.5 Ideal performance

The output of an ideal Lth-order ΣΔ modulator yields in the z-domain

$$Y(z) = z^{-L}X(z) + (1 - z^{-1})^L E(z) \tag{1.29}$$

where the input signal appears at the output with an Lth-order delay and the noise transfer function provides the quantization error of the embedded quantizer with a shaping of order L. If a B-bit quantizer is used and oversampling is applied, the dynamic range of the ΣΔM can be obtained from eq(1.19) as

$$DR \approx \frac{3}{2}(2^B - 1)^2 \cdot \frac{(2L+1)OSR^{(2L+1)}}{\pi^{2L}} \tag{1.30}$$

that, expressed in decibels, leads to

$$DR|_{\text{dB}} \approx 20\log_{10}(2^B - 1) + 1.76 + 10\log_{10}\left(\frac{2L+1}{\pi^{2L}}\right) + (2L+1)10\log_{10}(OSR) \tag{1.31}$$

Since most of ΣΔMs employ low-resolution internal quantizers ($B = 1 \sim 5$), the term $2^B - 1$ is not approximated in equations (1.30) and (1.31).

Note from eq(1.31) that the dynamic range of a ΣΔ modulator can be therefore increased if L, OSR, and/or B are increased. The pros and cons of each possibility are discussed next:

- ***Dependence on the modulator order, L.*** The performance of an oversampling ΣΔ converter can be considerably improved increasing the order of the noise-shaping, since quantization error will be more efficiently shaped and, hence, more attenuated at low frequencies. This is illustrated in Fig. 1.14, that compares the ideal NTF of a ΣΔM for L ranging from 1 to 5. The case $L = 0$ (no shaping) is also included for comparison purposes. For a given OSR, the increase in DR when increasing the modulator order L in one, leads from eq(1.31) to

$$\Delta DR|_{\mathrm{dB}} \approx 10\log_{10}\left[\frac{2L+3}{2L+1}\cdot\left(\frac{OSR}{\pi}\right)^2\right] \tag{1.32}$$

  This means, e.g., that the DR of a 4th-order ΣΔM with $OSR = 32$ is improved in 21.3dB (3.5bit) with respect to the 3rd-order one.
  However, as we show in Section 1.3, the use of high-order shaping gives rise to stability problems. Although these problems can be circumvented, the dynamic range of a high-order ΣΔM will be worse than predicted by eq(1.31).

- ***Dependence on the oversampling ratio, OSR.*** Note from eq(1.31) that the DR of an ideal Lth-order ΣΔ converter increases with OSR in $(L+1/2)$bit/octave. This is shown in Fig. 1.15, where the DR and $ENOB$ are plotted as a function of the oversampling ratio and the modulator order, in case of a single-bit internal quantizer. Note that for $OSR > 4$, the combined action of oversampling and noise-shaping considerably improves performance.

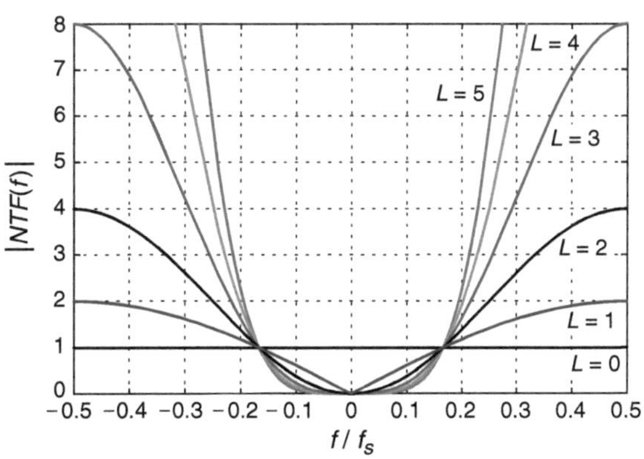

FIGURE 1.14 Illustration of $NTF(f)$ for different shaping orders (L) in the ΣΔ modulator.

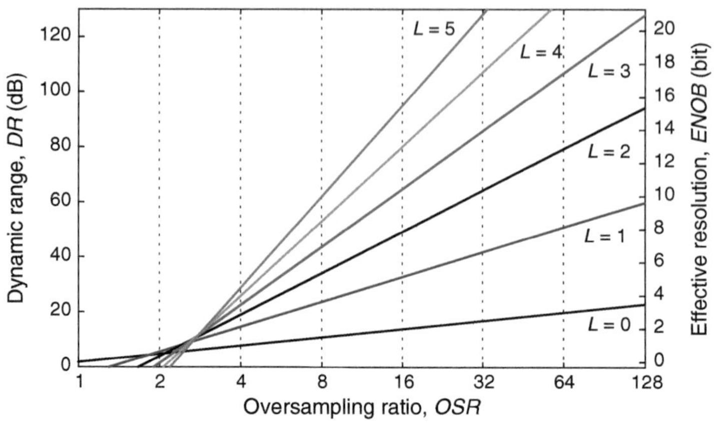

FIGURE 1.15 Ideal performance of oversampled ΣΔ converters. *DR* and *ENOB* versus *OSR* for different order *L* of the modulator. (A single-bit internal quantizer is assumed).

However, for a given signal band, a larger *OSR* implies a higher sampling frequency and a faster operation of the circuitry. The latter, if achievable in a given fabrication process, leads to a larger power dissipation.

- ***Dependence on the quantizer resolution, B.*** Increasing the resolution of the embedded quantizer also improves the performance of an ideal modulator, since the power associated to quantization error decreases. Table 1.1 shows the improvement in *DR*, obtained from eq(1.31), when using an multi-bit internal quantizer [Geer02]. These values are commonly approximated to consider that each extra bit in the quantizer leads to an increase of 6dB (1bit) in the modulator *DR*.

TABLE 1.1 *DR* improvement of an ideal ΣΔM if multi-bit quantization is used.

B	ΔDR compared to $B = 1$	ΔDR compared to $B-1$
2	9.5dB	9.5dB
3	16.9dB	7.4dB
4	23.5dB	6.6dB
5	29.8dB	6.3dB

This improvement is illustrated in Fig. 1.16, showing the performance of ideal multi-bit ΣΔMs versus *OSR*, for *B* ranging from 1 to 4.

However, multi-bit ΣΔMs require a multi-bit DAC in the feedback loop, that—on the contrary to a single-bit one, with only two levels—is not inherently linear. Note from Fig. 1.10*a* that any non-linearities in the multi-bit DAC will be directly added to the modulator input. Thus, in practice, the linearity required in the DAC equals that wanted for the ΣΔ modulator. This point will be further discussed in Section 1.5.

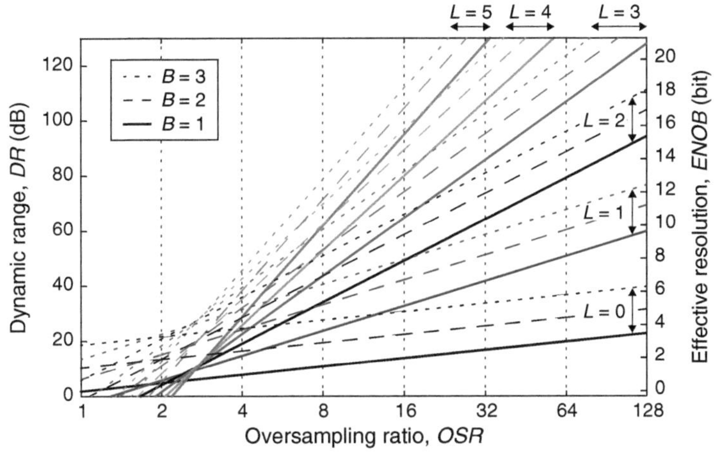

FIGURE 1.16 Ideal performance of multi-bit ΣΔMs. *DR* and *ENOB* versus *OSR* for different modulator orders (L) and resolutions of the internal quantizer (B).

1.3 Single-Loop ΣΔ Architectures

In the previous section, the operation principles and ideal performance of generic ΣΔMs have been introduced. This section presents ΣΔ topologies using a certain number of integrators and one quantizer, which are often referred to as *single-loop ΣΔMs*, but *single-stage* or *single-quantizer* architectures may also apply. Their linear performance will be discussed, as well as aspects that are not covered by the additive white noise approximation—such as pattern noise, idle tones, or instabilities. A single-bit quantizer will be assumed.

1.3.1 1st-order ΣΔ modulator

The simplest ΣΔ architecture—a 1st-order single-bit ΣΔM—can be built, as shown in Fig. 1.17*a*, using a discrete-time integrator as the loop filter and a single-bit quantizer (comparator). One may view the 'delta' and 'sigma' referring to the analog operations in the system loop: substraction (Δ) of the fed-back output signal and integration (Σ) of the difference. Note that the only difference between the representation Fig. 1.17*a* and the generic one in Fig. 1.10*a* is that two gain blocks—the integrator weights g_1 and g_1'—are included for the input and the feedback. Assuming a linear model for the comparator, as in Fig. 1.10*b*, the modulator output in the z-domain yields

$$Y(z) = \frac{g_1 g_q z^{-1} X(z) + (1 - z^{-1}) E(z)}{1 - (1 - g_1' g_q) z^{-1}} \tag{1.33}$$

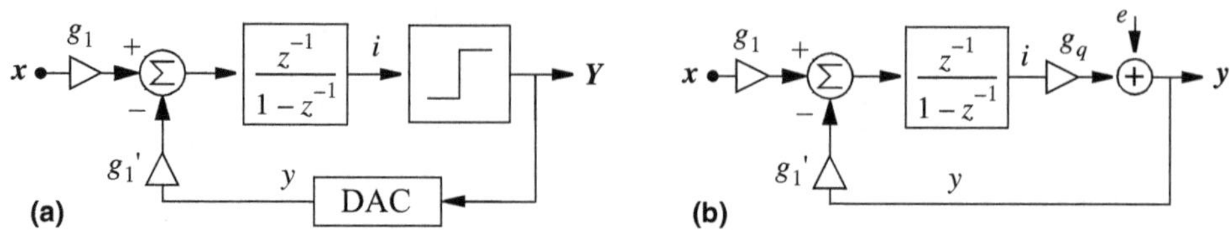

FIGURE 1.17 1st-order single-bit ΣΔM: (a) Generic scheme, (b) Corresponding linear model.

In order to achieve the 1st-order shaping of the quantization error, the following condition must be fulfilled

$$g_1'g_q = 1 \qquad \Rightarrow \qquad Y(z) = \frac{g_1}{g_1'} \cdot z^{-1}X(z) + (1 - z^{-1})E(z) \tag{1.34}$$

Note that $STF(z)$ can provide a gain $G = g_1/g_1'$ (see footnote 5). However, usually $g_1 \equiv g_1'$ and the input full-scale range X_{FS} equals the output Y_{FS}.

Using equations (1.18) and (1.30), the in-band quantization error and the dynamic range yield, respectively

$$P_Q \approx \frac{\Delta^2}{12} \cdot \frac{\pi^2}{3OSR^3} \tag{1.35}$$

$$DR|_{\mathrm{dB}} \approx 10\log_{10}\left(\frac{3}{2} \cdot \frac{3OSR^3}{\pi^2}\right) \tag{1.36}$$

The main disadvantage of this modulator is that very high oversampling ratios are needed to achieve medium resolutions, since DR increases only by 1.5bit/octave with OSR. This discards a priori its use in medium- to high-frequency applications. For example, $OSR \approx 960$ is needed to obtain 14-bit resolution. Therefore, $f_s \approx 2\mathrm{GHz}$ would be required for a 1-MHz signal band!

It must be mentioned that a two-level quantizer has no inherent gain (see Fig. 1.2*e*) and, therefore, there is no obvious value for g_q in eq(1.33). The linear model used henceforth for single-bit quantizers, proposed in [Will91], assumes that g_q is such that the product of the loop gain factors is forced to unity by the feedback loop. Thus, the ΣΔ loop can be viewed as an automatic gain control system (AGC), maintaining the loop gain at unity over most of the input range [Rebe97]. This model for comparators in a ΣΔM is entirely empirical [Will91]; its only justification is that analytical results obtained this way usually compare well to computer simulations using the true quantizer function

$$y(i) = \begin{cases} +\Delta/2, & i \ge 0 \\ -\Delta/2, & i < 0 \end{cases} \tag{1.37}$$

In general, the linear model of a ΣΔM—either for 1st-order single-bit modulators or for more complex architectures—, although not entirely correct, can be solved analytically and is used to gain insight of the system and predict the effect of different design parameters on its performance. However, the limitations of this linear representation of a truly non-linear system must be kept in mind—especially the white noise approximation of quantization error, that compromises the study of the stability of the non-linear system †7. Computer simulations using non-linear models are commonly performed to observe the true theoretical behavior of ΣΔMs [OptE90] [Rebe97].

In relation to the discussion above, there are also other important drawbacks for the use of 1st-order single-bit modulators; e.g., the presence of *pattern noise* and *idle tones* in the ΣΔ modulation that are not predicted by the white noise model [Candy81] [Candy97]. Equations (1.33) to (1.36) for the performance of the 1st-order ΣΔM assume that the quantization error is not correlated with the input signal and changes randomly. However, when the input to the 1st-order modulator is a DC signal, the output bounces between the two levels ($\pm\Delta/2$), trying to keep its mean value equal to the input under repetitive patterns. This leads to a *colored quantization error*, instead of white, and the in-band error power can be higher than predicted by the linear model if the repetition frequency lies within the signal band. This effect is illustrated in Fig. 1.18*a*, showing in-band quantization error power of a 1st-order single-bit ΣΔM for DC inputs in the range $[-\Delta/2, +\Delta/2]$. These type of representations are known as the pattern noise of the ΣΔ modulator. Note that the largest peaks exceed the prediction given by eq(1.35), -62.8dB in this example.

This correlation of the quantization error with the modulator input can also be observed in the *SNR* curves and in the output spectrum of the 1st-order ΣΔM. Fig. 1.18*b* shows the *SNR* as a function of the amplitude of three input sinewaves with different DC levels. Note that *SNR* can strongly deviate from the ideally expected performance. Fig. 1.18*c* illustrates the idle tones appearing in the modulator output spectrum due to this correlation. A possible solution to this problem is to include, usually at the quantizer input, a non-periodic signal as pseudo-random noise. With this technique, called *dithering* [Schu64] [Lips92] [Candy97], it is possible to partially decorrelate the quantization error and the input, but at the expense of larger complexity of the designs.

---

7. Like any feedback system, ΣΔMs may be unstable depending on the feedback dynamics. A ΣΔM is considered *stable* if, for bounded inputs and whatever integrator initial condition, the internal state variables remain also bounded over time. For instance, a 1st-order ΣΔM is stable for inputs in the range $[-\Delta/2, +\Delta/2]$. The stability of ΣΔMs is difficult to study analytically due to the actual non-linearity of the quantizer. For a better insight of the problem, readers can refer to [Agra83] [Adar87] [Stik88] [OptE90] [Baird94] [Enge99].

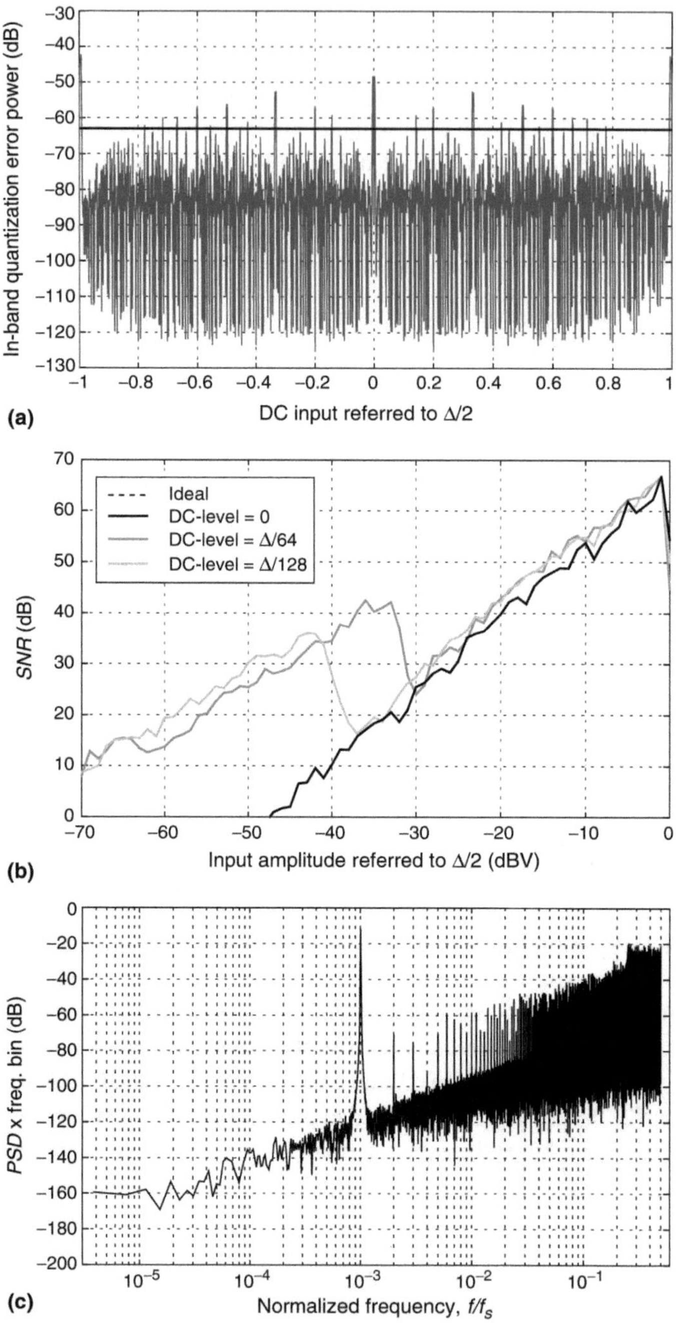

FIGURE 1.18 1st-order single-bit ΣΔM: (a) Pattern noise for DC inputs in the range $\pm\Delta/2$, (b) Effect on *SNR* curves for different DC levels, (c) Tonal behavior of the output spectrum for a half-scale input sinewave with DC-level $= 0$. ($g_1 = g_1' = 1$, $OSR = 128$, and $\Delta = 2$).

1.3.2 2nd-order ΣΔ modulator

If the quantizer in a 1st-order ΣΔM is replaced itself by a 1st-order modulator, the resulting architecture is the 2nd-order ΣΔM [Candy85], shown in Fig. 1.19 for the single-bit case. Assuming again a linear model for the comparator, the modulator output in the z-domain yields

$$Y(z) = \frac{g_1 g_2 g_q z^{-2} X(z) + (1 - z^{-1})^2 E(z)}{1 + (g_2' g_q - 2) z^{-1} + (1 + g_1' g_2 g_q - g_2' g_q) z^{-2}} \tag{1.38}$$

so that the following conditions must be fulfilled for a pure 2nd-order shaping

$$\left. \begin{aligned} g_1' g_2 g_q &= 1 \\ g_2' &= 2 g_1' g_2 \end{aligned} \right\} \Rightarrow Y(z) = \frac{g_1}{g_1'} \cdot z^{-2} X(z) + (1 - z^{-1})^2 E(z) \tag{1.39}$$

Using equations (1.18) and (1.30), the in-band quantization error and the dynamic range of the 2nd-order ΣΔM yield, respectively

$$P_Q \approx \frac{\Delta^2}{12} \cdot \frac{\pi^4}{5 OSR^5} \tag{1.40}$$

$$DR|_{\mathrm{dB}} \approx 10 \log_{10}\left(\frac{3}{2} \cdot \frac{5 OSR^5}{\pi^4} \right) \tag{1.41}$$

so that DR increases by 2.5bit/octave with OSR. For example, $OSR \approx 90$ is ideally enough to obtain 14-bit resolution; i.e., $f_s \approx 180\mathrm{MHz}$ for a 1-MHz signal band (instead of 2GHz with a 1st-order single-bit ΣΔM).

Note that, once the conditions in eq(1.39) are applied to the integrator weights of a 2nd-order ΣΔM, weights g_1 and g_2 still remain as free parameters. The proper selection of these free coefficients involves important design concerns, that are not covered by the linear model. The most important are:

- Keeping the state variables (integrator outputs) bounded to ensure the modulator stability. The 2nd-order ΣΔM is stable for inputs in the range $[-0.9\Delta/2, +0.9\Delta/2]$ if $g_2' > 1.25 g_1' g_2$, regardless the quantizer gain [Candy85]. This condition is met using $g_2' = 2 g_1' g_2$ —see eq(1.39).

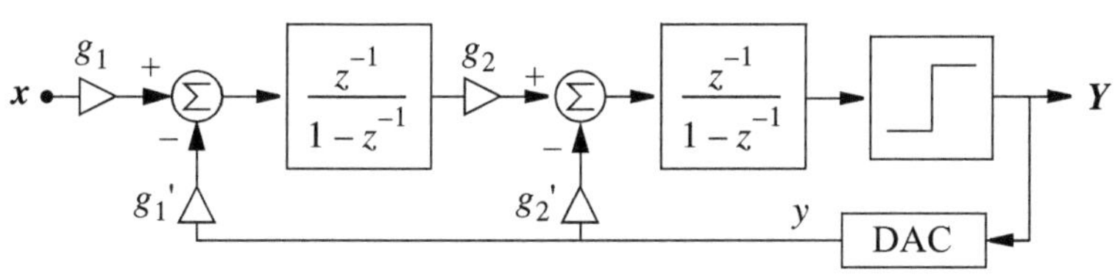

FIGURE 1.19 2nd-order ΣΔ modulator.

- Maximizing the overload level X_{OL} of the modulator to ensure a high SNR_{peak}—see Fig. 1.13.
- Minimizing the required signal range at the integrator outputs; i.e., the required integrator output swing OS must be, first, physically achievable within the intended supply, and second, as low as possible to reduce the power consumption and easy the design.
- Simplifying the circuit implementation of the final set of coefficients. Since integrator weights are realized in SC implementations as capacitor ratios using unit elements, a set of coefficients requiring a reduced number of unit capacitors leads to a silicon area saving.

In general, selecting the coefficients of a ΣΔM involves solving several trade-offs among architectural, circuital, and technological aspects of the practical implementation †8. Table 1.2 shows some sets of coefficients reported in literature. All exhibit an overload level $X_{OL} \approx -4\text{dBFS}$ (-4dB below the full-scale amplitude, $\Delta/2$). The required integrator output swing (relative to Δ) and the minimum number of unit capacitors are also included. Capacitor sharing between weights in the same integrator has been considered.

TABLE 1.2 Some of the reported coefficients for 2nd-order single-bit ΣΔMs.

Weights	[Boser88]	[Yin94b]	[Mede98a]	[Marq98b]
g_1, g_1'	0.5, 0.5	0.25, 0.25	0.25, 0.25	1/3, 1/3
g_2, g_2'	0.5, 0.5	0.5, 0.25	1, 0.5	0.6, 0.4
OS/Δ (at -4dBFS)	1.5	0.75	1.25	1.0
Unit capacitors	6 (3 + 3)	10 (4 + 6)	9 (5 + 4)	20 (4 + 16)

Besides the increased dynamic range, the use of two integrators in this modulator also contributes to a better decorrelation of the quantization error from the input signal. This is illustrated in the pattern noise in Fig. 1.20*a*, where the presence of peaks is notably reduced in comparison with a 1st-order ΣΔM. Also, quantization error in the stable input range is better approximated by the linear model (-97.2dB using eq(1.40) in this example). This makes SNR curves less dependent on the input DC-level and reduces the presence of idle tones in the output spectrum, as shown in Fig. 1.20*b* and Fig. 1.20*c*, respectively.

In fact, the decorrelation between the quantization error and the input signal increases with the modulator order. This, together with circuit noise acting as a dithering signal in practical implementations, greatly helps to palliate the coloration of quantization error.

---

8. The optimum selection for a given application may not apply in a different scenario.

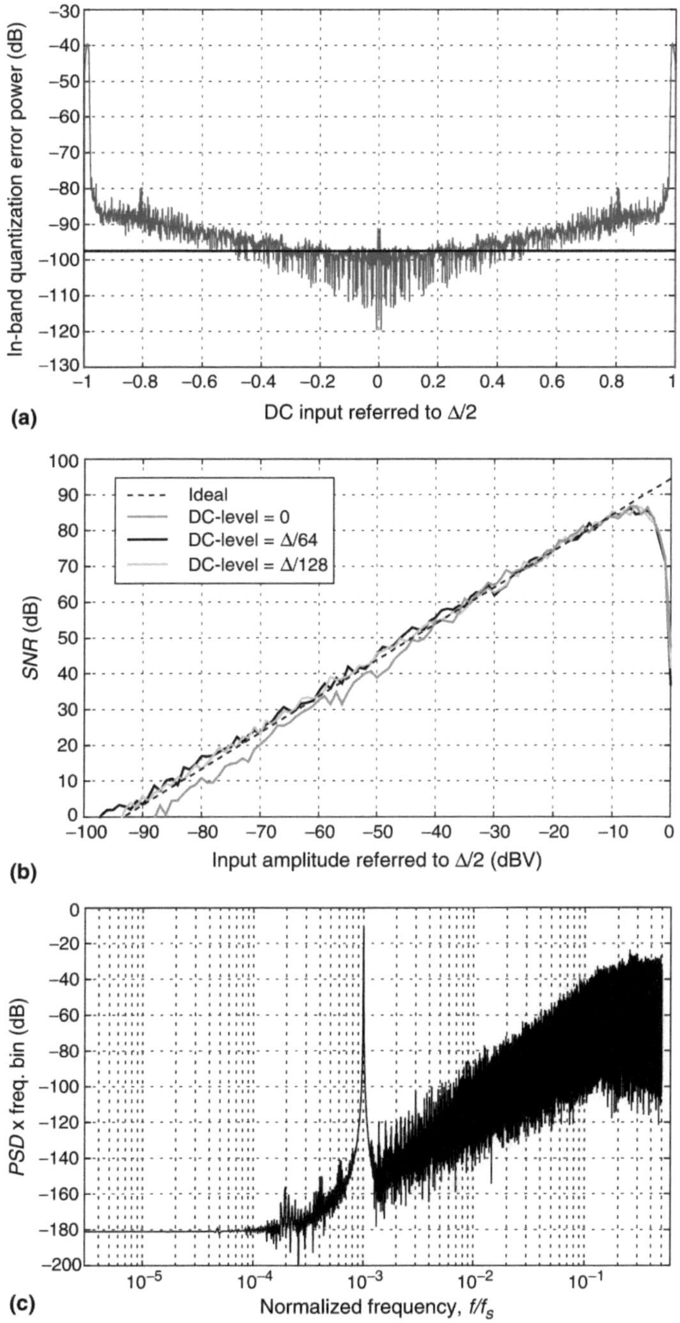

FIGURE 1.20 2nd-order single-bit ΣΔM: (a) Pattern noise for DC inputs in the range $\pm\Delta/2$, (b) Effect on *SNR* curves for different DC levels, (c) Tonal behavior of the output spectrum for a half-scale input sinewave with DC-level $= 0$. (Weights in [Boser88], $OSR = 128$, and $\Delta = 2$).

1.3.3 High-order ΣΔ modulators

Stability concerns

The simplest way to extend a ΣΔM towards an arbitrary Lth-order filtering consists of including L integrators before the quantizer [Ritc77]. Extending the 2nd-order ΣΔM in Fig. 1.19, the architecture in Fig. 1.21 can be obtained, which is often called *Lth-order single-loop ΣΔM with distributed feedback.* Using a linear model, its output would yield

$$Y(z) = \frac{g_1}{g_1'} \cdot z^{-L}X(z) + (1 - z^{-1})^L E(z) \tag{1.42}$$

if a set of relationships among the analog coefficients is fulfilled, as for 1st- and 2nd-order ΣΔMs—see equations (1.34) and (1.39). The in-band quantization error and the dynamic range would ideally yield

$$P_Q \approx \frac{\Delta^2}{12} \cdot \frac{\pi^{2L}}{(2L+1)OSR^{2L+1}} \tag{1.43}$$

$$DR|_{\text{dB}} \approx 10\log_{10}\left[\frac{3}{2} \cdot \frac{(2L+1)OSR^{2L+1}}{\pi^{2L}}\right] \tag{1.44}$$

achieving a high DR for large L, even for low OSR—see Fig. 1.15.

However, this performance is not achievable in practice because ΣΔMs with *pure differentiator* NTFs—i.e., FIR filters like $(1 - z^{-1})^L$—are prone to instability if $L > 2$, exhibiting unbounded states and poor SNR in comparison with that predicted by the linear model. In general, instability appears at the modulator output as a large-amplitude low-frequency oscillation, leading to long strings of alternating $+1$'s and -1 's [Adams97a].

This tendency to instability can be qualitatively explained as follows [Adams97a]. For a modulator to be stable, the input to the quantizer must not be allowed to become too large. Since the quantizer input is given by

$$I(z) = STF(z)X(z) + [NTF(z) - 1]E(z) \tag{1.45}$$

the gain of $NTF(z) - 1$, or simply $NTF(z)$, must not be too large. However, note from Fig. 1.14 that the gain of noise transfer functions of the form

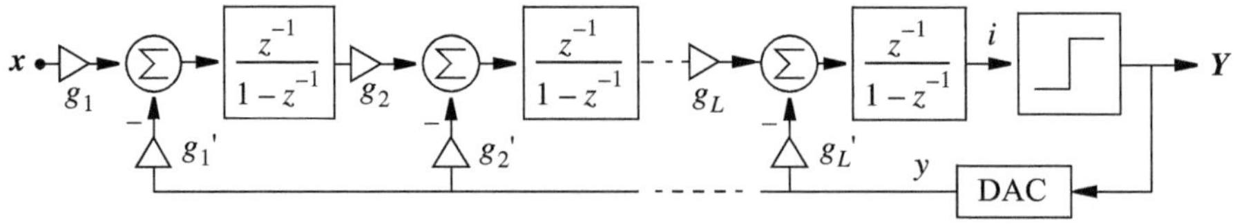

FIGURE 1.21 Lth-order single-loop ΣΔM with distributed feedback.

$(1-z^{-1})^L$ rapidly increases in the high-frequency region if $L > 2$, having a maximum $\|NTF\|_\infty = \max[NTF(z)] = 2^L$ at $z = -1$ ($f = f_s/2$).

Unlike 2nd-order ΣΔMs, for which a stability condition can be extracted, the determination of exact conditions to guarantee stable higher-order designs is still an open question. Some attempts have been reported to mathematically extract them [Good95], but they result in extremely complex expressions that cannot be generalized. In [OptE90] it is shown, using behavioral simulations, that high-order ΣΔMs are *conditionally stable*; i.e., with proper selection of the scaling coefficients, a stable operation can be obtained for inputs confined to a certain range and for certain initial conditions of the state variables. However, despite the absence of general stability conditions, high-order ΣΔMs have been successfully designed since the late 1980s.

Optimized *NTFs*

The first published work stating the viability of stable high-order ΣΔMs can possibly be found in [Lee87b]. Lee and Sodini proposed the architecture shown in Fig. 1.22, using multiple feedforward and feedback paths. With this topology—often called *interpolative*—, thanks to the large set of analog coefficients, more complex high-pass NTFs can be built with sufficiently low gain at the high-frequency region. Assuming a z^{-1} delay in the quantizer, the Lee-Sodini ΣΔM achieves the following noise transfer function

$$NTF(z) = \frac{z^{-1}\left[1 - \sum_{i=1}^{L} B_i\left(\frac{z^{-1}}{1-z^{-1}}\right)^i\right]}{1 - \sum_{i=1}^{L} B_i\left(\frac{z^{-1}}{1-z^{-1}}\right)^i + z^{-1}\sum_{i=0}^{L} A_i\left(\frac{z^{-1}}{1-z^{-1}}\right)^i}$$
$$= \frac{(z-1)^L - \sum_{i=1}^{L} B_i(z-1)^{L-i}}{z\left[(z-1)^L - \sum_{i=1}^{L} B_i(z-1)^{L-i}\right] + \sum_{i=0}^{L} A_i(z-1)^{L-i}} \quad (1.46)$$

To gain insight of the system it is useful to first consider that the feedback coefficients B_i are set to zero. In this case, the following IIR NTF is obtained

$$NTF(z) = \frac{(z-1)^L}{D(z)} \quad (1.47)$$

where $D(z)$ is a polynomial determined by the feedforward coefficients A_i. Note from eq(1.47) that NTF has all zeros located at $z = 1$ (DC). Therefore, coefficients A_i can be adjusted to build a high-pass Butterworth or

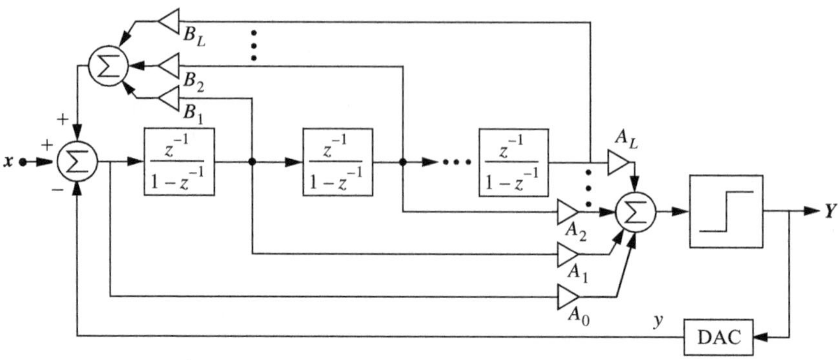

FIGURE 1.22 Lee-Sodini Lth-order ΣΔ modulator.

Chebyshev filter for NTF, with cutoff frequency beyond the signal band and approximately flat gain in the filter pass-band. For the ΣΔM to remain stable while providing a high SNR_{peak}, this gain must be adjusted to satisfy †9

$$\|NTF\|_\infty = \max[NTF(z)] \sim 1.5 \tag{1.48}$$

However, with all zeros at DC, NTF rises monotonically in the signal band like an Lth-order function, so that the PSD at the end of the signal band will practically determine the total in-band error power. If the feedback coefficients B_i are non-zero, but small in comparison with A_i, the position of the zeros of NTF can be controlled, whereas its poles will still be mostly controlled by coefficients A_i—see eq(1.46). Thus, NTF can be improved by placing notches in the signal band for further shaping of the quantization error, while preserving its flat out-of-band gain and, therefore, the modulator stability.

In [Lee87b] the zeros are fixed to obtain a NTF with equal-ripple response over the signal band, but other alternatives are also feasible. Indeed, the approach in [Schr93] leads to the optimal placement of the complex-conjugate

$$\min\left[\int_0^{f_b} |NTF(f)|^2 df\right] \Rightarrow \min\left[\begin{cases} \displaystyle\int_0^1 \prod_{i=1}^{L/2} (f^2 - f_{z_i}^2)^2 df, & L \text{ even} \\ \displaystyle\int_0^1 f^2 \prod_{i=1}^{(L-1)/2} (f^2 - f_{z_i}^2)^2 df, & L \text{ odd} \end{cases}\right] \tag{1.49}$$

9. This empirical condition has been derived from many stable high-order designs and is commonly accepted as a rule-of-thumb for flat-topped high-pass NTFs [Adams97a]. Although other stability criteria have been suggested—e.g., $\|NTF\|_\infty < 2$ [Lee87a], known as *Lee's rule*, and $\|NTF\|_2 = \text{rms}[NTF(z)] < \sqrt{3}$ [Agra83]—, as demonstrated in [Schr93], none of them can ensure the stability of an arbitrary high-order single-bit ΣΔM. For the time being, computer simulation is still the most reliable method to verify stability.

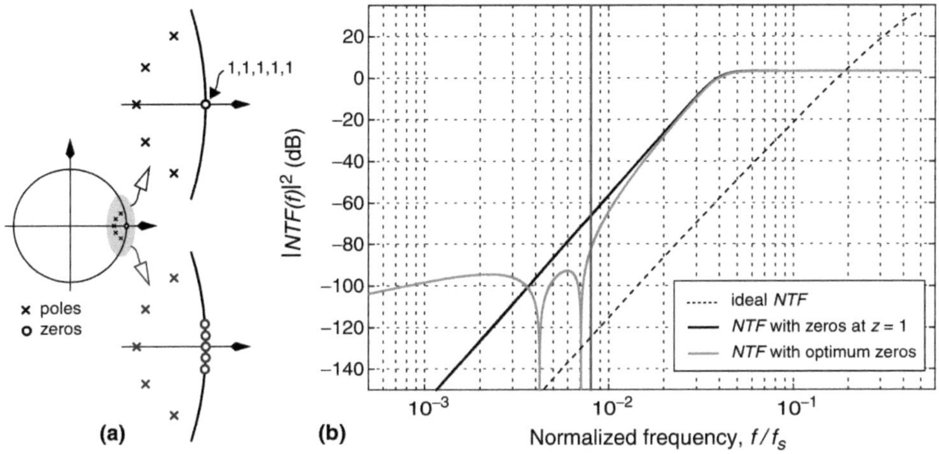

FIGURE 1.23 Comparison of different implementations of a 5th-order NTF: (a) Illustration of the pole-zero placement in the unit circle, (b) Magnitude response of the corresponding NTFs. ($OSR = 64$ is assumed for the design of the Butterworth poles).

zeros for minimizing the total in-band error power, by solving the problem as a function of f_{z_i}, which represents the location of the complex zeros times f_b. The solutions for eq(1.49) up to $L = 8$ can be found in [Schr93].

Fig. 1.23 illustrates these alternatives for implementing stable high-order NTFs. A 5th-order ΣΔM with $OSR = 64$ is considered in the example. Note from Fig. 1.23*a* that the position of the poles has been fixed following a Butterworth configuration, so that the cutoff frequency of NTF is beyond the signal band—vertical line in Fig. 1.23*b*—and its out-of-band gain is 1.5 (3.5dB) †10. Just by slightly moving four of the zeros at $z = 1$ along the unit circle into complex-conjugate positions, two notches are introduced in the signal band. As shown in Fig. 1.23*b*, this considerably reduces $|NTF(f)|$ at its upper edge and, therefore, the in-band error power. For the case considered, P_Q is reduced in 18dB by optimally spreading the zeros over the band.

Fig. 1.24 summarizes results obtained with these two approaches, showing the maximum SNR achievable by single-bit ΣΔMs of various orders against OSR. The theoretical performance for pure differentiator NTFs is included for comparison purposes. Note that in both cases the obtained SNR is much lower than that of the ideal case. Nevertheless, as shown in Fig. 1.24*b*, the optimal placement of zeros in the signal band can lead to stable high-order ΣΔMs with high SNR at moderate oversampling ratios ($OSR \sim 32, 64$).

---

10. At this point—i.e., in eq(1.47)—the noise transfer function must be designed to fulfill $NTF(\infty) = ntf(0) = 1$ for the practical realization of the corresponding IIR filter.

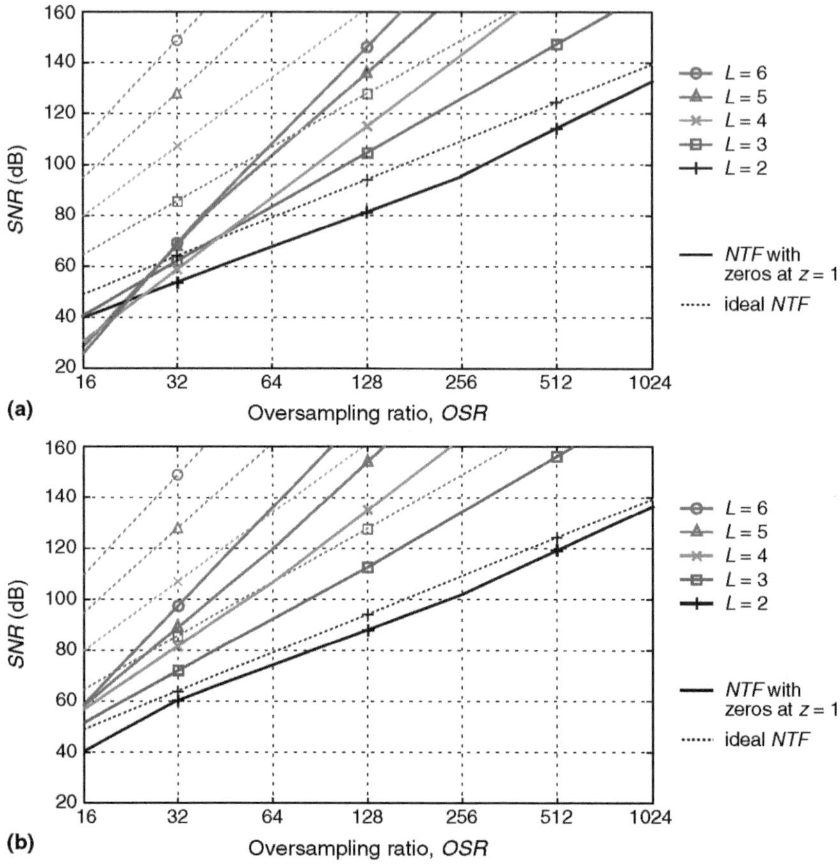

FIGURE 1.24 Maximum *SNR* achievable by *L*th-order single-loop ΣΔMs versus *OSR*: (a) *NTF* with all zeros at $z = 1$, (b) *NTF* with zeros optimally spread over the signal band. (Data taken from [Schr93]).

High-order topologies

Apart from the Lee-Sodini ΣΔM illustrated in Fig. 1.22, many other modulator topologies have been developed for implementing high-order *NTF*s with the characteristics of most common IIR filter families. A design procedure to obtain stable *NTF*s can be found in [Adams97a], whereas [Adams97b] offers a detailed overview of different topological alternatives for the modulator implementation. Most of them use multiple weighted feedback or feedforward paths (or both). The two modulators in Fig. 1.25 can be considered typical examples.

Fig. 1.25*a* illustrates a topology with feedforward summation of the integrator outputs before the quantizer. If $\gamma_i = 0$, the topology builds a high-pass *NTF* with all zeros at DC. By adding small negative feedback terms around pairs of integrators in the loop filter, pairs of zeros can be moved

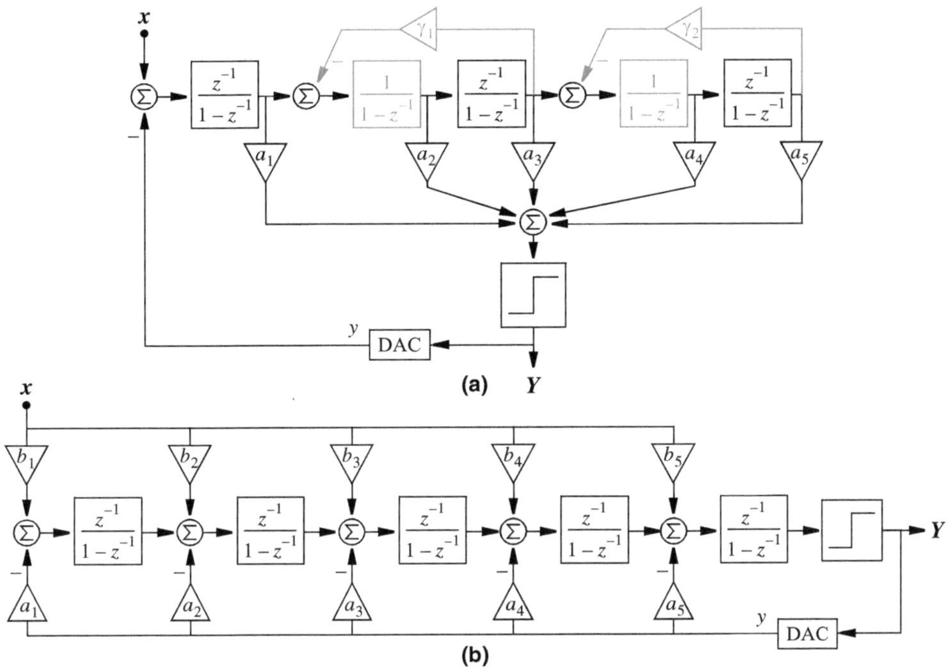

FIGURE 1.25 Illustration of high-order topologies: (a) 5th-order ΣΔM with feedforward summation and local resonator feedbacks, (b) 5th-order ΣΔM with distributed feedback and distributed feedforward input paths.

along the unit circle to create notches in $|NTF(f)|$ at frequencies †11

$$\Theta_i = \frac{2\pi f_i}{f_s} \approx \sqrt{\gamma_i} \tag{1.50}$$

Once NTF is set for the desired noise-shaping, this topology fixes the signal transfer function to $STF(z) = 1 - NTF(z)$. A drawback may arise if Butterworth poles are chosen, since STF will contain some peaking at high frequencies, what may jeopardize stability.

If a certain degree of freedom is desired in specifying both NTF and STF, the architecture in Fig. 1.25*b* can be used; it includes distributed feedback and feedforward paths. In this topology the zeros of STF can be fixed with coefficients b_i, without affecting the pole placement †12. Local resonator feedbacks can be also included to set notches in $|NTF(f)|$.

11. Note that one of the integrators has a z^{-1} delay in the numerator, whereas the other does not. A slightly less effective resonator would be built if both integrators had a z^{-1} delay, since zeros would be moved vertically from the $(1, j0)$ point.

12. Note from Fig. 1.25*b* that a modulator equivalent to that in Fig. 1.21 is obtained if all b_i coefficients are zero, except for b_1. In this case, NTF will be also given by eq(1.47), whereas STF will implement an all-pole low-pass filter—$STF(z) = b_1/D(z)$.

However, a common drawback of most high-order topologies implementing IIR filters for NTF is the increased circuit complexity due to the many feedback and/or feedforward coefficients required. Moreover, some of these coefficients can be considerably small [Chao90][Adams97b] [Kuo99], leading to large capacitor ratios in the SC implementation and, therefore, to increased area and power consumption.

Other alternatives exist for designing high-order ΣΔMs, but their underlying methodology is far from the design of optimized NTFs described above—i.e., the mapping of filter families onto particular modulator topologies. Procedures based on behavioral simulations have been presented for the selection of the scaling coefficients of high-order ΣΔMs with distributed feedback [OptE90] [Marq98b]. The approach basically consists of extensively exploring the design space searching for sets of integrator weights easy to implement that maximize the modulator SNR, while achieving a reasonable overload level. Obviously, high-pass IIR NTFs are obtained with zeros at DC, but the obtained placement of poles does not follow any particular configuration, because the selection of coefficients is not faced from a discrete filter design perspective.

Non-linear stabilization techniques

High-order modulators are only conditionally stable, so that instabilities may appear for inputs exceeding certain bounds or for certain initial conditions of the state variables, although having optimized the modulator NTF and STF. Two non-linear techniques are often used to ensure global stable operation:

- *State-variable clipping*: identifying the maximum amplitudes of the integrator outputs during stable operation of the modulator and including limiters at the integrators to preclude the outputs from being larger than these maxima [OptE90].
- *Integrator resetting*: forcing the integrators to zero or some other initial condition at power-up of the system or when unstable operation is detected [OptE91] [Mous94]. The detection of instability can be done at the integrator level, by placing comparators to determine if a state variable has surpassed a certain limit, or by monitoring the length of strings of consecutive pulses at the modulator output. The reset itself can be global (for all the integrators), or local (for some selected integrators).

In any case, as will be shown in Section 1.7, precluding instabilities in high-order single-loop ΣΔ modulators leads to attainable values of SNR that are considerably far from ideal.

1.4 Cascade ΣΔ Architectures

As stated in the previous section, high-order filtering using single-loop ΣΔMs suffers from instability problems and a considerable performance degradation compared to an ideal modulator. An alternative to circumvent instabilities while obtaining high-order shaping can be found in the so-called *MASH* (multi-stage noise-shaping) ΣΔ *modulators*, often referred to as *cascade* or *multi-stage* ΣΔ *modulators* [Mats87] [Longo88] [Chou89] [Rebe90].

Fig. 1.26 shows the generic architecture of a cascade ΣΔM. It consists of several stages of ΣΔ modulators, in which each stage re-modulates a signal containing the quantization error generated in the previous stage. Once in the digital domain, the outputs Y_i of the stages are properly processed and combined in order to cancel out the quantization errors of all the stages, but the last one in the cascade. This error appears at the overall modulator output shaped by a function of order equal to the summation of the order of all the stages. Furthermore, since all feedback loops are local and there is no interstage feedback, unconditionally stable high-order shaping can be obtained provided that only 1st- and/or 2nd-order ΣΔMs are cascaded. The performance of a multi-stage ΣΔM is therefore similar to that of an ideal high-order loop without instability problems.

The concept of cascade ΣΔMs is a priori extensible to whatever number of stages [†13] and also applies if the stages are simply quantizers; i.e., with no loop filter (0-order ΣΔMs) [Lesl90] [Dias93] [Broo97].

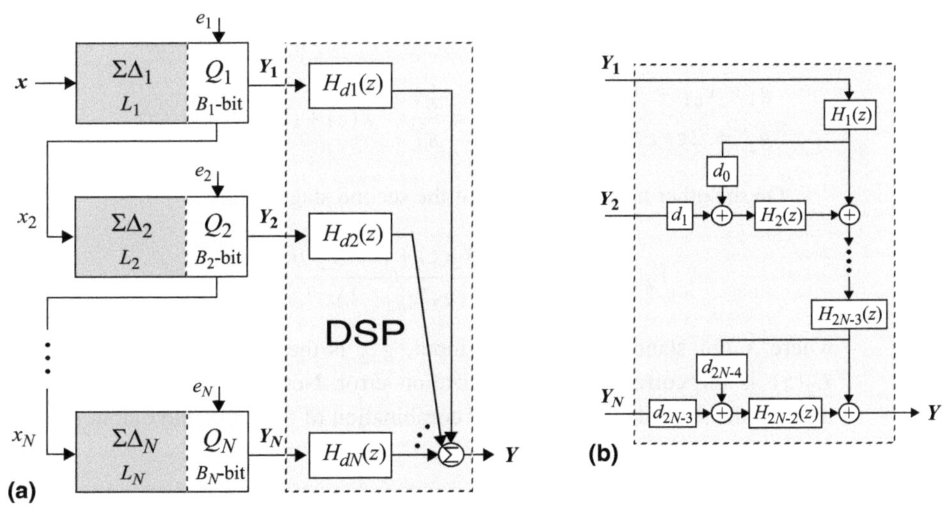

FIGURE 1.26 (a) Generic N-stage cascade ΣΔ modulator; (b) Digital signal processing (DSP) for the cancellation of quantization errors as it is often structured.

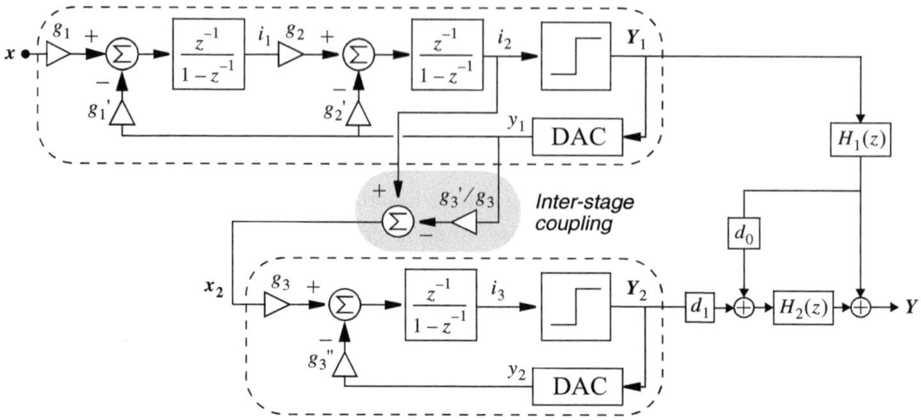

FIGURE 1.27 3rd-order 2-stage cascade ΣΔ modulator (2-1 ΣΔM).

The correct operation of cascade ΣΔMs relies in the adequate digital processing of the stages outputs. In order to illustrate its effect, let us consider the 3rd-order cascade in Fig. 1.27. The first stage is a 2nd-order ΣΔM and the second stage is a 1st-order one (both in dashed boxes), so that we can refer to the resulting topology as the 2-1 cascade ΣΔM [Longo88] [Ribn91a] [Will91].

Using linear analysis, the output of the first stage in the z-domain yields

$$Y_1(z) = \frac{g_1 g_2 g_{q1} z^{-2} X(z) + (1 - z^{-1})^2 E_1(z)}{1 + (g_2' g_{q1} - 2) z^{-1} + (1 + g_1' g_2 g_{q1} - g_2' g_{q1}) z^{-2}} \tag{1.51}$$

where g_{q1} is the quantizer gain and $E_1(z)$ its quantization error. The following relationships can be derived for a 2nd-order shaping at the 1st-stage output:

$$\left.\begin{aligned} g_1' g_2 g_{q1} &= 1 \\ g_2' &= 2 g_1' g_2 \end{aligned}\right\} \Rightarrow Y_1(z) = \frac{g_1}{g_1'} z^{-2} X(z) + (1 - z^{-1})^2 E_1(z) \tag{1.52}$$

On the other hand, the output of the second stage yields

$$Y_2(z) = \frac{g_3 g_{q2} z^{-1} X_2(z) + (1 - z^{-1}) E_2(z)}{1 + (g_3'' g_{q2} - 1) z^{-1}} \tag{1.53}$$

where $X_2(z)$ stands for the stage input, g_{q2} is the gain of its quantizer, and $E_2(z)$ is the corresponding quantization error. Note from Fig. 1.27, that the input to the second stage is a linear combination of the input and output signals of the 1st-stage quantizer, given by:

13. As we show in Chapter 2, the number of stages is limited in practice by circuit non-idealities (e.g., mismatch), that cause incomplete cancellation of low-order quantization errors at the modulator output. This effect is called *noise leakage*.

$$X_2(z) = I_2(z) - \frac{g_3'}{g_3} Y_1(z) \tag{1.54}$$

In order to obtain from eq(1.53) a 1st-order shaping of the 2nd-stage quantization error, the following condition must be fulfilled:

$$g_3'' g_{q2} = 1 \quad \Rightarrow \quad Y_2(z) = \frac{g_3}{g_3''} z^{-1} X_2(z) + (1 - z^{-1}) E_2(z) \tag{1.55}$$

Under conditions in equations (1.52) and (1.55), $X_2(z)$ can be written as

$$X_2(z) = \frac{Y_1(z) - E_1(z)}{g_{q1}} - \frac{g_3'}{g_3} Y_1(z) = \left(g_1' g_2 - \frac{g_3'}{g_3}\right) Y_1(z) - g_1' g_2 E_1(z) \tag{1.56}$$

and substituting the former equation in eq(1.55), $Y_2(z)$ yields

$$\begin{aligned} Y_2(z) &= \frac{g_1}{g_1'}\left(\frac{g_1' g_2 g_3}{g_3''} - \frac{g_3'}{g_3''}\right) z^{-3} X(z) + (1 - z^{-1}) E_2(z) \\ &+ \left[\left(\frac{g_1' g_2 g_3}{g_3''} - \frac{g_3'}{g_3''}\right)(1 - z^{-1})^2 - \frac{g_1' g_2 g_3}{g_3''}\right] z^{-1} E_1(z) \end{aligned} \tag{1.57}$$

involving the input signal, 1st-order shaped quantization error from the second stage, and both shaped and unshaped quantization error from the first stage.

The output of the stages can be processed in the digital domain in order to provide an overall output in which the presence of $E_1(z)$ is cancelled out. To that purpose, the modulator output is obtained as

$$Y(z) = H_{d1}(z) Y_1(z) + H_{d2}(z) Y_2(z) \tag{1.58}$$

with functions $H_{d1}(z)$ and $H_{d2}(z)$ given by:

$$\begin{aligned} H_{d1}(z) &= H_1(z)[1 + d_0 H_2(z)] = z^{-1}\left[1 + \left(\frac{g_3'}{g_1' g_2 g_3} - 1\right)(1 - z^{-1})^2\right] \\ H_{d2}(z) &= d_1 H_2(z) = \frac{g_3''}{g_1' g_2 g_3}(1 - z^{-1})^2 \end{aligned} \tag{1.59}$$

The blocks of the cancellation logic are then given by

$$d_0 = \frac{g_3'}{g_1' g_2 g_3} - 1 \qquad d_1 = \frac{g_3''}{g_1' g_2 g_3} \qquad H_1(z) = z^{-1} \qquad H_2(z) = (1 - z^{-1})^2 \tag{1.60}$$

in order to provide the following modulator output:

$$Y(z) = \frac{g_1}{g_1'} z^{-3} X(z) + d_1 (1 - z^{-1})^3 E_2(z) \tag{1.61}$$

Note from eq(1.61) that $Y(z)$ contains only a delayed version of the modulator input and a 3rd-order shaped version of the 2nd-stage quantization error, whereas the transfer function of the 1st-stage quantization error is nulled thanks to the cancellation logic. Thus, the output of the 2-1 cascade ΣΔM yields

$$Y(z) = STF(z)X(z) + NTF_2(z)E_2(z) \Rightarrow \begin{cases} STF(z) \propto z^{-3} \\ NTF_1(z) = 0 \\ NTF_2(z) = d_1(1-z^{-1})^3 \end{cases} \tag{1.62}$$

so that its performance is similar to that of an ideal ΣΔM with a 3rd-order FIR noise transfer function, but unconditionally stable by construction. The in-band quantization error power of the 2-1 cascade ΣΔM is therefore

$$P_Q \approx d_1^2 \cdot \frac{\Delta_2^2}{12} \cdot \frac{\pi^6}{7OSR^7} \tag{1.63}$$

where Δ_2 is the quantization step of the 2nd-stage quantizer. Note that the performance would equal that of an ideal 3rd-order ΣΔM if $d_1 = 1$, but it will be lowered if $d_1 > 1$, because of the amplification of the quantization error by the factor d_1^2. Unfortunately, d_1 equals $g_3''/g_1'g_2g_3$, so that reducing d_1 involves increasing the integrator weights, what can lead to an excessively large swing of the internal state variables and/or a premature overload of the quantizers. Thus, a trade-off must be established among minimizing the excess of in-band quantization error and maximizing the overload level and SNR_{peak} of the cascade ΣΔM.

In general, for a N-stage cascade ΣΔM like the one shown in Fig. 1.26, a stable Lth-order ΣΔM can be built by cascading stages of order $L_i \leq 2$. If the stages outputs are adequately processed in the digital domain, only the modulator input signal $X(z)$ and the last-stage quantization error $E_N(z)$ remain in the z-domain modulator output, yielding

$$\begin{aligned} Y(z) &= STF(z)X(z) + NTF_N(z)E_N(z) \\ &= z^{-L}X(z) + d_{2N-3}(1-z^{-1})^L E_N(z) \end{aligned} \tag{1.64}$$

where $L = L_1 + L_2 + \ldots + L_N$ and d_{2N-3} is the *scaling factor* related to the integrator weights that amplifies the last-stage quantization error. The in-band quantization error power of a N-stage cascade is then given by

$$P_Q \cong d_{2N-3}^2 \cdot \frac{\Delta_N^2}{12} \cdot \frac{\pi^{2L}}{(2L+1)OSR^{(2L+1)}} \tag{1.65}$$

with Δ_N being the level spacing in the B_N-bit quantizer of the Nth stage.

Hence, the performance corresponds to that of an ideal Lth-order B_N-bit ΣΔM, except for the scalar d_{2N-3} that causes a systematic loss of performance. Common values for this amplifying factor are 2 and 4, which lead to decreases in the ideal DR of 6dB (1bit) and 12dB (2bit), respectively. However, these performance degradations, inherent to multi-stage ΣΔMs, are considerably lower than the ones shown in Section 1.3 for high-order single-loop architectures—see Fig. 1.24. Moreover, in the case of multi-stage ΣΔMs, they are independent of OSR.

This fact has favoured the development of a great number of cascade ΣΔMs. Some of these topologies will be shown next, including in each case the relationships among integrator weights (analog coefficients) and *error cancellation logic* coefficients (digital blocks) that must be fulfilled for a proper performance of the architecture.

Fig. 1.28 shows an alternative representation of the 2-1 cascade ΣΔM [Longo88] [Ribn91a] [Will91] already considered, which is more compact than the one in Fig. 1.27. The relationships obtained during the former analysis are summarized in Table 1.3. Note that other 3rd-order cascade ΣΔMs can be constructed using two stages (1-2 ΣΔM) or three stages (1-1-1 ΣΔM). In fact, the 1-1-1 cascade ΣΔM [Mats87] [Rebe90] was the first multi-stage modulator reported in literature. However, several reasons preclude the use of a 1st-order modulator as first stage of the cascade [Rebe97]. First, the quantization error from the first stage, to be cancelled out, will be 1st-order shaped, instead of 2nd-order shaped. This means that more in-band error needs to be annulled. Second, if the cancellation of the 1st-stage quantization error is not complete, the

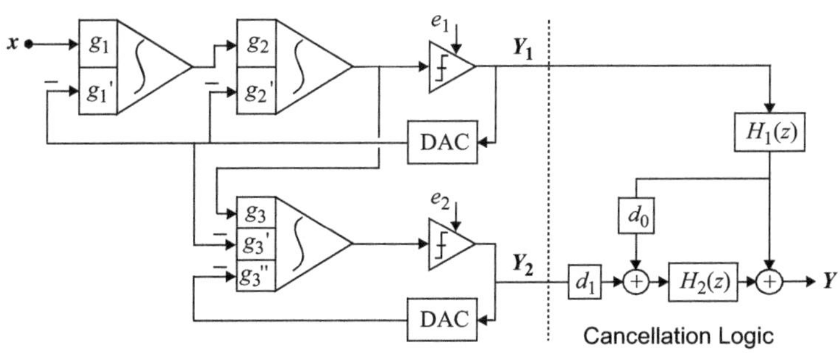

FIGURE 1.28 Compact representation of the 3rd-order 2-stage 2-1 ΣΔM.

TABLE 1.3 Relationships to be fulfilled for error cancellation in the 2-1 ΣΔM.

Analog	Digital	
$g_2' = 2g_1'g_2$	$d_0 = \frac{g_3'}{g_1'g_2g_3} - 1$	$H_1(z) = z^{-1}$
	$d_1 = \frac{g_3''}{g_1'g_2g_3}$	$H_2(z) = (1-z^{-1})^2$

uncancelled error leaking to the output (*noise leakage*) will be larger. Finally, the tonal behavior of the 1st-order first stage menaces the performance of the cascade. Note that the latter does not apply if a 1st-order stage is used as second or successive stages. In that case, the input to the 1st-order stage will contain quantization error from the previous one, what helps to prevent tones, because the input to the quantizer will be much more random.

Bearing this in mind, practical 4th-order cascade ΣΔMs can be built. Fig. 1.29*a* shows a 2-stage topology (2-2 ΣΔM) [Kare90] [Baher92], whereas Fig. 1.29*b* shows a 3-stage one (2-1-1 ΣΔM) [Yin94b].

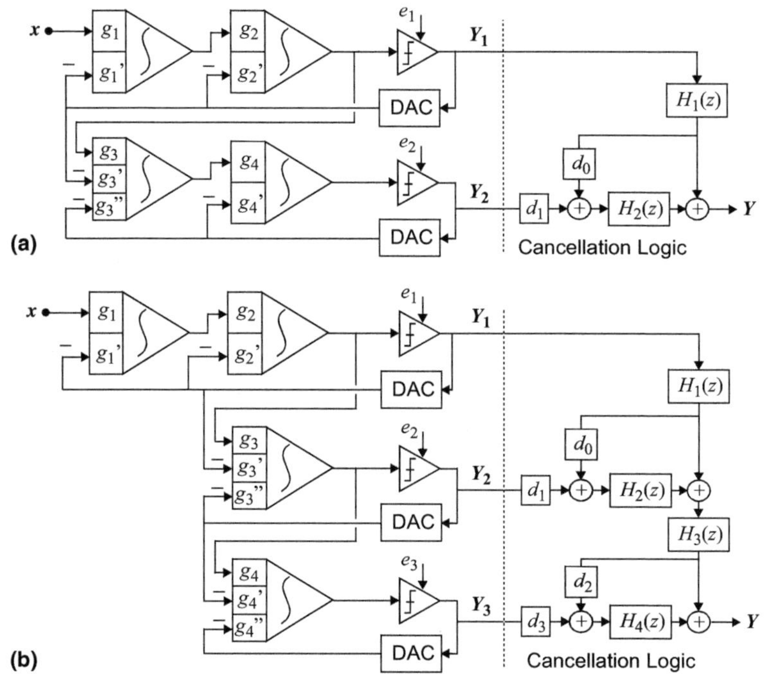

FIGURE 1.29 4th-order cascade ΣΔ modulators: (a) 2-2 ΣΔM, (b) 2-1-1 ΣΔM.

TABLE 1.4 Relationships to be fulfilled for error cancellation in the 2-2 and the 2-1-1 ΣΔM.

2-2 ΣΔM		
Analog	Digital	
$g_2' = 2g_1'g_2$	$d_0 = \frac{g_3'}{g_1'g_2g_3} - 1$	$H_1(z) = z^{-2}$
$g_4' = 2g_3''g_4$	$d_1 = \frac{g_3''}{g_1'g_2g_3}$	$H_2(z) = (1-z^{-1})^2$

2-1-1 ΣΔM		
Analog	Digital	
$g_2' = 2g_1'g_2$	$d_0 = \frac{g_3'}{g_1'g_2g_3} - 1$	$H_1(z) = z^{-1}$
$g_4' = g_3''g_4$	$d_1 = \frac{g_3''}{g_1'g_2g_3}$	$H_2(z) = (1-z^{-1})^2$
	$d_2 = 0$	$H_3(z) = z^{-1}$
	$d_3 = \frac{g_4''}{g_1'g_2g_3g_4}$	$H_4(z) = (1-z^{-1})^3$

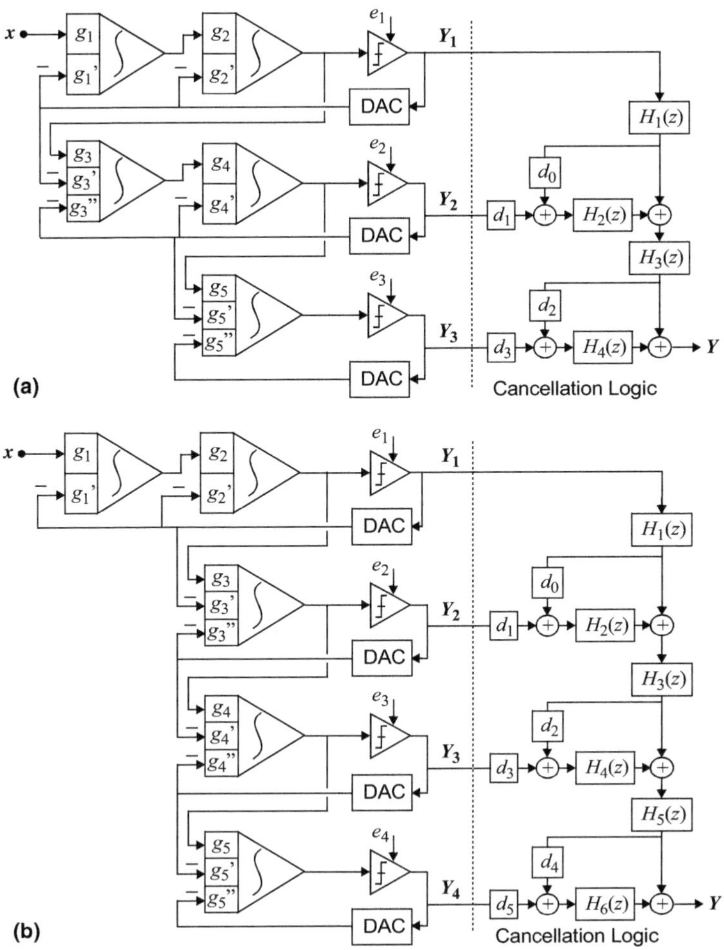

FIGURE 1.30 5th-order cascade ΣΔ modulators: (a) 2-2-1 ΣΔM, (b) 2-1-1-1 ΣΔM.

TABLE 1.5 Relationships to be fulfilled for error cancellation in the 2-2-1 and the 2-1-1-1 ΣΔM.

2-2-1 ΣΔM		
Analog	Digital	
$g_2' = 2g_1'g_2$	$d_0 = \frac{g_3'}{g_1'g_2g_3} - 1$	$H_1(z) = z^{-2}$
$g_4' = 2g_3''g_4$	$d_1 = \frac{g_3''}{g_1'g_2g_3}$	$H_2(z) = (1-z^{-1})^2$
$g_5' = g_3''g_4g_5$	$d_2 = 0$	$H_3(z) = z^{-1}$
	$d_3 = \frac{g_5''}{g_1'g_2g_3g_4g_5}$	$H_4(z) = (1-z^{-1})^4$

2-1-1-1 ΣΔM		
Analog	Digital	
$g_2' = 2g_1'g_2$	$d_0 = \frac{g_3'}{g_1'g_2g_3} - 1$	$H_1(z) = z^{-1}$
$g_4' = g_3''g_4$	$d_1 = \frac{g_3''}{g_1'g_2g_3}$	$H_2(z) = (1-z^{-1})^2$
$g_5' = g_4''g_5$	$d_2 = 0$	$H_3(z) = z^{-1}$
	$d_3 = \frac{g_4''}{g_1'g_2g_3g_4}$	$H_4(z) = (1-z^{-1})^3$
	$d_4 = 0$	$H_5(z) = z^{-1}$
	$d_5 = \frac{g_5''}{g_1'g_2g_3g_4g_5}$	$H_6(z) = (1-z^{-1})^4$

The former topologies can be easily extended to higher orders. Fig. 1.30 shows two realizations of 5th-order cascades: the 2-2-1 ΣΔM [Vleu01] and the 2-1-1-1 ΣΔM [Rio00]. Finally, the 6th-order 2-2-2 cascade [Dedic94] [Feld98] is illustrated in Fig.1.31.

Note that the expressions in Tables 1.3 to 1.6 for the digital blocks assume that the error cancellation logic is structured according to Fig. 1.26*b*. For other kind of structures, the expressions may be different, but anyhow the digital part will have to fulfil certain relationships with the analog one (cascade of stages).

Once these relationships are fulfilled, the value of the free analog coefficients is fixed for minimum loss of performance and simpler circuit implementation of the cascade ΣΔM. Proper concerns for their selection are:

- Minimizing the scaling factor that determines the loss of resolution.
- Maximizing the overload level X_{OL} to ensure a high peak SNR.
- Minimizing the required output swing of the integrators, especially in low-voltage implementations.
- Simplifying the set of analog coefficients to be easily implemented as capacitor ratios using unit elements.

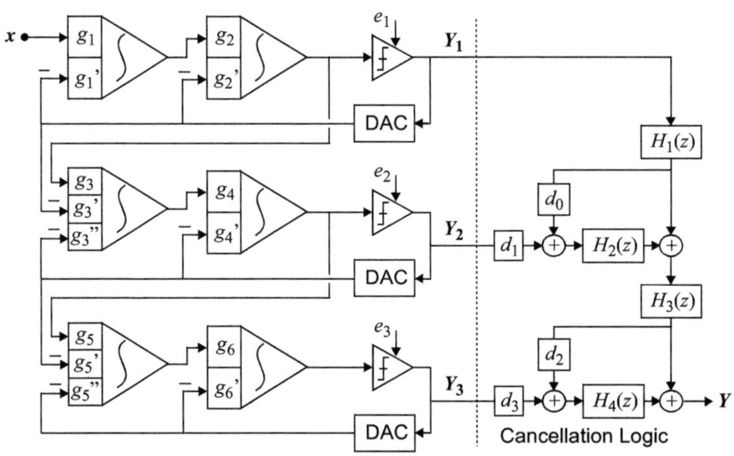

FIGURE 1.31 6th-order 3-stage cascade ΣΔ modulator (2-2-2 ΣΔM).

TABLE 1.6 Relationships to be fulfilled for error cancellation in the 2-2-2 ΣΔM.

Analog	Digital	
$g_2' = 2g_1'g_2$	$d_0 = \frac{g_3'}{g_1'g_2g_3} - 1$	$H_1(z) = z^{-2}$
$g_4' = 2g_3''g_4$	$d_1 = \frac{g_3''}{g_1'g_2g_3}$	$H_2(z) = (1-z^{-1})^2$
$g_5' = g_3''g_4g_5$	$d_2 = 0$	$H_3(z) = z^{-2}$
$g_6' = 2g_5''g_6$	$d_3 = \frac{g_5''}{g_1'g_2g_3g_4g_5}$	$H_4(z) = (1-z^{-1})^4$

- Reducing the total number of unit capacitors to save silicon area.
- Easing the implementation of the error cancellation logic. To that purpose, power-of-two digital coefficients should be chosen, so that the digital scaling can be implemented with simple shift registers.

Obviously, many trade-offs exist among the former issues. Table 1.7 shows some reported coefficients for the 4th-order 2-2 cascade ΣΔM. Related aspects—like nominal performance loss, overload level, and minimum number of unit capacitors—are also included in each case.

TABLE 1.7 Some reported coefficients for 4th-order 2-2 cascade ΣΔMs.

Weights	[Miao98]	[Marq98b]-A	[Marq98b]-B
g_1, g_1'	0.25, 0.25	0.5, 0.5	0.5, 0.5
g_2, g_2'	0.5, 0.25	0.5, 0.5	0.5, 0.5
g_3, g_3', g_3''	0.5, 0.125, 0.25	1, 0.5, 0.5	0.5, 0.25, 0.5
g_4, g_4'	0.5, 0.25	0.5, 0.5	0.5, 0.5
d_0, d_1	1, 4	1, 2	1, 4
ΔDR due to scaling	–12dB (–2bit)	–6dB (–1bit)	–12dB (–2bit)
$X_{OL}/(\Delta/2)$	–2dBFS	–5dBFS	–2dBFS
Unit capacitors	28 (4 + 6 + 12 + 6)	13 (3 + 3 + 4 + 3)	16 (3 + 3 + 7 + 3)

Note that the set of weights due to [Marq98b]-A lead to $d_1 = 2$ and, therefore, to a nominal reduction of the dynamic range of only 6dB (1bit). However, this is done at the expense of a larger coupling between the two stages, what results in a smaller overload level in comparison with [Marq98b]-B and [Miao98]. On the other hand, the implementation of the coefficients requires only 13 unit capacitors, thanks to the larger value of the weights.

Fig. 1.32 shows the *SNR* curves of the three 2-2 ΣΔMs for an oversampling ratio of 32. Note that peak *SNR*s of approximately 91dB are obtained. If we compare this value with the peak *SNR* achieved by a stable 4th-order single-loop ΣΔM with all *NTF*-zeros at DC (Fig. 1.24*a*), the performance is improved in 30dB. The improvement is still around 10dB in comparison with Fig. 1.24*b*, where the four zeros are optimally spread over the signal band [†14].

The potentialities of cascade ΣΔMs have led us to propose a family of cascades that can be easily expanded to any order, while preserving a low systematic loss of resolution (only 1bit) and a high overload level. It comprises

14. Larger improvements can also be obtained in the 2-2 cascade if a local resonator feedback is added in the 2nd-order second stage to move the two zeros of its *NTF* to optimal positions in the baseband [Rito94] [Olia02], but the corresponding weight will be considerably small.

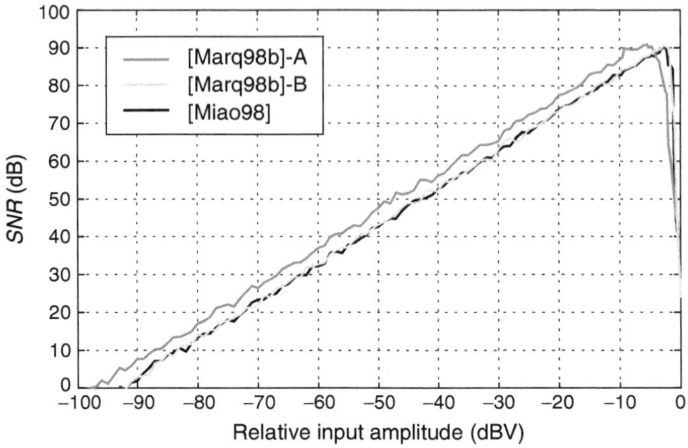

FIGURE 1.32 *SNR* versus relative input amplitude (times the full-scale amplitude $\Delta/2$) of the different 2-2 cascade ΣΔMs in Table 1.7 for $OSR = 32$.

a 2nd-order loop followed by identical 1st-order stages, thanks to an adequate selection of coefficients. Appendix A discusses in detail the proposed topology.

1.5 Multi-Bit ΣΔ Architectures

As stated in previous sections, the dynamic range achieved by a ΣΔM with single-bit internal quantization for given *OSR* can be enhanced by increasing the modulator order. However, these improvements rapidly decrease due to instabilities in single-loop ΣΔ architectures (see Section 1.3.3) or noise leakage in cascade ΣΔ topologies (see Section 1.4). According to eq(1.19), an alternative way to further increase the modulator *DR* is to use multi-bit internal quantization. The main advantages of multi-bit ΣΔ modulators are:

- Smaller internal quantization step and hence smaller in-band error power than in single-bit counterparts. As shown in Section 1.2.5, a 6-dB reduction in the in-band error power is roughly obtained per additional bit in the internal quantizer.
- Weaker internal non-linearity than for single-bit quantizers, so that those phenomena caused by the non-linear dynamics—such as idle tones and noise patterns—are less notorious.
- Better fitting to the additive white noise approximation of the quantization error than for single-bit quantizers.
- Better stability properties than single-bit ΣΔ architectures for a given loop filter order.

In general, the larger the resolution of the multi-bit quantizer, the more noticeable the pros and the closer the modulator SNR to its ideal value.

The latter is illustrated in Fig. 1.33 for high-order single-loop ΣΔMs, showing the attainable SNR_{peak} versus the number of bits in the internal quantizer. Note that if the NTF is optimized for the single-bit case, the SNR_{peak} improves at 6-dB/bit rate. Even so, this choice is far from optimum. In fact, larger improvements are feasible, because the tendency of the internal quantizer to overload decreases as B increases and more aggressive NTFs can be obtained without jeopardizing stability. For instance, for $B > 3$ the optimized 2nd-order ΣΔM features larger SNR_{peak} than the 3rd-order loop optimized for $B = 1$. By re-optimizing the latter around 30-dB SNR enhancement is achieved for $B = 4$. Note from Fig. 1.33 that by re-optimizing the NTFs for each B value, improvements larger than 20dB per extra bit are obtained in certain cases, simply due to the improved stability properties.

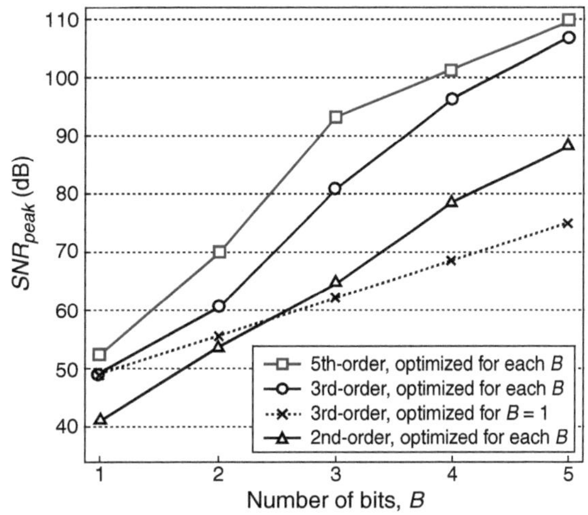

FIGURE 1.33 SNR_{peak} vs. B for different optimized NTFs with $OSR = 16$. (Data taken from [Broo02]).

These considerations suggest that, for a given modulator DR, multi-bit quantization enables the reduction of the oversampling ratio. Hence, multi-bit ΣΔMs are better suited for wideband applications, since the required clock frequency is lower. This helps to reduce the power consumption, not only in the ΣΔ modulator itself, but also in the decimation filter.

Unfortunately, multi-bit quantizers have also important drawbacks that counter the former advantages, among them:

- Multi-bit quantizers require much more analog circuitry and are more difficult to design than single-bit ones.
- On the contrary to 1-bit quantizers—which are intrinsically linear because of the two-level quantization—multi-bit quantizers exhibit in practice some non-linearities, mostly due to component mismatch. As we show immediately on, these errors have a significant impact on modulator performance.

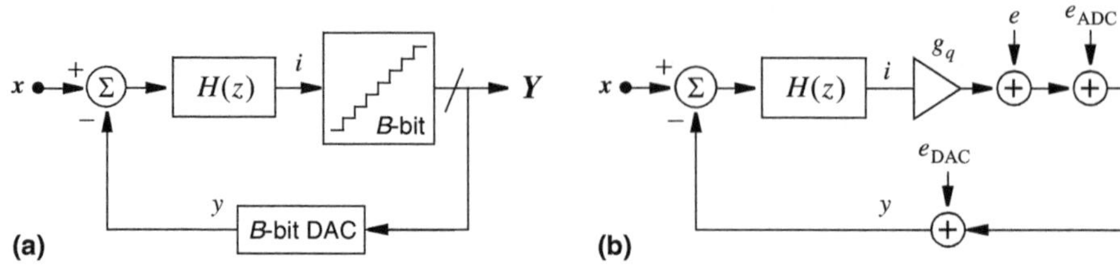

FIGURE 1.34 Multi-bit ΣΔM architecture: (a) Basic scheme, (b) Corresponding linear model considering errors in the internal ADC and DAC.

Influence of DAC errors

The former issue can be clearly understood with the help of Fig. 1.34. In the linear model of the multi-bit ΣΔM in Fig. 1.34*b*, new errors related to the multi-bit conversion are added to the quantization error e: an error e_{ADC} associated to the A-to-D conversion process and an error e_{DAC} in the subsequent D-to-A conversion needed to reconstruct the analog feedback signal. Note that e_{ADC} is injected in the same path as the quantization error e, and, therefore, is also attenuated by the action of negative feedback. However, DAC errors are injected in the feedback path, so that non-linearities in the DAC introduce distortion directly at the modulator input that is not be mitigated by feedback. Consequently, the linearity of a multi-bit ΣΔM will be no better than the linearity of the multi-bit internal DAC [Carl97]. In other words, the DAC must be designed to reach the linearity targeted for the whole ΣΔ ADC, what is quite challenging due to the impact of component mismatch.

Fig. 1.35 shows the architecture typically employed for the multi-bit ADC and DAC in ΣΔ modulators. Since normally $B \le 5$, full-parallel topologies are used. The B-bit ADC consists of a bank of $2^B - 1$ comparators that digitizes the output of the loop filter into thermometer code, which is decoded to binary at the back-end. On the other hand, the

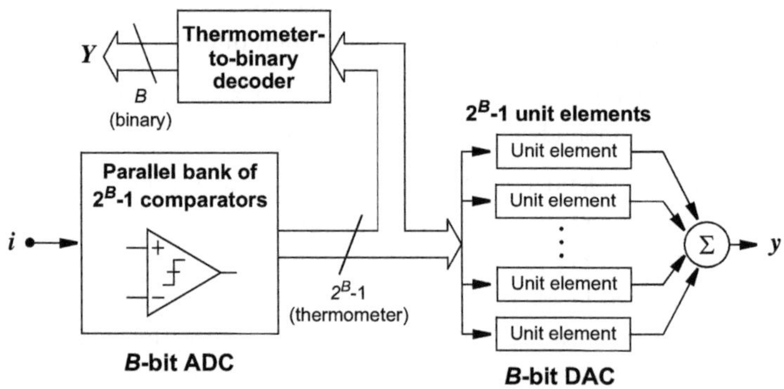

FIGURE 1.35 Typical architecture of the B-bit internal ADC and DAC.

DAC uses $2^B - 1$ unit elements (capacitors, resistors, current sources, etc.) to reconstruct the analog feedback signal in 2^B levels (numbered from 0 to $2^B - 1$). The ith analog output level is generated by activating i unit elements and summing up their outputs (charges or currents). DAC errors are due to mismatch among the unit elements, which deviates the output levels from their ideal values. Assuming that the actual value of each unit element follows a Gaussian distribution, the worst-case relative error in the DAC output can be estimated as [Raza95] [Carl97]

$$\sigma\left(\frac{\Delta y}{y}\right) \approx \frac{1}{2\sqrt{2^B}}\sigma\left(\frac{\Delta U_e}{U_e}\right) \tag{1.66}$$

where $\sigma(\Delta U_e / U_e)$ is the relative error in the unit element. Note that, thanks to the parallel topology, the accuracy of the DAC will be larger than that of the unit elements. However, for a 4-bit ΣΔM with 16-bit linearity, the required matching of the unit elements in the DAC should be better than 0.01% (13 bits). Unfortunately, component matching that can be achieved in present-day standard CMOS processes is in the range of 0.1% (10 bits), so that the required accuracy in the unit elements could only be obtained through the parallel connection of many (over 64) large components. Obviously, this means that obtaining multi-bit ΣΔMs with linearities larger than 12 or 13 bits, while relying only on standard component matching, leads to prohibitive area occupation.

Several alternative approaches to achieve high-linearity multi-bit ΣΔMs will be now outlined, some of them requiring modest component matching.

1.5.1 Element trimming and analog calibration

A straight-forward method to increase the accuracy of multi-bit DACs is to improve the matching of the unit elements by *trimming*. Different methods can be applied depending on the kind of elements [Carl97]. Laser trimming can be used for resistors. Trimming of capacitors can be done by switching on or off small capacitors in parallel with the unit element and the settings of the switches can be stored in a PROM. These trimmings can be done at the foundry, but at the expense of additional fabrication and/or measurement steps, so that the cost of the IC will be significantly increased. Moreover, they cannot compensate drift with temperature or aging.

Trimming methods can also be periodically applied during operation, but precision measurement hardware must be added on-chip to determine how to trim the elements. Moreover, the interruption of normal operation for *periodical calibration* is not allowed in many applications. There, *background calibration* can be used, but at the cost of a significant increase of the circuit complexity.

1.5.2 Digital correction

A different strategy consists of converting DAC errors into digital form and correcting them in the digital domain using look-up tables [Lars88] [Cata89]. Fig. 1.36*a* illustrates the concept, which consists basically of mapping the low-accuracy digital output codes of the ΣΔM (due to the non-ideal DAC) into high-accuracy ones. For each B-bit output of the ΣΔM, a digital correction block provides an N-bit output that represents the corresponding DAC output level with the accuracy targeted for the ΣΔM (N, with $N \gg B$). The operation relies in the action of negative feedback. The spectrum of $y(n)$ will follow very closely that of the input signal $x(n)$ within the baseband, where the loop gain provided by $H(z)$ is very high. At the same time $Y_c(n)$ is, by assumption, an accurate digital replica of the DAC output, so that its baseband spectrum must also correspond very accurately to that of the input signal.

Fig. 1.36*b* depicts a scheme for accurately acquiring the DAC outputs in the digital correction block, which is usually a RAM [Lars88]. A B-bit digital counter successively generates all possible 2^B input codes of the DAC. Each analog output code acts as the static input signal to the same ΣΔM, but reconfigured to single-bit for intrinsical linearity †[15], that converts it into an N-bit length digital sequence. A digital filter (usually a counter) finds the digital equivalent of the DAC output code as the mean value of the bit stream and it is stored in the RAM at the address defined by the B-bit counter. Since the total calibration requires 2^{N+B} clock periods, it is performed at power-up.

Improved schemes for digital correction can be found in literature [Wald90] [Sarh93] [Carl97].

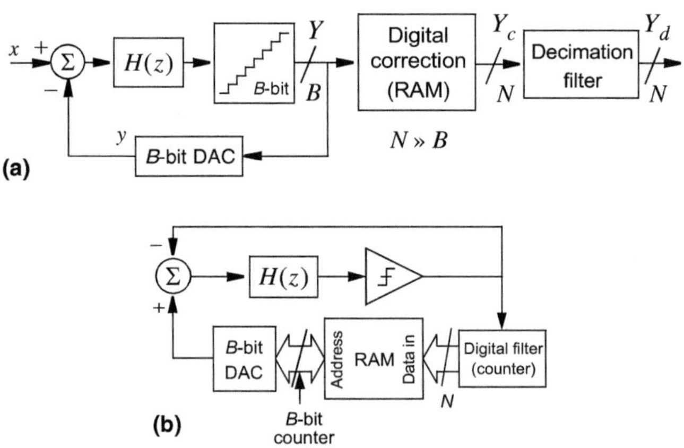

FIGURE 1.36 Digital correction: (a) General scheme, (b) Calibration scheme.

15. Keeping in mind that optimized multi-bit ΣΔMs can result into unstable single-bit ones.

1.5.3 Dynamic element matching

As previously discussed, mismatches among the DAC unit elements cause linearity errors and harmonic distortion in the ΣΔM. For a given integrated DAC with the structure in Fig. 1.35, there is a direct correspondence between the thermometer input code and the respective DAC error, because the same unit elements are used to generate the DAC output level every time a certain input code is active. The basic idea behind *dynamic element matching* (DEM) techniques consists of breaking this direct correspondence, so that the unit elements selected to generate a given DAC level output vary over time. This way, the fixed error of that level will be transformed into a time-varying error.

To that purpose, a block is added to control the selection of elements at each clock cycle (see Fig. 1.37). The selection is done according to simple algorithms that try to drive the average error in each DAC level to zero over time. Thus, part of the DAC error power that laid in the low-frequency range will be moved to higher frequencies, where it can be removed by the decimation filter.

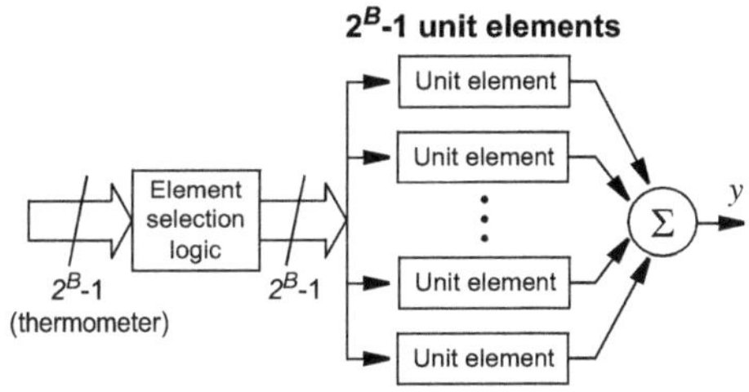

FIGURE 1.37 Architectural concept of DEM.

A detailed overview of the different DEM techniques can be found in [Carl97] [Geer02]. They can be grossly grouped in the following categories:

- *Randomization algorithms*: selection of the unit elements using pseudo-randomly configured networks [Carl88]—e.g., butterfly structures similar to those in FFTs. Harmonic distortion induced by the DAC is transformed into white noise, whose out-of-band power will be removed by digital filtering. The remaining power will however increase the noise floor.
- *Rotation algorithms*: selection of the unit elements in a periodic fashion to shift harmonic distortion out of the signal band, as in *clocked averaging* (CLA) [Klaa75]. It does not increase the noise floor, but mixed components can be originated that fold back into the baseband.
- *Mismatch-shaping algorithms*: selection of the unit elements according to algorithms that shape the mismatch error to reduce its in-band power: *individual level averaging* (ILA) [Leung92], *data weighted averaging* (DWA) [Baird95], *data directed scrambling* (DDS) [Adams95], etc. The shaping is limited to first and second order.
- *Vector-quantizer structures*: incorporation of a digital ΣΔ converter with error-feedback topology in order to achieve high-order shaping [Schr95].

1.5.4 Dual-quantization

The techniques previously presented for correcting non-linearities in multi-bit ΣΔMs involve a large increase in the circuit complexity (DEM, digital correction, and analog calibration) or have a clear impact on the production cost (trimming). An alternative approach are *dual-quantization* techniques, which combine single-bit and multi-bit quantization in the same ΣΔM. The idea behind is to combine their benefits: the intrinsic linearity of 1-bit quantization and the reduced quantization error of B-bit quantizers. This can be done with little extra circuitry. Architectural examples of these techniques will be discussed next.

Leslie-Singh architecture

The general scheme of the dual-quantizer architecture proposed by Leslie and Singh [Lesl90] is illustrated in Fig. 1.38. Single-bit quantization is used in the ΣΔ loop, with intrinsic linearity due to the 1-bit feedback. A path containing a multi-bit quantizer is cascaded. The two quantizer outputs are then properly combined in the digital domain to reduce the quantization error at the output to that of the multi-bit quantizer. Note that, in practice, the scheme in Fig. 1.38 can be simplified by using the MSB of the B-bit quantizer to feed the 1-bit signal back to the ΣΔ loop.

This topology can be viewed as an L-0 cascade ΣΔM, where the first stage is an Lth-order single-bit ΣΔM and the second one is a 0-order B-bit ΣΔM. Thus, the Leslie-Singh architecture requires perfect cancellation of the 1st-stage 1-bit quantization error at the modulator output and also suffers from noise leakage problems.

Note also that, although the modulator output ideally contains only multi-bit quantization error, the architecture does not benefit from the better stability performance of standard multi-bit ΣΔMs, since the loop is closed through single-bit feedback. Therefore, the NTF cannot be optimized to aggressive high-order shapings without jeopardizing stability.

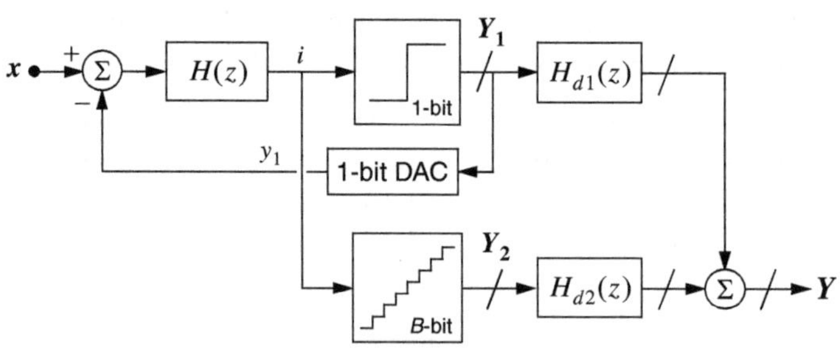

FIGURE 1.38 Leslie-Singh ΣΔ modulator.

Single-loop ΣΔMs

Fig. 1.38 illustrates an alternative topology for high-order ΣΔMs employing dual-quantization. In the 3rd-order ΣΔM shown [Hair91] [Hair94], the first two integrators are fed through a 1-bit DAC, whereas the third integrator is fed through a multi-bit DAC. The modulator linearity will not be menaced, since DAC non-linearities will be suppressed by the gain of two preceding integrators in the ΣΔ loop. At the same time, the topology benefits from improved stability performance, thanks to the multi-bit feedback in the last integrator. In practice, the MSB of the B-bit quantizer can be used to establish the 1-bit feedback. Linear analysis shows that, under ideal conditions, the modulator output yields

$$Y(z) = z^{-1}X(z) + 2(1-z^{-1})^3E_2(z) - 2z^{-1}(1-z^{-1})^2E_D(z) \qquad (1.67)$$

providing 3rd-order shaping for the B-bit quantization error and 2nd-order shaping for the DAC error, because it is injected after the second integrator. Note that this topology also suffers in practice from noise leakage of the 1-bit quantization error to the output.

The architecture in Fig. 1.38 can be generalized to higher-order shaping, leading to the so-called *dual-quantization* or *dual-feedback single-loop ΣΔMs*. As the order of the loop increases, the number of back-end integrators with multi-bit feedback can be traded off with aggressive noise-shaping (improved stability) and linearity demands for the multi-bit DAC.

Cascade ΣΔMs

Dual-quantization can be easily applied in cascade ΣΔMs [Bran91b] [Tan93] [Mede98a]. As shown in Section 1.4, the output of a cascade ΣΔM ideally contains only the input signal and the last-stage quantization error, whereas quantization errors from the rest of stages are cancelled out in the digital domain. Thus, for a reduction of the in-band quantization error power, only the last-stage quantizer needs to be multi-bit. The remaining quantizers are usually single-bit for inherent linear feedback in that stages. In such a case, the multi-bit DAC errors are injected in the last stage of the

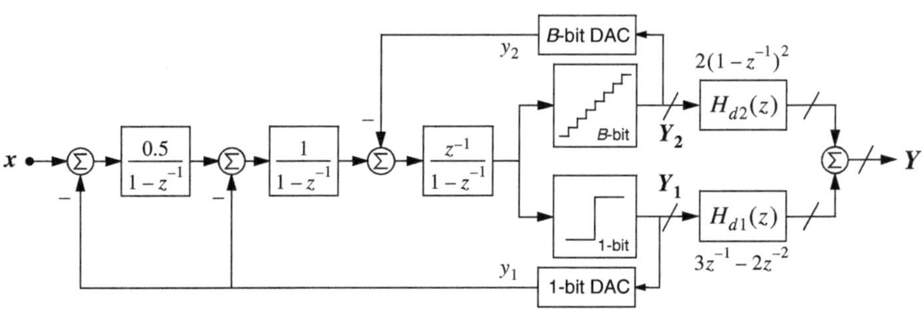

FIGURE 1.39 A 3rd-order single-loop ΣΔM with dual-quantization.

cascade, so that the linearity requirements will not be demanding, thanks to the in-band attenuation provided by the preceding stages.

The resulting topology is that illustrated in Fig. 1.26, considering $B_i = 1$ for $i = 1, \ldots, N-1$ and $B_N = B$. Under ideal linear analysis, the output of a generic N-stage cascade ΣΔM with dual-quantization yields

$$\begin{aligned} Y(z) &= STF(z)X(z) + NTF_N(z)E_N(z) + NTF_D(z)E_D(z) \\ &= z^{-L}X(z) + d_{2N-3}(1-z^{-1})^L E_N(z) \\ &\quad + d_{2N-3}(1-z^{-1})^{(L-L_N)}E_D(z) \end{aligned} \tag{1.68}$$

where L is the summation of the stage orders, d_{2N-3} is the scaling factor due to inter-stage coupling (usually 2 or 4), $E_N(z)$ is the last-stage quantization error, and $E_D(z)$ is the error in the multi-bit DAC. In comparison with eq(1.64), an extra term appears in eq(1.68), showing that the DAC error will be $(L-L_N)$th-order shaped—i.e., the overall modulator order minus that of the last stage. For instance, if the 2-1 ΣΔM in Fig. 1.28 is considered to include a multi-bit quantizer in its last stage, errors in the DAC will be 2nd-order shaped, so that eq(1.63) is written as

$$P_Q \approx d_1^2 \cdot \frac{\Delta_2^2}{12} \cdot \frac{\pi^6}{7OSR^7} + d_1^2 \cdot \sigma_D^2 \cdot \frac{\pi^4}{5OSR^5} \tag{1.69}$$

where σ_D^2 stands for the DAC error power.

Many IC implementations of cascade ΣΔMs using this dual-quantization scheme have been reported; e.g., a 2-1 cascade with 3-bit quantization in the last stage [Bran91b], a 2-1-1 cascade with 3 bits in the third stage [Mede99b], and a 2-2 cascade with 5 bits in the last stage [Mori00]. We can refer to these cascades as 2-1(5b) ΣΔM, 2-1-1(3b) ΣΔM, and 2-2(5b) ΣΔM, respectively, explicitly including the resolution of the stage if the corresponding quantizer is multi-bit. This compact notation for multi-bit cascades will be used henceforth.

Cascade ΣΔMs employing multi-bit quantization in all stages have been also reported; e.g., a 2(4b)-1(4b)-1(4b) ΣΔM [Fuji00] and 2(5b)-2(3b)-1(3b) ΣΔM [Vleu01]. Note that, under ideal conditions, the quantization error of the first two stages is cancelled out in the digital domain. Multi-bit quantization in these stages is used for the only purpose of reducing the corresponding quantization errors that will, in practice, leak to the output [†16]. In [Vleu01]

16. Digital compensation techniques for cascade ΣΔMs have been also proposed for reducing noise leakage caused by non-idealities in the analog circuitry [Wies96] [Kiss00]. Adaptive schemes are presented for reducing the 1st-stage quantization error leaking to the output, by means of injecting a test signal at the quantizer and digitally adapting the error cancellation logic to reduce its in-band power at the modulator output.

DEM techniques are incorporated in the multi-bit first stage to improve the modulator linearity, while simply relying in the attenuation of the DAC errors in the multi-bit second stage provided by the first two integrators. In [Fuji00] DEM techniques are applied to the three modulator stages. At a first glance, one could argue that using multi-bit quantization in all stages may not worth the increased circuit complexity. However, it leads to an additional appealing feature: the inter-stage coupling can be increased in comparison with a single-bit approach without overloading the quantizers. This way, the scaling factor d_{2N-3} that multiplies the last-stage quantization error can be smaller than unity, leading to an improvement of the global performance, instead of a reduction. For example, in [Fuji00], $d_3 = 1/32$, what results in a *DR* improved by 30dB, in comparison with an ideal 4th-order 4-bit ΣΔM.

Cascades using tri-level (1.5-bit) quantizers can also be found in literature; e.g., a 2(1.5b)-2(1.5b)-2(1.5b) ΣΔM [Dedic94] and a 2-2-2(1.5b) ΣΔM [Feld98]. As reported in [Paul87], tri-level coding helps to reduce quantization errors in comparison with 1-bit quantization—over 3-dB *SNR* improvement is measured for a 2nd-order ΣΔM. Although this coding is not inherently linear, 1.5-bit quantizers are often used in fully-differential SC ΣΔMs, since highly-linear tri-level DACs can be easily implemented using extra feedback switches, thus avoiding DEM techniques [Lewis92] [Balm00] [Reut02].

1.6 Parallel ΣΔ Architectures

Conventional ΣΔMs basically offer three degrees of freedom in order to achieve certain specifications in the A-to-D conversion; i.e., the loop order L, the oversampling ratio OSR, and the internal quantizer resolution B. An extra degree of freedom can be obtained through the use of parallelism; that is, several ΣΔ ADCs working together on the input signal to increase the effective conversion rate. Although the increase in hardware is obvious, these architectures represent an alternative to achieve wideband A-to-D conversion for those applications in which conventional ΣΔ ADCs would require unaffordable sampling rates.

A generic parallel ΣΔ ADC is composed of N channels, each one containing a ΣΔM and digital filter. The analog input signal is multiplexed at the input of each channel, processed by the corresponding ΣΔ ADC, and digitally demultiplexed. The overall output of the parallel ΣΔ ADC is finally obtained as the summation of the demultiplexed channel outputs.

The three basic approaches for multiplexing the input and output signals in parallel ΣΔ ADCs will be briefly presented here [Eshr96] [Kozak00].

1.6.1 Frequency division multiplexing

Fig. 1.40 illustrates the scheme of a parallel ΣΔ ADC with frequency multiplexing, often called a *multi-band* ΣΔ *ADC*. It uses N band-pass ΣΔMs (BPΣΔMs) operating in parallel to digitize the input signal. Each BPΣΔM has a different center frequency and operates on a sub-band of the original signal. A bank of digital band-pass filters decimates each of the N channels to finally reconstruct the frequency decomposed signal [Aziz93] [Corm94].

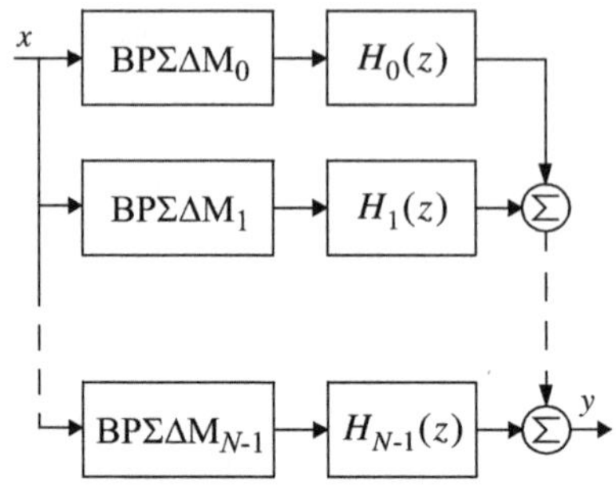

FIGURE 1.40 Block diagram of a multi-band ΣΔ ADC.

The global performance achieved by N parallel BPΣΔMs with equal OSR is similar to that of a single low-pass ΣΔM (LPΣΔM) with oversampling ratio $N \times OSR$, since the oversampling of each channel only applies to a sub-band.

Hardware complexity can make this scheme unpractical, because the N channels must be different for proper operation.

1.6.2 Time division multiplexing

Fig. 1.41 illustrates the basic scheme of a parallel ΣΔ ADC with time multiplexing, often referred to as a *time-interleaved* ΣΔ *ADC* (TI ΣΔ ADC). The input signal is sampled at frequency $N \times f_s$ and distributed through a multiplexer over N channels. Each channel operates at frequency f_s and their outputs are demultiplexed to construct the final output. The global performance is similar to sampling at clock rate $N \times f_s$, although each individual channel indeed operates at f_s.

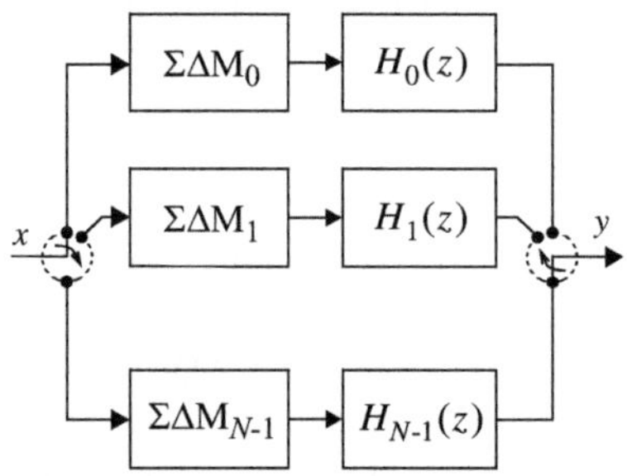

FIGURE 1.41 Block diagram of a time-interleaved ΣΔ ADC.

Unlike Nyquist-rate ADCs [Black80], this technique is not directly applicable to ΣΔ ADCs, due to oversampling. In practice, the principles of *multi-rate systems* [Croc83] need to be applied to find the time-interleaved equivalent of a given ΣΔ topology, what implies the cross-coupling of channels [Khoi93] [Khoi97] [Kozak00] [Wang00]. An alternative approach has been recently proposed to overcome this increase in circuit complexity [Eshr03].

1.6.3 Code division multiplexing

The scheme of parallel ΣΔ ADCs using code multiplexing is shown in Fig. 1.42. It consists of N channels, in which the input signal is multiplied by an analog signal, ΣΔ modulated, decimated and filtered, and multiplied by a digital signal. The outputs of the channels are finally summed to obtain the overall output. The multiplying signals are channel-specific and correspond to Hadamard sequences containing ± 1 's. Thus, only sign inversion is needed, avoiding real multipliers.

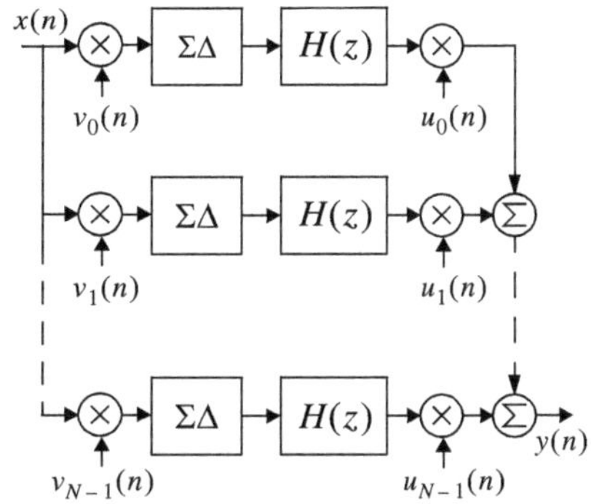

FIGURE 1.42 Block diagram of a ΠΣΔ ADC.

This architecture is often referred to as *parallel ΣΔ ADC* (ΠΣΔ ADC) [King94] [Galt95] [Galt96] [King98]. The performance of N identical channels with oversampling OSR can be made close to that of a conventional ΣΔADC with oversampling $N \times OSR$.

Many works on these three types of parallel ΣΔ ADCs have been presented at architectural level, but these strategies seem not yet mature, since very few IC prototypes have been reported [King94] [King98].

1.7 State of the Art in ΣΔ ADCs

From the previous sections it can be concluded that many different alternatives exist for the realization of low-pass ΣΔ modulators. This section presents a detailed review of those that have led to practical ICs and have been reported in open literature. The purpose is to illustrate the possibilities and existing tendencies for achieving common ADC specifications, taking into account architectural, electrical, and technological aspects.

Most of the reported LPΣΔM ICs are discrete-time (DT) implementations. They have been classified from an architectural point-of-view, separately considering single-loop and cascade ΣΔMs, as well as single-bit or multi-bit quantization (Tables 1.8 to 1.11). Although not specifically treated in this work, continuous-time (CT) implementations are covered in Table 1.12 for the sake of completeness. The performance of each reported IC is summarized in terms of:

- Effective resolution achieved by the ΣΔM ($ENOB$)—see eq(1.28).
- Bandwidth of the input signal, represented by the *digital output rate* of the converter ($DOR = 2f_b$); i.e., the Nyquist rate after decimation.
- Oversampling ratio used in the ΣΔM (OSR).
- Employed architecture, including the order of the loop filter and the number of bits in the internal quantizer.
- Technology and supply voltage used for the implementation.
- Power consumption of the ΣΔM.
- Quantifications of the 'quality' of the IC performance, in terms of two figures-of-merit (FOMs) typically employed for LPΣΔMs, namely:

$$FOM_1\big|_{\text{pJ/conv}} = \frac{\text{Power(W)}}{2^{ENOB(\text{bit})} \cdot DOR(\text{S/s})} \times 10^{12} \tag{1.70}$$

$$FOM_2 = 2kT \cdot \frac{3 \cdot 2^{2ENOB(\text{bit})} \cdot DOR(\text{S/s})}{\text{Power(W)}} \times 10^{5} \tag{1.71}$$

proposed in [Good96] and [Rabii97], respectively. FOM_1 emphasizes power consumption, whereas FOM_2 emphasizes resolution. The smaller the FOM_1 value and the larger the FOM_2 value, the 'better' the ΣΔM.

Most of the reported ΣΔM ICs correspond to DT switched-capacitor (SC) implementations in CMOS technologies. That must be assumed in Tables 1.8 to 1.12 if not specifically written. For high-order single-loops, a distributed feedback architecture must be also assumed if not specified. The presented power dissipation accounts only for the ΣΔ modulator, excluding the consumption of the decimation filter. In the tables '–' means that the datum is not specified in the corresponding reference.

On top of the former items for summarizing the IC performance, some extra information is displayed in between brackets, regarding architectural and electrical techniques and especial characteristics of technology used. The corresponding abbreviations are summarized before Table 1.8.

Abbreviations					
Architecture	DFB	Distributed Feedback	Technique	2S	Double Sampling
	DFF	Distributed Feedforward		AES	Additive-Error Switching
	FB	Feedback		CT-FE	Continuous-Time Front-End
	FF	Feedforward		DC	Digital Correction
	FFS	Feedforward Summation		DD	Dynamic Dither
	LFB	Local Feedback		MOS-O	MOSFET-Only
	LR	Local Resonator		RO	Reset-Opamp
	SLL	Stabilizing Local Loops		SWO	Switched-Opamp
DEM	Bi-DWA	Bi-Directional DWA	Technology	2P	Double-Poly
	DDS	Data Directed Scrambling		MCM	Multi-Chip Module
	DWA	Data Weighted Averaging		MiM	Metal-insulator-Metal
	I-DWA	Incremental DWA		MS	Mixed-Signal Process
	ILA	Individual Level Averaging		STD	Standard Digital Process
	mDWA	modified DWA			
	P-DWA	Partitioned DWA			

TABLE 1.8 DT single-loop single-bit LPΣΔM CMOS ICs.

REFs	*ENOB* (bit)	*DOR* (S/s)	*OSR*	Architecture	Technology	Supply (V)	Power (W)	FOM_1	FOM_2
[Boser88]	14.5	16k	256	2nd-or	3μm –	5	12m	32.37	1.79
[Bran91a]	16	50k	256	2nd-or	1μm STD	5	13.8m	4.21	38.84
[Burm96]	14.16	19.53k	256	2nd-or [2S, AES]	2μm –	5	13m	36.36	1.26
[Grilo96]	15.32	7k	286	2nd-or	0.6μm STD	1.8	2m	6.98	14.62
[Mede97]	16.4	9.6k	256	2nd-or	0.7μm STD	5	1.71m	2.06	104.79
[Pelu97]	12	6.8k	74	2nd-or [SWO]	0.7μm STD	1.5	101μ	3.63	2.82
[Send97]	14.33	6k	128	2nd-or [2S]	0.5μm MS [2P]	1.5	550μ	4.46	11.55
[Thanh97]	13.4	195.3k	128	2nd-or [2S, ILA]	1.2μm MS [2P]	5	25.9m	12.27	2.20
[Tille01]	13	16k	64	2nd-or [MOS-O]	0.25μm STD	1.8	1m	7.63	2.68
[Kesk02]	13	40k	256	2nd-or [RO]	0.35μm MS [2P]	1	5.6m	17.09	1.20
	12	100k	102.4					13.67	0.75
[Saue02]	12.17	16k	64	2nd-or [SWO]	0.18μm STD	0.7	80μ	1.09	10.54
[Saue03]	13	16k	64	2nd-or [SWO]	0.18μm MS [MiM]	0.65	45.5μ	0.35	58.90
	11	32k	32					0.69	7.36
[Nade94]	13.83	2k	250	3rd-or [LR]	2μm MS [2P]	5	940μ	32.27	1.13
[Au97]	12	16k	64	3rd-or [SLL]	1.2μm MS [2P]	2	340μ	5.19	1.97
[Pelu98]	12.5	32k	48	3rd-or [SWO]	0.5μm STD	0.9	40μ	0.22	66.85
[Gero03]	8.5	500	16	3rd-or [SWO, LR]	0.8μm MS [2P]	1.8	2.2μ	12.15	0.07
[OptE91]	13.5	500k	64	4th-or [DFB]	1.5μm MS [2P]	5	160m	27.62	1.05
[Kerth94]	21	800	320	4th-or [FFS, LR]	3μm MS [2P]	10	25m	14.90	351.28
[Yama94]	18	20	1600	4th-or [FFS, LR]	1.2μm MS [2P]	5	1.3m	247.96	2.64
[Coban99]	16	40k	64	4th-or [FB-FF, LR]	0.5μm STD	1.5	1m	0.38	428.81
[Kasha99]	20	800	320	4th-or [FFS, LR]	0.6μm MS [2P]	5	16m	19.07	137.22
[Snoe01]	16.65	22k	64	4th-or [DFB-DFF, LR]	0.5μm MS [2P]	2.5	2.5m	1.11	232.29
[Bajd02]	13	22k	64	4th-or [DFB-DFF, CT-FE]	0.5μm MS [2P]	1.8	1.7m	9.43	2.17
[Brig02]	17.1	800	320	4th-or [FFS, LR]	0.6μm MS [2P]	5	50m	444.90	0.79
[Maul00]	15.32	500k	64	5th-or [FFS, LRs]	0.6μm STD	5	210m	10.27	9.94

TABLE 1.9 DT single-loop multi-bit LPΣΔM CMOS ICs.

REFs	*ENOB* (bit)	*DOR* (S/s)	*OSR*	Architecture	Technology	Supply (V)	Power (W)	FOM_1	FOM_2
[Sarh93]	15.66	41k	128	2nd-or(4b) [DC]	2μm MS [2P]	5	–	–	–
[Chen95]	15.65	40k	64	2nd-or(3b) [ILA]	1.2μm –	5	67.5m	32.82	3.91
[Nys97]	19	800	512	2nd-or(3b) [DWA]	2μm –	5	2.175m	5.19	252.36
[Fogl00]	16.22	48k	64	2nd-or(5b) [1st-or DEM]	0.5μm STD	3.3	68.6m	18.72	10.18
[Fogl01]	16.65	40k	64	2nd-or(5b) [2nd-or DEM]	0.5μm STD	3.3	70.4m	17.11	15.00
[Grilo02]	13	1M	32	2nd-or(4b) [1st-or DEM]	0.35μm BiCMOS	2.7	11.88m	1.45	14.10
[Gomez02]	12.83	400k	65	2nd-or(5b) [ILA]	0.13μm STD	1.5	2.4m	0.82	22.07
	12					1.2	1.4m	0.85	11.96
	8.01	4M	12			1.5	2.9m	2.81	0.23
[Gaggl03]	13.83	600k	96	2nd-or(3b) [w/o DEM]	0.18μm MS [dual-gate, MiM]	1.8 (3.3)	15m	1.72	21.09
[Mill03]	15.32	36k	639	2nd-or(6b) [mDWA]	0.18μm MS [dual-gate, MiM]	2.7	30m	20.37	5.01
	13.50	400k	57.5					6.47	4.47
	12.83	1.25M	18					3.30	5.51
	11.67	3.84M	12				50m	4.00	2.04
[Hair94]	16	39k	128	3rd-or(1b,5b) [dual-quant.]	2μm MS [2P]	5	58m	33.26	4.92
[Geer00]	15.8	2.5M	24	3rd-or(4b) [DWA]	0.65μm MS [2P]	5	295m	2.07	68.85
	11.5	12.5M	8				380m	10.50	0.69
[Baird96]	13.66	500k	16	4th-or(4b) [DFB, DC]	1.2μm MS [2P]	5	58m	8.96	3.61
[Kuo02]	13.7	1.25M	12	4th-or(4b) [FB-FF, I-DWA]	0.25μm MS [MiM]	2.5	100m	6.01	5.53
	13.0	2M					105m	6.41	3.19
[Jiang02]	13.8	4M	8	5th-or(4b) [hybrid FIR-IIR, DWA]	0.18μm STD	1.8	149m	2.61	13.63
[Reut02]	14	2.5M	32	5th-or(1.5b) [FFS, LRs]	0.25μm STD	2.5	24m	0.59	69.79
[Leung97]	19.3	96k	64	7th-or(1.5b) [FFS, LRs]	0.8μm MS [2P]	5	760m	12.26	131.36

TABLE 1.10 DT cascade single-bit LPΣΔM CMOS ICs.

REFs	*ENOB* (bit)	*DOR* (S/s)	*OSR*	Archit.	Technology	Supply (V)	Power (W)	FOM_1	FOM_2
[Yin93]	15.7	320k	64	2-1	1.2μm STD	5	65m	3.82	34.82
[Will94]	17	50k	128	2-1	1μm STD	5	47m	7.17	45.62
[Rabii97]	16.1	50k	80	2-1	0.8μm STD	1.8	2.5m	0.71	246.29
[Gomez00]	16.65	44k	128	2-1 [DD]	0.6-0.3μm MCM	3	22m	4.86	52.88
[Saue03]	13	16k	64	2-1 [SWO]	0.18μm MS [MiM]	0.65	61.75μ	0.47	43.40
	12.17	32k	32					0.42	27.47
[Rito94]	16.15	44k	64	2-2 [LR]	1.2μm BiCMOS	5	102m	31.82	5.72
	15.49					3	55m	27.19	4.22
[Fuji97]	18.15	48k	128	2-2 [LFB]	0.7μm MS [2P, low-Vt]	5	500m	35.91	20.17
[Miao98]	14.82	50k	64	2-2	3μm –	5	74m	51.03	1.42
[Wang01]	18.1	25k	64	2-2	0.6μm MS [2P]	5	75m	10.68	65.68
[Olia02]	13.5	360k	36	2-2 [LR]	0.4μm BiCMOS [2P, low-Vt]	1.8 (2.4)	5m	1.20	24.12
[Lee03]	14.16	1M	64	2-2	0.35μm MS [2P]	1.8	150m	8.20	5.57
	12	2M	32					18.31	0.56
[Rebe90]	15	180k	64	1-1-1	1.5μm MS [2P]	5	76m	12.89	6.35
[Yin94b]	15.82	1.5M	64	2-1-1	2μm BiCMOS	5	180m	2.07	69.61
[Marq98a]	14.8	2M	24	2-1-1	1μm MS [2P]	5	230m	4.03	17.66
[Geer99]	15	2.2M	24	2-1-1	0.5μm MS [2P]	3.3	200m	2.77	29.48
[Mori00]	13	2.2M	24	2-2-2	0.35μm MS [2P]	3.3	150m	8.32	2.46
[Yoon98]	15.3	64k	16	2-1-1-2	2μm MS [2P]	6.6	79m	30.60	3.29

TABLE 1.11 DT cascade multi-bit LPΣΔM CMOS ICs.

REFs	*ENOB* (bit)	*DOR* (S/s)	*OSR*	Architecture	Technology	Supply (V)	Power (W)	FOM_1	FOM_2
[Broo97]	14.5	2.5M	8	2(5b)-0(12b) [5-b flash with DDS, 12-b pipeline]	0.6μm MS [2P]	5 & 3	500m	8.63	6.70
[Bran91b]	12	2.1M	24	2-1(3b)	1μm STD	5	41m	4.77	2.14
[Reve03]	13.33	20k	64	2-1(1.5b) [2S]	0.35μm MS [2P, low-Vt]	0.8	60μ	0.29	88.09
[Mori00]	12	2.2M	24	2-2(5b)	0.35μm MS [2P]	3.3	99m	10.99	0.93
[Lamp01]	13.5	1.5625M	32	2-2(3b) [LR]	0.35μm –	2.5	50m	2.76	10.47
[Mede99b]	13	2.2M	16	2-1-1(3b)	0.7μm STD	5	55m	3.05	6.70
[Fuji00]	15	2.5M	8	2(4b)-1(4b)-1(4b) [Bi-DWA in all stages]	0.5μm MS [2P, low-Vt]	5	105m	1.28	63.81
[Rio01b]	13	2.2M	16	2-1-1(4b)	0.35μm STD	3.3	73.7m	4.09	5.00
	12	4M					78.3m	4.78	2.14
[Gupta02]	14.6	2.2M	29	2-1-1(2b) [2S]	0.35μm STD	3.3	180m	3.29	18.81
[Rio04]	13.8	2.2M	32	2-1-1(3b)	0.25μm MS [MiM]	2.5	65.8m	2.10	16.98
	12.7	4.4M	16					2.25	7.39
[Vleu01]	15.5	4M	16	2(5b)-2(3b)-1(3b) [2S, P-DWA]	0.5μm MS [2P]	2.5	150m	0.81	142.94
[Dedic94]	14.65	200k	16	2(1.5b)-2(1.5b)-2(1.5b) [Dither]	1.2μm MS [2P]	5	40m	7.78	8.25
[Feld98]	12.5	1.4M	16	2-2-2(1.5b)	0.7μm MS [2P]	3.3	81m	9.99	1.45

TABLE 1.12 CT LPΣΔM CMOS ICs.

REFs	ENOB (bit)	DOR (S/s)	OSR	Architecture	Technology	Supply (V)	Power (W)	FOM_1	FOM_2
[Hall92]	10	1.17M	128	2nd-or [GmC]	2μm –	5	–	–	–
[Luh98a]	8.0	2M	25	2nd-or [GmC]	2μm MS [2P]	5	15m	29.30	0.02
	11.3	200k	250					29.75	0.21
[Luh98b]	9.7	2M	25	2nd-or [GmC]	2μm MS [2P]	5	16.6m	9.98	0.21
[Lin99]	10.5	5M	16	2nd-or [GmC]	1.2μm –	3	12m	1.66	2.18
[Zwan99]	10.4	1M	10	2nd-or(5b) [GmC, w/o DEM]	0.5μm STD	5	7.2m	5.33	0.63
[Dorr03]	10.84	4M	64	2nd-or [RC-active]	0.12μm –	1.2	4m	0.55	8.35
[Gerf03]	13	50k	48	3rd-or [RC-active]	0.5μm –	1.5	135μ	0.33	62.04
[Hall93]	10	–	–	2-2 [GmC]	2μm –	5	–	–	–
[Redm94]	13	44k	64	4th-or [RC-active, Interpolat.]	1.6μm –	5	–	–	–
[Zwan96]	13	8k	64	4th-or [GmC, FFS]	0.5μm STD	2.2	200μ	3.05	6.70
[Blan02]	11.3	16k	62.5	4th-or [GmC, FFS]	0.35μm STD	2.5	75μ	1.86	3.39
[Veld02]	11.3	4M	40	4th-or(1.5b) [RC-active & GmC, FFS]	0.18μm STD	1.8	6.6m	0.65	9.62
[Luh00]	10	6.2M	64	5th-or [GmC, FFS]	0.6μm STD	3.3	16m	2.52	1.01

Several observations can be made from the information in the tables above:

- There is an extensive use of SC techniques in the reported LPΣΔM ICs. Some CT implementations have more recently appeared, using either GmC or RC-active techniques. DT modulators using switched-current (SI) techniques have been also reported, but they are usually intended for band-pass applications and are therefore not included in the tables.
- Many of the presented ICs are implemented in mixed-signal (MS) oriented CMOS technologies that provide high-quality capacitors. Extra process steps, such as a second poly layer or thin-oxide between metals, are commonly used (for poly-poly and MiM capacitors, respectively). Also many of the ICs employ standard capacitors structures available in pure digital (STD) CMOS technologies. Among them, there is only one IC implemented with just transistors [Tille01].
- Some ICs are reported in technologies with 'more special' features. Dual-gate processes are used in [Gaggl03] and [Mill03] to improve the linearity of the switches by employing thicker transistors with higher gate voltage. Low-Vt transistors are used for a more comfortable design of the amplifiers within the given supply range [Fuji97] [Fuji00] [Olia02] [Reve03]. Also BiCMOS implementations are reported [Grilo02] [Rito94] [Yin94b], the latter oriented for high-speed applications.

- Most of the reported ICs are still operated with supplies larger than 2.5V. As a consequence of technology scaling, the number of low-voltage implementations ($V_{DD} \leq 1.5\text{V}$) is increasing in recent years, but most are restricted to medium frequency applications ($DOR \leq 100\text{kS/s}$). The most extended low-voltage technique is switched-opamp (SWO) [Pelu97] [Pelu98] [Saue02] [Saue03] [Gero03].
- There is a large number of alternative topologies for high-order single-loops, but the most popular one is the feedforward summation (FFS) architecture—up to 7th-order in [Leung97]. The 'standard' distributed feedback scheme is usually restricted to 3rd-order loops, except for [OptE91] and [Baird96] in which 4th-order modulators are reported. Local resonators (LRs) are also extensively used for the placement of some of the NTF zeros within the signal band.
- In single-loop multi-bit ΣΔMs, DEM techniques have clearly gained ground over other correction or calibration methods. The most extended algorithm is DWA and its many modifications. Digital correction is used in only two of the reported ICs [Sarh93] [Baird96]. Special mention must be made to the modulators in [Zwan99] and [Gaggl03], which employ multi-bit quantization with no use of calibration, correction, or DEM techniques. Both designs rely only in the matching provided by the process. On the other hand, dual-quantization schemes are common practice to obtain multi-bit quantization in cascade ΣΔMs, given the relaxed linearity requirements of the multi-bit DAC.
- Clearly influenced by the growing market in xDSL applications, many medium-resolution wideband ICs have been reported (12~15bit at rates higher than 1MS/s). Given the high input frequency, SC ΣΔMs must be operated at low oversampling ratios (8~32), so that the required resolution is usually achieved by incorporating multi-bit quantization to high-order modulators. Examples of single-loop realizations are found in [Geer00] [Jiang02] [Kuo02] [Reut02]. On the other hand, almost all the reported cascade multi-bit ICs are oriented to these applications. Special attention is lately being paid on CT implementations, in which OSR can be increased without compromising the integrators dynamics (on the contrary to the settling requirements in SC ICs). For example, [Luh00] and [Dorr03] use $OSR = 64$ and clock frequencies of 256MHz and 400MHz, respectively, what is unaffordable in DT implementations. In spite of this, the reported resolutions are limited for the moment to 11~12bit.

For comparison purposes, much of the information in the former tables is graphically displayed in the following figures.

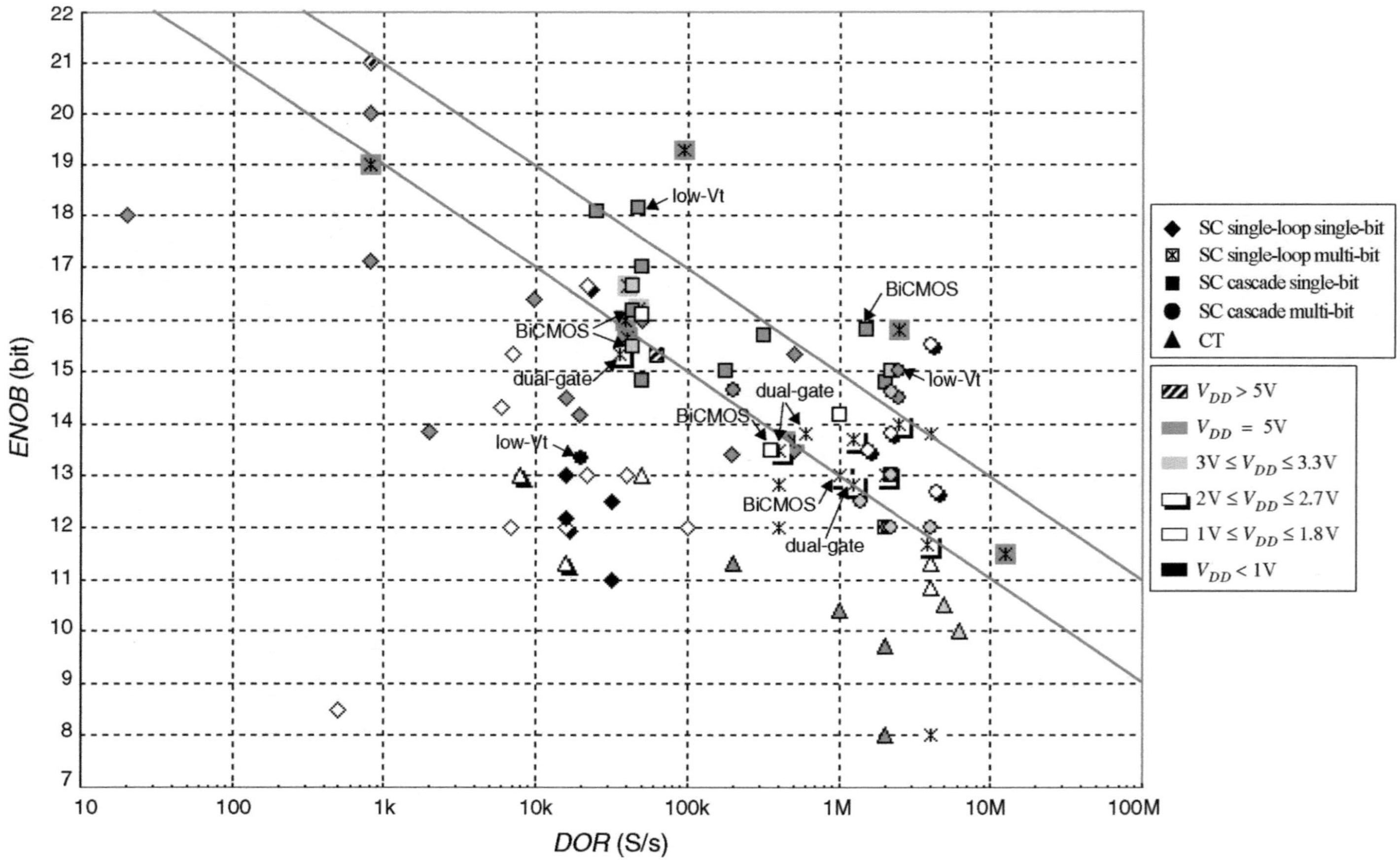

FIGURE 1.43 Location of the reported LPΣΔM ICs in the *ENOB—DOR* plane.

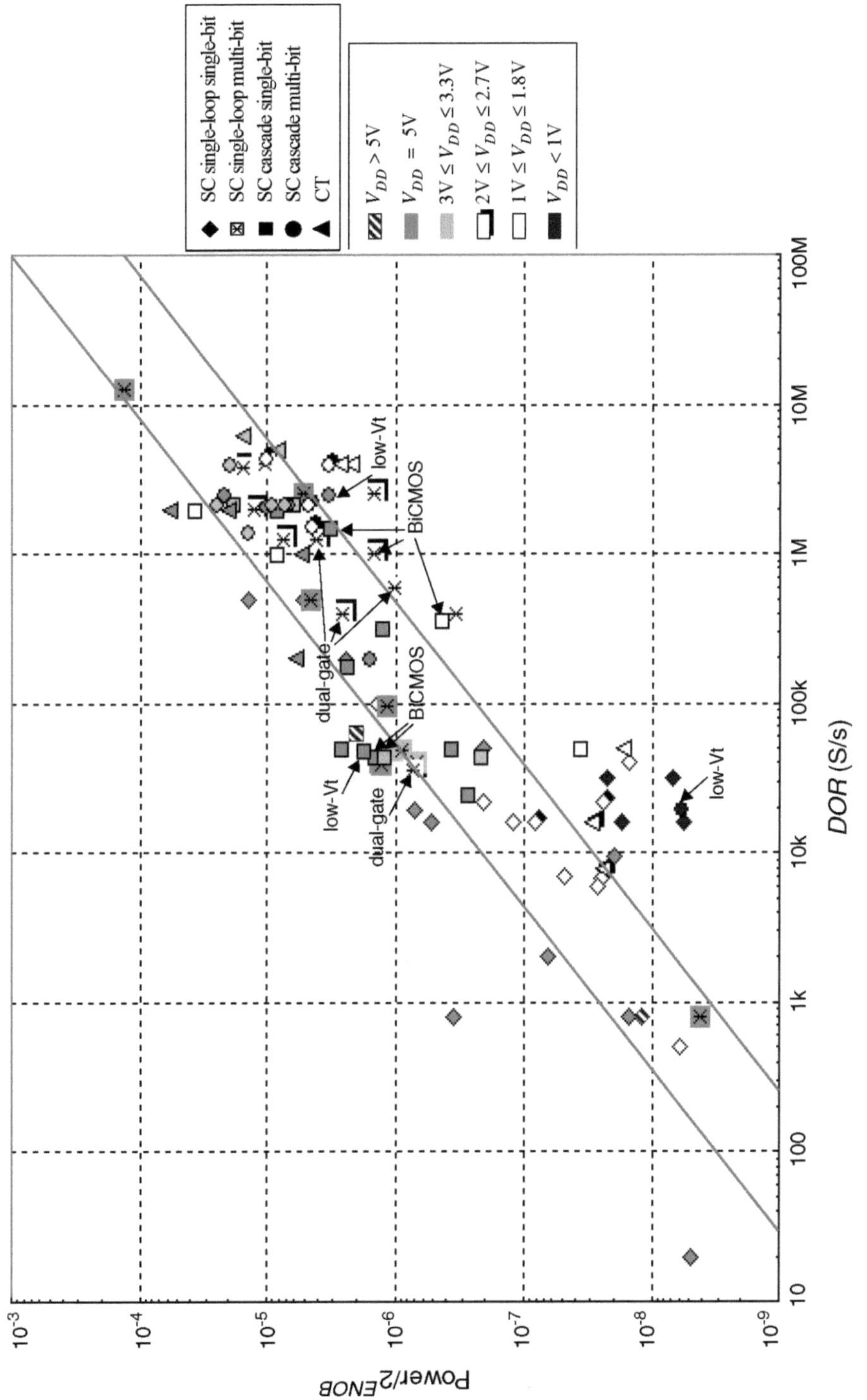

FIGURE 1.44 Normalized power consumption per conversion level versus *DOR* of the reported LPΣΔM ICs.

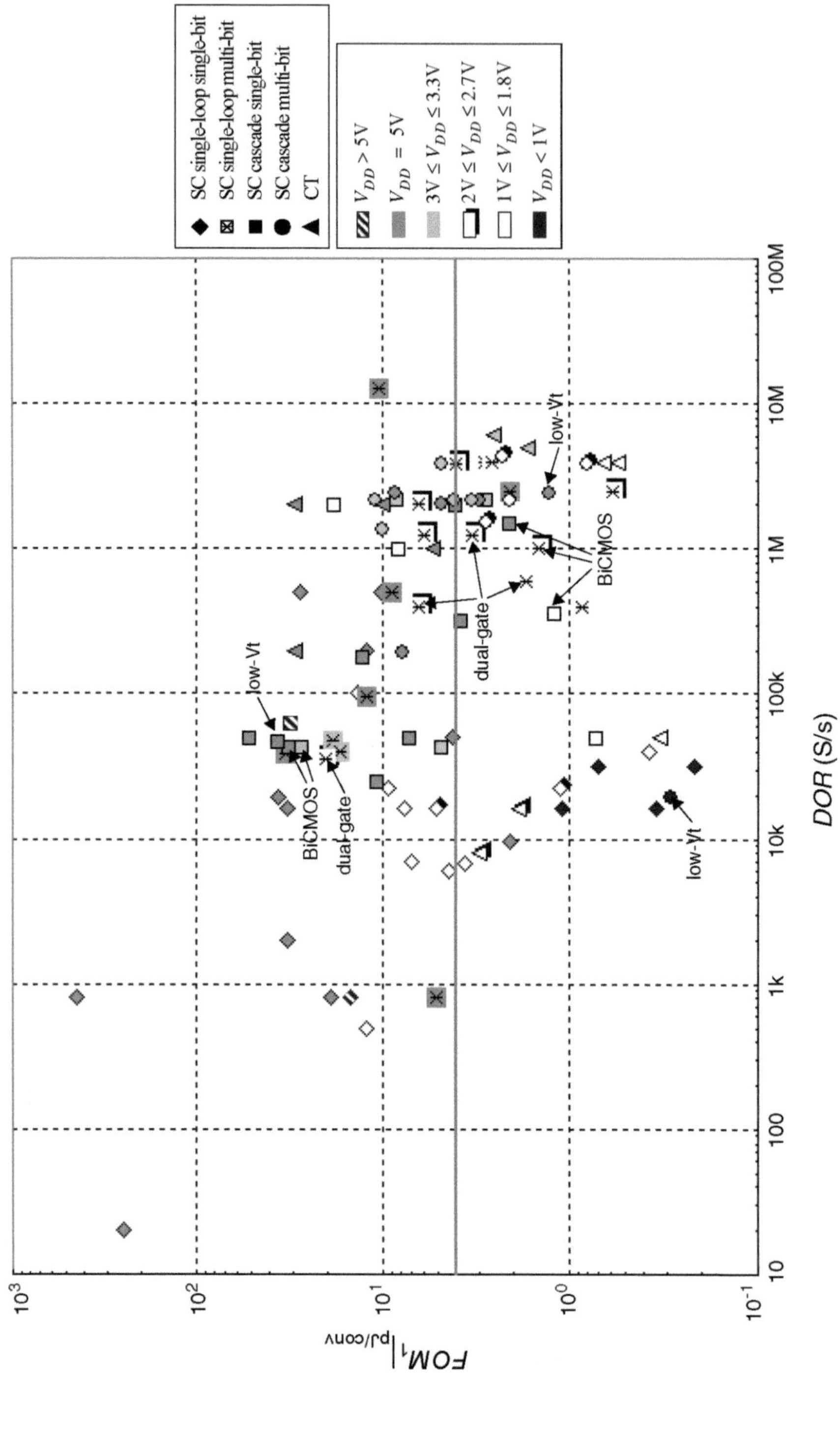

FIGURE 1.45 FOM_1 versus DOR of the reported LPΣΔM ICs.

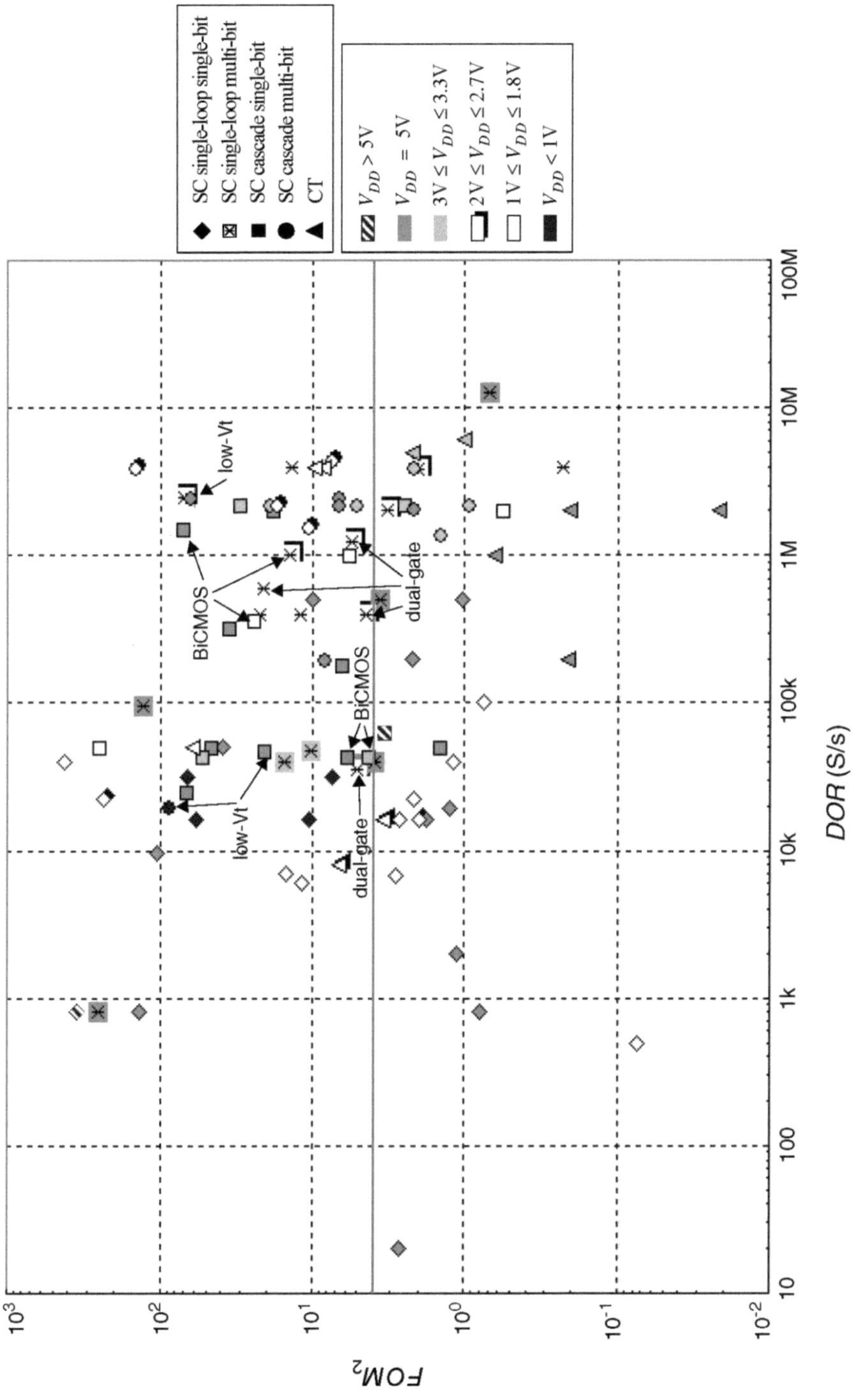

FIGURE 1.46 FOM_2 versus DOR of the reported LPΣΔM ICs.

Fig. 1.43 shows the effective resolution of the reported ICs versus their corresponding digital output rate. It is noticeable that most of the ΣΔ designs are concentrated in the frequency ranges of 20kS/s-60kS/s and 1MS/s-4MS/s, oriented for audio and xDSL applications, respectively. The former is dominated by single-loop single-bit designs, whereas the latter is clearly dominated by cascade multi-bit ΣΔMs. On the other hand, low-voltage implementations are more concentrated in the low-frequency range. Except for particular low-voltage ICs—e.g., those in [Rabii97] and [Lee03]—, it is visible that the achieved resolutions are not yet comparable to those provided by ICs operating at higher supply.

Fig. 1.44 depicts the power consumption of each IC, times the number of effective conversion levels (Power/2^{ENOB}), against the corresponding output rate. Note that the normalized power dissipation is considerably smaller for the low-voltage ICs in the medium-frequency range. On the contrary, the spread in this figure is much lower for high frequencies.

Fig. 1.45 shows the FOM_1 value achieved by each IC versus the corresponding DOR. As previously stated, this figure-of-merit emphasizes power consumption and the smaller its value, the 'better' the ΣΔM. Note that most of the ICs result into $FOM_1 \leq 10$. Also, it can be grossly derived from Fig. 1.45 that FOM_1 decreases as the supply does. Note that low-voltage implementations lead to the smaller FOM_1 values—especially single-loops for medium frequencies and CT modulators for high frequencies.

Fig. 1.46 depicts the FOM_2 value of each reported ΣΔM versus its output rate. In this case, the larger the FOM_2, the 'better' the design, with emphasis in the effective resolution. Note that most of the ICs achieve $FOM_2 \geq 1$.

The performance of the disclosed ΣΔMs can be compared from the figures presented above. For instance, globally speaking, designs like those in [Kerth94], [Leung97], and [Vleu01] can be considered to outperform in the low-, medium-, and high-frequency ranges, respectively.

1.8 Summary

In this chapter the basic principles of ΣΔ modulation have been analyzed. The effects of oversampling and noise-shaping on the ADC performance have been presented and the general structure, ideal performance, and metrics of ΣΔ converters have been defined and compared with Nyquist-rate ADCs.

Alternative topologies for the practical implementation of ΣΔ modulators have been discussed. The limitations on the linear analysis of ΣΔ loops have been addressed, considering the problems associated to colored quantization error in low-order single-loop ΣΔMs and unstable non-linear dynamics of high-order loops. Techniques for the realization of stable high-order ΣΔMs have been presented, such as optimization of the noise transfer function and non-linear mechanisms for integrator clipping or resetting.

Cascade topologies have been presented as an alternative to obtain high-order shaping with unconditional stability and reduced performance degradation. Their inherent problem of noise leakage under non-ideal operation conditions has been foreseen and will be covered in detail in Chapter 2.

The use of multi-bit internal quantization has also been presented as an alternative to enhance the effective resolution of the A-to-D conversion. Besides reducing quantization error, it provides better stability properties to single-loop architectures, but jeopardizes linearity. Non-linearity error in the multi-bit DAC due to component mismatch has been discussed, together with techniques to palliate its impact on the modulator performance, such as digital correction, DEM, and dual-quantization schemes.

The many design alternatives for the practical implementation of ΣΔ ADCs are finally summarized in the state of the art of reported low-pass ICs.

CHAPTER 2

Non-Ideal Performance of ΣΔ Modulators

THE PRINCIPLES OF ΣΔ MODULATION were studied in the previous chapter and alternative modulator topologies were presented. The achievable performance of different architectures was addressed, taking into account the quantization error. Besides this error—inherent to the A-to-D conversion—, only the additional effect of DAC errors was considered, in order to establish a comparison at the architectural level between modulators employing single-bit quantization and those using multi-bit quantizers.

This chapter presents a detailed study of the main non-ideal mechanisms affecting the performance of SC ΣΔ modulators. Although it is commonly accepted that ΣΔ conversion is less sensitive to non-idealities in the analog building blocks than other data conversion techniques [Nors97a], their influence will be large the more demanding the specifications for the A/D converter. Therefore, the impact of these errors on the modulator behavior must be carefully considered during the design phase.

The first part of this chapter is devoted to errors that alter the noise transfer function of the modulator, such as the amplifier finite DC gain (Section 2.1), capacitor mismatch (Section 2.2), and defective integrator settling (Section 2.3). As we will see, the sensitivity of ΣΔ converters to this sort of errors is heavily dependent on the modulator topology.

The second part of the chapter is dedicated to non-idealities that can be modeled as additive error sources at the modulator input, since they are not attenuated in the converter baseband. Circuit noise (Section 2.4), clock jitter (Section 2.5), and non-linear non-idealities generating distortion (Section 2.6) are considered.

System-level considerations, behavioral models, and closed-form expressions are obtained for the effect of each non-ideality. From them, estimable guidelines for the design of ΣΔ modulators can be extracted.

2.1 Integrator Leakage

In the previous chapter, the following integrator transfer function was used in the study of the ΣΔM topologies under ideal conditions:

$$H(z) = \frac{z^{-1}}{1 - z^{-1}} \tag{2.1}$$

The expression above assumes an infinite gain for the SC integrator at DC ($z = 1$). Obviously, this can not be achieved in practice due to the limited DC gain of the amplifier in the integrator. This non-ideality basically affects the noise transfer function $NTF(z)$ of the modulator, so that the assumption $NTF(z)|_{z \to 1} \to 0$ is no longer valid. As we will see in this section, this leads to an increase of the in-band error power that heavily depends on the modulator topology.

Leaky integrator

Let us consider the SC integrator in Fig. 2.1, where the operational amplifier is modeled by a simple voltage-controlled voltage source with gain $A_{DC} \gg 1$. The integrator is considered to have n_b input branches to gain generality. Using this model, the difference equation for the integrator can be written as

$$v_{o,n} = \frac{\left(1 + \frac{1 + C_P/C_I}{A_{DC}}\right) \cdot v_{o,n-1}}{1 + \left(1 + \sum_{i=1}^{n_b} \frac{C_{Si}}{C_I} + \frac{C_P}{C_I}\right) / A_{DC}} + \frac{\sum_{i=1}^{n_b} C_{Si}/C_I \cdot v_{i,n-1}}{1 + \left(1 + \sum_{i=1}^{n_b} \frac{C_{Si}}{C_I} + \frac{C_P}{C_I}\right) / A_{DC}} \tag{2.2}$$

where we can see that the finite DC gain yields to a memory error and a gain error—first and second term in the equation, respectively.

Transforming eq(2.2) to the z-domain, we get

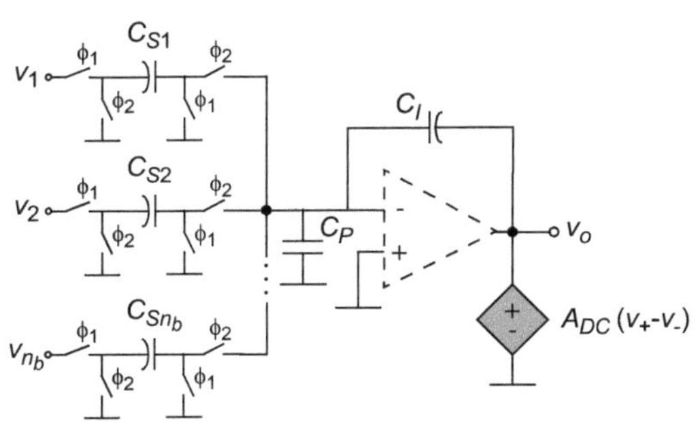

FIGURE 2.1 SC integrator model considering the finite amplifier DC gain.

$$v_o(z) = \frac{(1 + k_{fbs}/A_{DC}) \cdot z^{-1} v_o(z)}{1 + k_{fbi}/A_{DC}} + \frac{z^{-1} \sum_{i=1}^{n} C_{Si}/C_I v_i(z)}{1 + k_{fbi}/A_{DC}} \quad (2.3)$$

with $k_{fbs} = 1 + C_P/C_I$ being the inverse of the capacitive feedback factor during sampling, $k_{fbi} = 1 + g + C_P/C_I$ the inverse of the capacitive feedback factor during integration, and $g = \sum C_{Si}/C_I$. Operating and identifying terms as in the following expression

$$v_o(z) = \left[\sum_{i=1}^{n_b} \frac{C_{Si}}{C_I} v_i(z) \right] \cdot H(z) \quad (2.4)$$

we get that the transfer function of the leaky integrator is given by

$$H(z) = \frac{1}{1 + k_{fbi}\mu} \cdot \frac{z^{-1}}{1 - z^{-1}\left[\frac{1 + k_{fbs}\mu}{1 + k_{fbi}\mu}\right]} \quad (2.5)$$

where $\mu = 1/A_{DC}$. Considering $\mu \ll 1$ and neglecting the error in the gain, eq(2.5) can be approximated to

$$H(z) \cong \frac{z^{-1}}{1 - z^{-1}[1 + (k_{fbs} - k_{fbi})\mu]} = \frac{z^{-1}}{1 - z^{-1}(1 - g\mu)} \quad (2.6)$$

showing that the basic effect is a shift of the pole by an amount equal to $g\mu$.

2.1.1 Single-loop ΣΔ modulators

1st-order loop

Considering the integrator leakage in a 1st-order loop, the modulator output is given by

$$Y(z) = STF(z)X(z) + NTF(z)E(z)$$

$$STF(z) = \frac{z^{-1}}{1 + \mu z^{-1}} \cong z^{-1} \qquad NTF(z) = \frac{1 - z^{-1}(1 - \mu)}{1 + \mu z^{-1}} \cong (1 - z^{-1}) + \mu z^{-1} \quad (2.7)$$

where it has been assumed that $g = \sum C_{Si}/C_I \sim 1$ [†1].

Note from eq(2.7) that the leakage affects the signal transfer function introducing a gain error that, in general, will be negligible. On the other hand,

1. As we will see, this approximation is made to enable a comparison of the sensitivity to leakage of different ΣΔ architectures, regardless of the specific selection and implementation of integrator weights. Since the comparison will be made in terms of powers of μ, the approximation does not affect the validity of the obtained results.

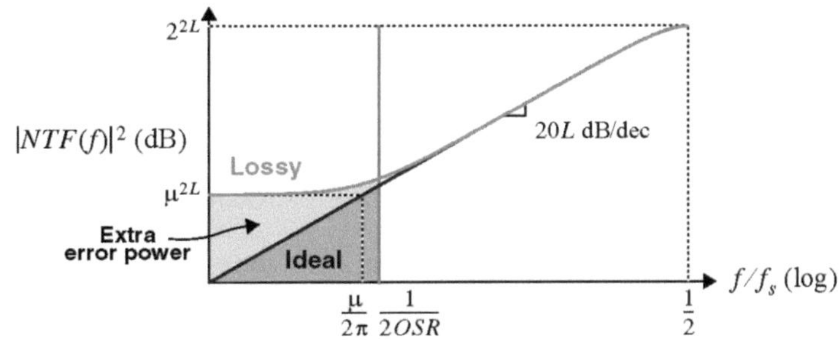

FIGURE 2.2 Magnitude response of NTF and extra in-band error power considering integrator leakage. L stands for the loop order.

it leads to a change in the noise-shaping function due to the shift of the zero from its ideal location at $z = 1$ (DC). Note that the first term in $NTF(z)$ corresponds to the ideal 1st-order shaping, whereas the leakage adds a second term that is non-shaped.

By doing the transformation $z \to \exp(j2\pi f/f_s)$, the in-band error power at the modulator output under this degradation of the quantization error shaping (see Fig. 2.2) can be calculated as

$$\begin{aligned} P_Q(\mu) &= \int_{-f_b}^{+f_b} S_E(f)|NTF(f)|^2 df = \frac{\Delta^2}{12 f_s}\int_{-f_b}^{+f_b} |NTF(f)|^2 df \\ &\cong \frac{\Delta^2}{12}\left(\frac{\pi^2}{3OSR^3} + \frac{\mu^2}{OSR}\right) \end{aligned} \tag{2.8}$$

with f_b being the signal bandwidth, f_s the modulator sampling frequency, $OSR = f_s/(2f_b)$ the oversampling ratio, and Δ the spacing between adjacent levels in the quantizer.

Note from eq(2.8), that the term inversely proportional to OSR^3 corresponds to the ideal 1st-order shaped quantization error, whereas the term introduced by the leakage is proportional to μ^2 and inversely proportional to OSR. This second term may dominate the in-band error power for low amplifier DC gains and/or high oversampling ratios.

2nd-order loop

Proceeding in a similar way for a 2nd-order single-loop ΣΔM, the noise transfer function can be calculated as

$$NTF(z) = \left[\frac{1 - z^{-1}(1-\mu)}{1 + \mu z^{-1}}\right]^2 \cong (1 - z^{-1})^2 + 2\mu z^{-1}(1 - z^{-1}) + \mu^2 z^{-2} \tag{2.9}$$

so that the output-referred in-band error power considering the leakages of the two integrators in the loop is given by

$$P_Q(\mu) = \frac{\Delta^2}{12 f_s} \int_{-f_b}^{+f_b} |NTF(f)|^2 df \cong \frac{\Delta^2}{12}\left(\frac{\pi^4}{5 OSR^5} + \frac{2\mu^2}{3 OSR^3} + \frac{\mu^4}{OSR}\right) \tag{2.10}$$

Note that the first term in eq(2.10) corresponds to the ideal 2nd-order shaped quantization error, whereas the leakages introduce terms that are 1st-order shaped ($\propto \mu^2$) and non-shaped terms ($\propto \mu^4$).

Lth-order loops

These latter expressions can be generalized for an Lth-order single-loop modulator as follows

$$NTF(z) = \left[\frac{1 - z^{-1}(1-\mu)}{1 + \mu z^{-1}}\right]^L \cong [1 - z^{-1}(1-\mu)]^L \tag{2.11}$$

the output-referred in-band error power being given by

$$P_Q(\mu) \cong \frac{\Delta^2}{12}\left[\frac{\pi^{2L}}{(2L+1) OSR^{(2L+1)}} + \frac{\mu^{2L}}{OSR} + \sum_{m=1}^{L-1} \frac{L(L-1)\ldots(L-m+1)}{m!} \frac{\mu^{2(L-m)}\pi^{2m}}{(2m+1) OSR^{(2m+1)}}\right] \tag{2.12}$$

where the second and third terms are due to integrator leakage.

Note that in an Lth-order loop, L terms corresponding to shapings of order $0, 1, 2, \ldots, L-1$ are added. These extra terms are proportional to decreasing powers of μ—i.e., $\mu^{2L}, \mu^{2(L-1)}, \ldots, \mu^4, \mu^2$, respectively— so that, for usual values of the oversampling ratio and the amplifier DC gain, the dominant extra term is that with a shaping of order $L-1$. Taking this into account, eq(2.12) can be approximated to

$$P_Q(\mu) \cong \frac{\Delta^2}{12}\left[\frac{\pi^{2L}}{(2L+1) OSR^{(2L+1)}} + \frac{L\mu^2\pi^{2L-2}}{(2L-1) OSR^{(2L-1)}}\right] \tag{2.13}$$

Expressing eq(2.13) in relative terms to the ideal quantization error, we get that the increase of the in-band error power due to leakages is

$$\Delta P_Q(\mu)\big|_{\mathrm{dB}} = 10\log_{10}\left(1 + L \cdot \frac{2L+1}{2L-1} \cdot \frac{\mu^2 OSR^2}{\pi^2}\right) \tag{2.14}$$

Note that the increase of the in-band error power depends on OSR^2 and will exhibit a smooth increase when increasing the order of the loop.

2.1.2 Cascade ΣΔ modulators

Let us consider the generic N-stage cascade ΣΔ modulator shown in Fig. 2.3. As stated in the previous chapter, in this kind of topologies an Lth-order modulator is built by cascading unconditionally stable ΣΔMs of order $L_i \leq 2$, so that the quantization error $E_i(z)$ of one stage is re-shaped by the following ones. Then, all but the last-stage error are cancelled out in the digital domain, so that the modulator output after digital cancellation is

$$\begin{aligned} Y(z) &= STF(z)X(z) + NTF_N(z)E_N(z) \\ &= z^{-L}X(z) + d_{2N-3}(1 - z^{-1})^L E_N(z) \end{aligned} \tag{2.15}$$

where $L = L_1 + L_2 + \ldots + L_N$ and d_{2N-3} is a scaling coefficient larger than unity—2 or 4 are the most common values—resulting from the analog-domain scaling required to prevent premature overloading (see Section 1.4).

Given eq(2.15), the ideal quantization error power is given by

$$P_Q \cong \frac{\Delta_N^2}{12} d_{2N-3}^2 \left[\frac{\pi^{2L}}{(2L+1)OSR^{(2L+1)}} \right] \tag{2.16}$$

with Δ_N being the level spacing in the B_N-bit quantizer of the Nth stage.

If the integrator leakage are considered, the noise transfer functions of the different stages are modified and their quantization errors are incompletely cancelled in the digital domain, so that the modulator output becomes

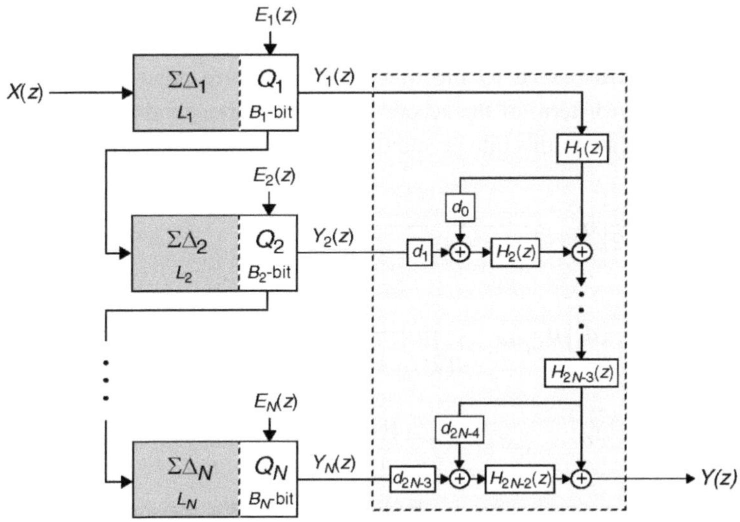

FIGURE 2.3 Generic N-stage cascade ΣΔ modulator; cancellation logic in dashed box.

$$Y(z) = STF(z)X(z) + \sum_{i=1}^{N} NTF_i(z)E_i(z) \quad (2.17)$$

leading to the following expression for the output-referred in-band error power

$$\begin{aligned} P_Q(\mu) \cong & \frac{\Delta_1^2}{12}\left[\sum_{m=0}^{L_1-1} \alpha_m \frac{\mu^{2(L_1-m)}\pi^{2m}}{(2m+1)OSR^{(2m+1)}}\right] \\ & + \frac{\Delta_2^2}{12}d_1^2\left[\sum_{m=L_1}^{L_1+L_2-1} \alpha_m \frac{\mu^{2(L_1+L_2-m)}\pi^{2m}}{(2m+1)OSR^{(2m+1)}}\right] + \dots \\ & \frac{\Delta_N^2}{12}d_{2N-3}^2\left[\sum_{m=L-L_N}^{L-1} \alpha_m \frac{\mu^{2(L-m)}\pi^{2m}}{(2m+1)OSR^{(2m+1)}} + \frac{\pi^{2L}}{(2L+1)OSR^{(2L+1)}}\right] \end{aligned} \quad (2.18)$$

where Δ_i is the level spacing in the quantizer of the ith stage, α_m are scalars related to the order of the stages, and $d_1, \dots, d_{2N-3}$ are the digital coefficients in the direct path from the output of the stages to $Y(z)$ (see Fig. 2.3).

As for Lth-order loops, note from eq(2.18) that, in an Lth-order cascade, L terms corresponding to shapings of order $0, 1, 2, \dots, L-1$ are added, but each stage of order L_i contributes with L_i terms with increasing shaping and decreasing powers of μ. For instance, in a 2-2 cascade the leakages in the first stage will introduce terms proportional to μ^4/OSR and μ^2/OSR^3, whereas the second stage will add terms proportional to μ^4/OSR^5 and μ^2/OSR^7 to the ideal one ($\propto 1/OSR^9$).

For usual values of the oversampling ratio and the amplifier DC gain, the dominant term added by the first stage will correspond to $m = L_1 - 1$, the dominant error term of the second stage will correspond to $m = L_1 + L_2 - 1$, and so on. Taking this into account, eq(2.18) can be approximated to

$$\begin{aligned} P_Q(\mu) \cong & \frac{\Delta_1^2}{12}\left[\alpha_{L_1-1}\frac{\mu^2\pi^{2L_1-2}}{(2L_1-1)OSR^{(2L_1-1)}}\right] \\ & + \frac{\Delta_2^2}{12}d_1^2\left[\alpha_{L_1+L_2-1}\frac{\mu^2\pi^{2(L_1+L_2)-2}}{[2(L_1+L_2)-1]OSR^{[2(L_1+L_2)-1]}}\right] + \dots \\ & \frac{\Delta_N^2}{12}d_{2N-3}^2\left[\alpha_{L-1}\frac{\mu^2\pi^{2L-2}}{(2L-1)OSR^{(2L-1)}} + \frac{\pi^{2L}}{(2L+1)OSR^{(2L+1)}}\right] \end{aligned} \quad (2.19)$$

The former equation can be particularized for the most common cascade topologies. Results are summarized in Table 2.1, where the value of the scalar α_m is also included.

TABLE 2.1 In-band error power considering leakages in cascade ΣΔMs.

	$P_Q(\mu)$
1-1-1	$\frac{\Delta_1^2}{12}\left(\frac{\mu^2}{OSR}\right)+\frac{\Delta_2^2}{12}d_1^2\left(\frac{\mu^2\pi^2}{3OSR^3}\right)+\frac{\Delta_3^2}{12}d_3^2\left(\frac{\mu^2\pi^4}{5OSR^5}+\frac{\pi^6}{7OSR^7}\right)$
1-1-1-1	$\frac{\Delta_1^2}{12}\left(\frac{\mu^2}{OSR}\right)+\frac{\Delta_2^2}{12}d_1^2\left(\frac{\mu^2\pi^2}{3OSR^3}\right)+\frac{\Delta_3^2}{12}d_3^2\left(\frac{\mu^2\pi^4}{5OSR^5}\right)+\frac{\Delta_4^2}{12}d_5^2\left(\frac{\mu^2\pi^6}{7OSR^7}+\frac{\pi^8}{9OSR^9}\right)$
2-2	$\frac{\Delta_1^2}{12}\left(\frac{4\mu^2\pi^2}{3OSR^3}\right)+\frac{\Delta_2^2}{12}d_1^2\left(\frac{2\mu^2\pi^6}{7OSR^7}+\frac{\pi^8}{9OSR^9}\right)$
2-2-2	$\frac{\Delta_1^2}{12}\left(\frac{4\mu^2\pi^2}{3OSR^3}\right)+\frac{\Delta_2^2}{12}d_1^2\left(\frac{4\mu^2\pi^6}{7OSR^7}\right)+\frac{\Delta_3^2}{12}d_3^2\left(\frac{2\mu^2\pi^{10}}{11OSR^{11}}+\frac{\pi^{12}}{13OSR^{13}}\right)$
2-1	$\frac{\Delta_1^2}{12}\left(\frac{4\mu^2\pi^2}{3OSR^3}\right)+\frac{\Delta_2^2}{12}d_1^2\left(\frac{\mu^2\pi^4}{5OSR^5}+\frac{\pi^6}{7OSR^7}\right)$
2-1-1	$\frac{\Delta_1^2}{12}\left(\frac{4\mu^2\pi^2}{3OSR^3}\right)+\frac{\Delta_2^2}{12}d_1^2\left(\frac{\mu^2\pi^4}{5OSR^5}\right)+\frac{\Delta_3^2}{12}d_3^2\left(\frac{\mu^2\pi^6}{7OSR^7}+\frac{\pi^8}{9OSR^9}\right)$
2-1-1-1	$\frac{\Delta_1^2}{12}\left(\frac{4\mu^2\pi^2}{3OSR^3}\right)+\frac{\Delta_2^2}{12}d_1^2\left(\frac{\mu^2\pi^4}{5OSR^5}\right)+\frac{\Delta_3^2}{12}d_3^2\left(\frac{\mu^2\pi^6}{7OSR^7}\right)+\frac{\Delta_4^2}{12}d_5^2\left(\frac{\mu^2\pi^8}{9OSR^9}+\frac{\pi^{10}}{11OSR^{11}}\right)$

If the noise leakage from the first stage is considered to be dominant, eq(2.19) can be further simplified to

$$P_Q(\mu) \cong \frac{\Delta^2}{12}\left[\frac{\alpha_{L_1-1}\mu^2\pi^{2L_1-2}}{(2L_1-1)OSR^{(2L_1-1)}}\right]+\frac{\Delta_N^2}{12}d_{2N-3}^2\left[\frac{\pi^{2L}}{(2L+1)OSR^{(2L+1)}}\right] \tag{2.20}$$

where scalar α_{L_1-1} equals 1 if $L_1 = 1$, or 4 if $L_1 = 2$.

Expressing eq(2.20) in relative terms to the ideal quantization error, we get that the increase of the in-band error power due to leakages is

$$\Delta P_Q(\mu)\big|_{\text{dB}} \cong 10\log_{10}\left[1+(2^{B_N}-1)^2\frac{\alpha_{L_1-1}}{d_{2N-3}^2}\cdot\frac{2L+1}{2L_1-1}\cdot\frac{\mu^2 OSR^{2(L-L_1+1)}}{\pi^{2(L-L_1+1)}}\right] \tag{2.21}$$

given that $\Delta_N = \Delta/(2^{B_N}-1)$, with B_N being the resolution of the last-stage quantizer.

Note from eq(2.21) that:

- Cascade topologies are always more sensitive to leakages than single-loop topologies.
- Unlike single-loop topologies, the sensitivity to leakages rapidly increases when increasing the order of the modulator.

- For a given order of the cascade, using a 1st-order front-end stage leads to more sensitivity to leakages than using a 2nd-order stage.
- For a given cascade topology, the impact of leakages increases as the resolution in the multi-bit last stage does, because the ideal quantization error is reduced, while the error power due to leakages remains the same.

Finally, the above statements are illustrated in the following figures. The curves for which the modulator order is specified correspond to 'ideal' single-loop topologies †2, whereas the rest correspond to different cascade architectures.

Fig. 2.4 shows the obtained dynamic range *DR* versus the amplifier DC gain for several 4th-order architectures and different oversampling ratios. For given L and OSR, the ideal DR of cascade architectures differs from that of a single-loop implementation due to the scaling factor d_{2N-3}—see eq(2.16).

Fig. 2.5 shows the required amplifier DC gain versus *OSR* in order to lose only 1-bit in DR. Single-loop and cascade architectures of order 3 to 6 have been considered.

Fig. 2.6 illustrates the increased sensitivity to leakages of multi-bit cascades. For a 2-1-1 cascade, the required DC gain to have 1-bit loss in *DR* is plotted, for resolutions in the last-stage quantizer ranging from 1 to 6 bits.

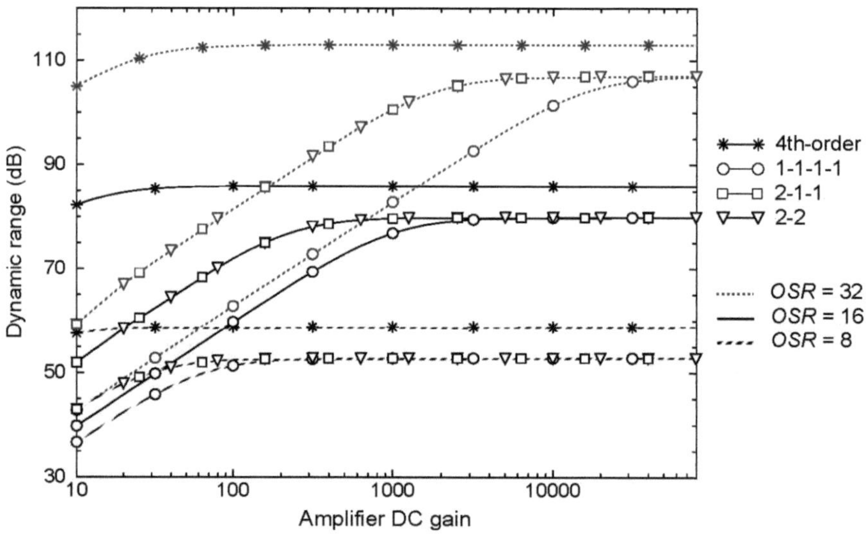

FIGURE 2.4 Dynamic range vs. amplifier DC gain for several 4th-order ΣΔ modulators.

2. With 'ideal' we refer to single-loop topologies implementing pure-differentiator (FIR) *NTF* s, assuming their stable operation, regardless the loop order. That is, the performance degradation introduced to ensure conditional stability is not considered (see Section 1.3.3).

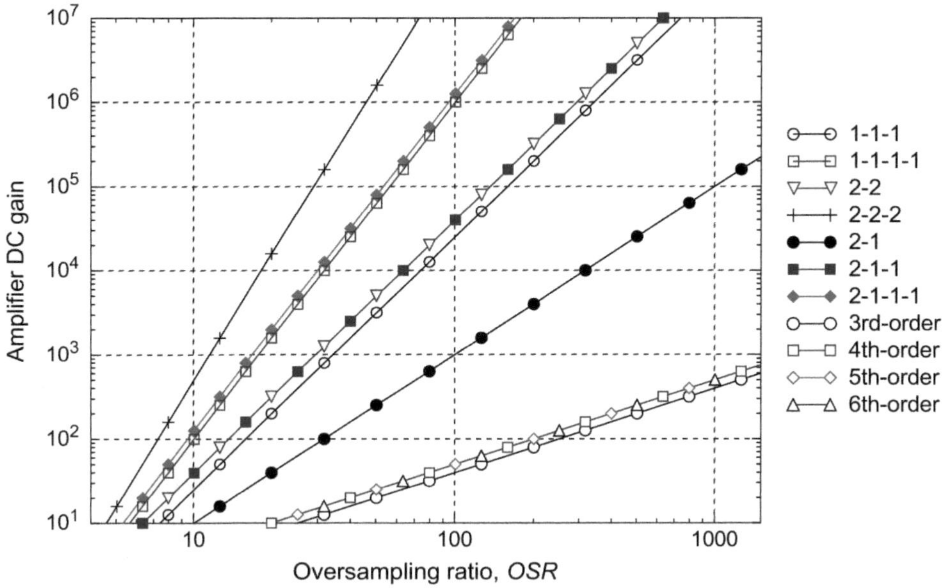

FIGURE 2.5 Sensitivity to integrator leakage of single-loop and cascade modulators. Required DC gain vs. oversampling ratio for 1-bit loss in the modulator dynamic range.

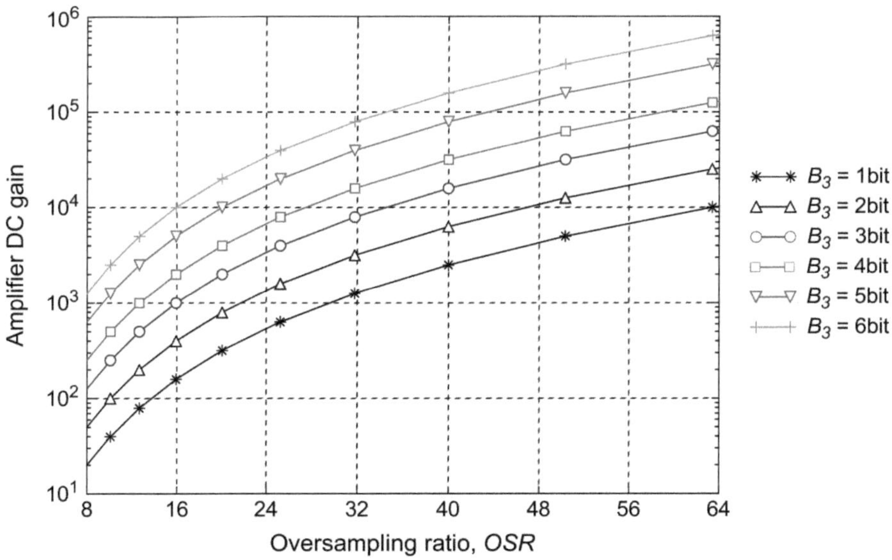

FIGURE 2.6 Sensitivity to integrator leakage in the 3-stage 2-1-1 multi-bit cascade. Required DC gain vs. oversampling ratio for 1-bit loss in dynamic range.

2.2 Capacitor Mismatch

In SC ΣΔ modulators, integrator weights are implemented by means of capacitor ratios, so that the variation of the process parameters and sizes cause in practice a deviation of the implemented weights from their nominal values. For instance, if a weight g_i is implemented as the ratio of m_i to n_i unit capacitors, the actual implemented weight $g_i{*}$ will present an error ε_{g_i} given by

$$\left.\begin{aligned} g_i{*} &= g_i(1 \pm \varepsilon_{g_i}) \\ g_i &= \frac{m_i C_u}{n_i C_u} \end{aligned}\right\} \Rightarrow \quad \varepsilon_{g_i} = 3\frac{\sigma_{g_i}}{g_i} = 3\sqrt{\frac{1}{m_i} + \frac{1}{n_i}}\sigma_C \tag{2.22}$$

where the worst-case error in the weight has been estimated as three times its relative standard deviation and $\sigma_C = \sigma_{C_u}/C_u$ is the relative sigma of the unit capacitor C_u used. Note from eq(2.22) that, for the same unit capacitor, ε_{g_i} will be divided by $\sqrt{2}$ in a fully-differential implementation.

These weight errors alter the relationships among coefficients required for the correct functioning of ΣΔMs, so that they usually map into modifications of the ideal noise transfer function. In this section we show that—as for integrator leakage—capacitor mismatch causes an increase of the in-band error power that is heavily dependent on the modulator architecture.

2.2.1 Single-loop ΣΔ modulators

2nd-order loop

Let us consider the 2nd-order ΣΔM illustrated in Fig. 2.7*a*. Assuming that the actual integrator weights deviate from their nominal values following $g_i{*} = g_i(1 \pm \varepsilon_{g_i})$, the modulator output in the z-domain is [†3]

$$\begin{aligned} Y(z) &= STF(z)X(z) + NTF(z)E(z) \\ STF(z) &\cong (1 - |\varepsilon_{g_1'} - \varepsilon_{g_1}|)z^{-2} \cong z^{-2} \\ NTF(z) &\cong (1 + \varepsilon_g)(1 - z^{-1})^2 \quad , \quad \varepsilon_g = \varepsilon_{g_2} + \varepsilon_{g_1'} \end{aligned} \tag{2.23}$$

where terms in $\varepsilon_{g_i}^2$ have been neglected.

Note from eq(2.23) that the errors in the weights lead to a small

3. It is assumed that the nominal weights verify $g_1 = g_1'$ and $g_2' = 2g_1'g_2$. In case $g_1 \neq g_1'$ the signal transfer function $STF(z)$ will be multiplied by a factor g_1/g_1'.

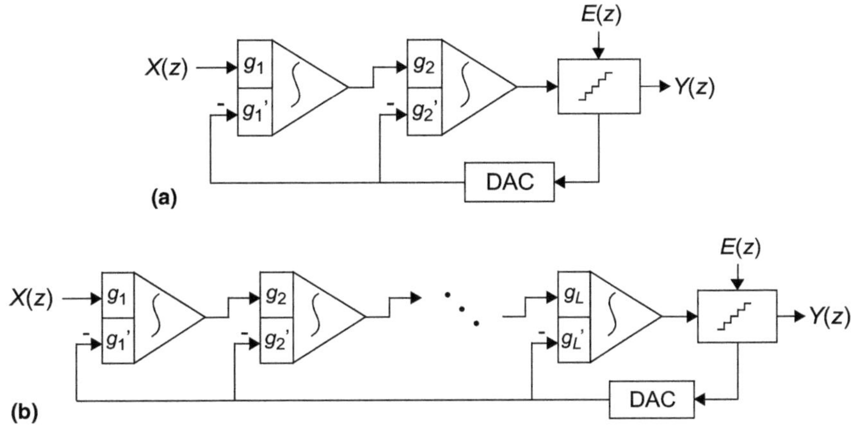

FIGURE 2.7 Single-loop ΣΔMs: (a) 2nd-order modulator, (b) Lth -order modulator.

decrease in the gain of STF, which can be neglected in practice. On the other hand, the noise transfer function still provides a 2nd-order shaping of the quantization error, but the mismatch causes an increase of the modulus of NTF due to the errors in g_1' and g_2.

The output-referred in-band error power under capacitor mismatch is

$$P_Q(\varepsilon_g) \cong \frac{\Delta^2}{12}\left[\frac{(1+\varepsilon_g)^2\pi^4}{5OSR^5}\right] \tag{2.24}$$

where eq(2.22) can be used to calculate the worst-case errors in g_1' and g_2 considering their actual implementation in terms of unit capacitors.

Lth-order loops

Likewise, the output of an Lth -order single-loop ΣΔM, like the one illustrated in Fig. 2.7b, can be written as

$$Y(z) \cong z^{-L}X(z) + (1+\varepsilon_g)(1-z^{-1})^L E(z),\ \ \varepsilon_g = \varepsilon_{g_L} + \ldots + \varepsilon_{g_2} + \varepsilon_{g_1'} \tag{2.25}$$

where the Lth -order shaping of $E(z)$ is maintained and the increase of the modulus of NTF leads to

$$P_Q(\varepsilon_g) \cong \frac{\Delta^2}{12}\left[\frac{(1+\varepsilon_g)^2\pi^{2L}}{(2L+1)OSR^{(2L+1)}}\right] \tag{2.26}$$

Expressing eq(2.26) in relative terms to the ideal quantization error, the increase of the in-band error power due to capacitor mismatch is obtained as

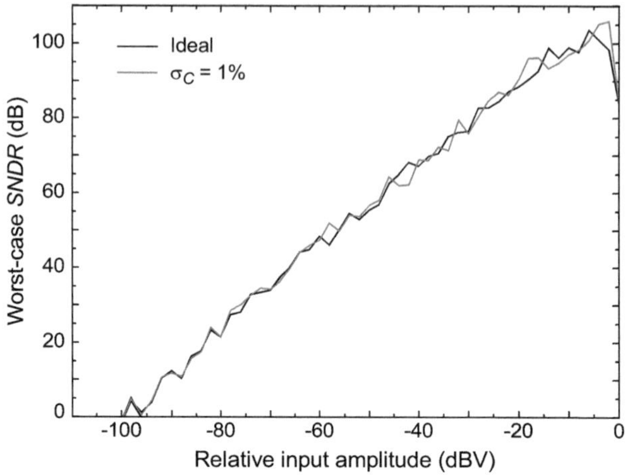

FIGURE 2.8 *SNDR* vs. relative input amplitude for a 2nd-order ΣΔ modulator.

$$\Delta P_Q(\varepsilon_g)\big|_{\text{dB}} = 20\log_{10}(1+\varepsilon_g) \tag{2.27}$$

Note that it is independent of the oversampling ratio and only exhibits a smooth increase with the loop order, because the more weights to be implemented, the more error terms will be appear in ε_g—see eq(2.25).

This renders single-loop topologies relatively insensitive to capacitor mismatch, and the performance of those modulators is not degraded for typical values of σ_C attainable in nowadays technologies ($0.05\% \sim 0.1\%$). This is illustrated in Fig. 2.8, that shows the signal-to-(noise + distortion) ratio *SNDR* of a 2nd-order loop versus the input amplitude, obtained through behavioral simulation. Note that considering a capacitor mismatch as large as 1% does not affect much the modulator *SNDR*.

2.2.2 Cascade ΣΔ modulators

Let us consider an Lth-order N-stage cascade, like that in Fig. 2.3. If the actual integrator weights are given by $g_i^* = g_i(1 \pm \varepsilon_{g_i})$, the modulator output—ideally given by eq(2.15)—is now re-calculated as

$$\begin{aligned} Y(z) \cong {} & z^{-L}X(z) + \varepsilon_1 z^{-(L-L_1)}(1-z^{-1})^{L_1}E_1(z) \\ & + d_1\varepsilon_2 z^{-(L-L_1-L_2)}(1-z^{-1})^{(L_1+L_2)}E_2(z) + \ldots \\ & + d_{2N-3}(1+\varepsilon_N)(1-z^{-1})^{L}E_N(z) \end{aligned} \tag{2.28}$$

where errors $\varepsilon_1, \varepsilon_2, \ldots, \varepsilon_N$ are related to the errors in the integrator weights.

Note from eq(2.28) that, besides an increase of the modulus of the last-stage quantization error—similar to what happens in single-loop topologies—, mismatch causes incomplete cancellation of the quantization error of the remaining stages, so that they 'leak' to the modulator output. This makes cascade topologies more sensitive to mismatch than single-loop modulators, since part of the quantization error of the first stage, $E_1(z)$, is present at the output with a shaping of order L_1, a portion of $E_2(z)$ appears with a shaping of order $L_1 + L_2$, etc.

The output-referred in-band error power considering mismatch in the weights becomes

$$P_Q(\varepsilon_g) \cong \frac{\Delta_1^2}{12}\left[\frac{\varepsilon_1^2 \pi^{2L_1}}{(2L_1+1)OSR^{(2L_1+1)}}\right] + \frac{\Delta_2^2}{12} d_1^2 \left[\frac{\varepsilon_2^2 \pi^{2(L_1+L_2)}}{[2(L_1+L_2)+1]OSR^{[2(L_1+L_2)+1]}}\right] + \ldots + \frac{\Delta_N^2}{12} d_{2N-3}^2 \left[\frac{(1+\varepsilon_N)^2 \pi^{2L}}{(2L+1)OSR^{(2L+1)}}\right] \tag{2.29}$$

with Δ_i being the spacing between adjacent levels in the quantizer of the ith stage.

Eq(2.29) can be particularized for most common cascade topologies. Results are summarized in Table 2.2, whereas the expressions for $\varepsilon_1, \varepsilon_2, \ldots, \varepsilon_N$ as functions of the errors in the integrator weights are shown in Table 2.3.

Note from Table 2.3 that some errors contain contributions like $|\varepsilon_{g_i''} - \varepsilon_{g_i}|$, relating weights of the same integrator, in which the effect of $\varepsilon_{g_i''}$ and ε_{g_i} will be partially cancelled. For instance, if the weight g_i is implemented as the ratio of a sampling capacitor formed by m_i unit elements to a feedback capacitor formed by n_i unit elements and g_i'' is the ratio of m_i'' unit elements to the same n_i unit elements above, the partial cancellation of their errors yields †4

$$g_i = \frac{m_i C_u}{n_i C_u}, \quad g_i'' = \frac{m_i'' C_u}{n_i C_u} \quad \Rightarrow \quad |\varepsilon_{g_i''} - \varepsilon_{g_i}| = 3\sqrt{\frac{1}{m_i''} + \frac{1}{m_i}}\sigma_C \tag{2.30}$$

---

4. In order to reduce the number of sampling unit capacitors and save area, it is common practice to distribute the weights of a given integrator so that some unit capacitors are used in more than one weight. In such a case, the errors may be further cancelled. For instance, if $g_i' = m_i'/n_i$, $g_i'' = m_i''/n_i$, and $g_i = (m_i' + m_i'')/n_i$, the error cancellation will lead to

$$|\varepsilon_{g_i''} - \varepsilon_{g_i}| = 3\sqrt{\frac{1}{m_i''} - \frac{1}{m_i' + m_i''}}\sigma_C$$

TABLE 2.2 In-band error power considering weights mismatch in cascade ΣΔMs.

	$P_Q(\varepsilon_g)$
1-1-1	$\frac{\Delta_1^2}{12}\left(\frac{\varepsilon_1^2\pi^2}{3OSR^3}\right)+\frac{\Delta_2^2}{12}d_1^2\left(\frac{\varepsilon_2^2\pi^4}{5OSR^5}\right)+\frac{\Delta_3^2}{12}d_3^2\left[\frac{(1+\varepsilon_3)^2\pi^6}{7OSR^7}\right]$
1-1-1-1	$\frac{\Delta_1^2}{12}\left(\frac{\varepsilon_1^2\pi^2}{3OSR^3}\right)+\frac{\Delta_2^2}{12}d_1^2\left(\frac{\varepsilon_2^2\pi^4}{5OSR^5}\right)+\frac{\Delta_3^2}{12}d_3^2\left(\frac{\varepsilon_3^2\pi^6}{7OSR^7}\right)+\frac{\Delta_4^2}{12}d_5^2\left[\frac{(1+\varepsilon_4)^2\pi^8}{9OSR^9}\right]$
2-2	$\frac{\Delta_1^2}{12}\left(\frac{\varepsilon_1^2\pi^4}{5OSR^5}\right)+\frac{\Delta_2^2}{12}d_1^2\left[\frac{(1+\varepsilon_2)^2\pi^8}{9OSR^9}\right]$
2-2-2	$\frac{\Delta_1^2}{12}\left(\frac{\varepsilon_1^2\pi^4}{5OSR^5}\right)+\frac{\Delta_2^2}{12}d_1^2\left(\frac{\varepsilon_2^2\pi^8}{9OSR^9}\right)+\frac{\Delta_3^2}{12}d_3^2\left[\frac{(1+\varepsilon_3)^2\pi^{12}}{13OSR^{13}}\right]$
2-1	$\frac{\Delta_1^2}{12}\left(\frac{\varepsilon_1^2\pi^4}{5OSR^5}\right)+\frac{\Delta_2^2}{12}d_1^2\left[\frac{(1+\varepsilon_2)^2\pi^6}{7OSR^7}\right]$
2-1-1	$\frac{\Delta_1^2}{12}\left(\frac{\varepsilon_1^2\pi^4}{5OSR^5}\right)+\frac{\Delta_2^2}{12}d_1^2\left(\frac{\varepsilon_2^2\pi^6}{7OSR^7}\right)+\frac{\Delta_3^2}{12}d_3^2\left[\frac{(1+\varepsilon_3)^2\pi^8}{9OSR^9}\right]$
2-1-1-1	$\frac{\Delta_1^2}{12}\left(\frac{\varepsilon_1^2\pi^4}{5OSR^5}\right)+\frac{\Delta_2^2}{12}d_1^2\left(\frac{\varepsilon_2^2\pi^6}{7OSR^7}\right)+\frac{\Delta_3^2}{12}d_3^2\left(\frac{\varepsilon_3^2\pi^8}{9OSR^9}\right)+\frac{\Delta_4^2}{12}d_5^2\left[\frac{(1+\varepsilon_4)^2\pi^{10}}{11OSR^{11}}\right]$

TABLE 2.3 Errors in the noise leakage of the different stages for cascade ΣΔMs.

<table>
<tr><th></th><th>ε_1</th><th>ε_2</th><th>ε_3</th><th>ε_4</th></tr>
<tr><td>1-1-1</td><td rowspan="2">$\lvert\varepsilon_{g_2''}-\varepsilon_{g_2}\rvert+\varepsilon_{g_1'}$</td><td rowspan="2">$\lvert\varepsilon_{g_3''}-\varepsilon_{g_3}\rvert+\varepsilon_{g_2''}$</td><td>$\varepsilon_{g_3''}$</td><td>-</td></tr>
<tr><td>1-1-1-1</td><td>$\lvert\varepsilon_{g_4''}-\varepsilon_{g_4}\rvert+\varepsilon_{g_3''}$</td><td>$\varepsilon_{g_4''}$</td></tr>
<tr><td>2-2</td><td rowspan="5">$\lvert\varepsilon_{g_3''}-\varepsilon_{g_3}\rvert+$ $+\varepsilon_{g_2}+\varepsilon_{g_1'}$</td><td>$\varepsilon_{g_4}+\varepsilon_{g_3''}$</td><td>-</td><td>-</td></tr>
<tr><td>2-2-2</td><td>$\lvert\varepsilon_{g_5''}-\varepsilon_{g_5}\rvert+\varepsilon_{g_4}+\varepsilon_{g_3''}$</td><td>$\varepsilon_{g_6}+\varepsilon_{g_5''}$</td><td>-</td></tr>
<tr><td>2-1</td><td>$\varepsilon_{g_3''}$</td><td>-</td><td>-</td></tr>
<tr><td>2-1-1</td><td rowspan="2">$\lvert\varepsilon_{g_4''}-\varepsilon_{g_4}\rvert+\varepsilon_{g_3''}$</td><td>$\varepsilon_{g_4''}$</td><td>-</td></tr>
<tr><td>2-1-1-1</td><td>$\lvert\varepsilon_{g_5''}-\varepsilon_{g_5}\rvert+\varepsilon_{g_4''}$</td><td>$\varepsilon_{g_5''}$</td></tr>
</table>

If the noise leakage from the first stage is considered to be dominant, eq(2.29) can be simplified to

$$P_Q(\varepsilon_g)\cong\frac{\Delta^2}{12}\left[\frac{\varepsilon_1^2\pi^{2L_1}}{(2L_1+1)OSR^{(2L_1+1)}}\right]+\frac{\Delta_N^2}{12}d_{2N-3}^2\left[\frac{(1+\varepsilon_N)^2\pi^{2L}}{(2L+1)OSR^{(2L+1)}}\right] \quad (2.31)$$

so that, by comparing it with eq(1.65), the increase of the in-band error power due to mismatch is

$$\Delta P_Q(\varepsilon_g)\big|_{\mathrm{dB}} = 10\log_{10}\left[(1+\varepsilon_N)^2 + \frac{(2^{B_N}-1)^2}{d_{2N-3}^2}\cdot\frac{2L+1}{2L_1+1}\cdot\frac{\varepsilon_1^2 OSR^{2(L-L_1)}}{\pi^{2(L-L_1)}}\right] \quad (2.32)$$

where $\Delta_N = \Delta/(2^{B_N}-1)$ and B_N is the resolution of the multi-bit quantizer.

Note from eq(2.32) that:

- Unlike single-loop topologies, the sensitivity to mismatch rapidly increases with the oversampling ratio and the modulator order.
- For a given order, using a 1st-order front-end stage entails a larger sensitivity to mismatch than using a 2nd-order front-end stage.
- For a given cascade topology, the impact of mismatch increases when increasing B_N, because the last-stage quantization error is reduced, while the error power due to mismatch remains the same.

Fig. 2.9 shows the required capacitor matching versus OSR in order to lose only 1bit in DR for several cascade ΣΔ modulators.

Fig. 2.10*a* shows $SNDR$ curves of a 2-1-1 cascade, obtained through Monte Carlo behavioral simulation with $\sigma_C = 0.5\%$. Note that the modulator performance is considerably degraded. Fortunately, present CMOS technologies provide capacitor matching as good as $0.05\% \sim 0.1\%$ [Lin00], what greatly limits the degradation due to mismatch (see Fig. 2.10*b*).

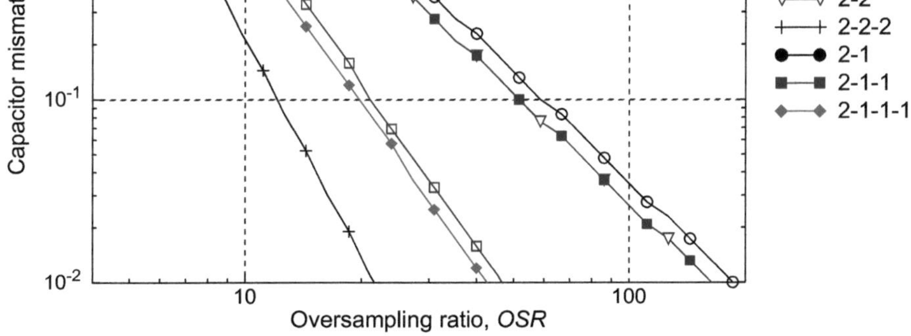

FIGURE 2.9 Sensitivity to capacitor mismatch of cascade modulators. Required standard deviation vs. oversampling ratio for 1-bit loss in the modulator dynamic range.

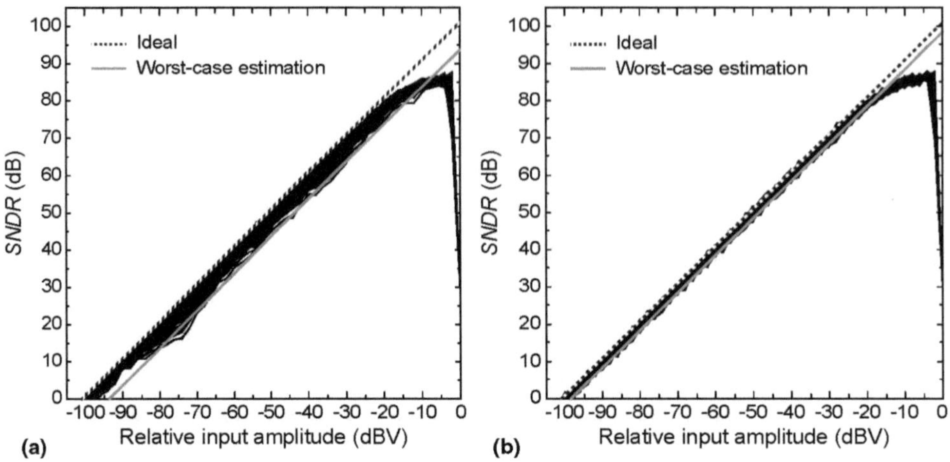

FIGURE 2.10 *SNDR* curves of a 2-1-1 ΣΔM operating with $OSR = 32$. Behavioral simulation results for a 100-run Monte Carlo and worst-case estimation—eq(2.31)—for: (a) $\sigma_C = 0.5\%$, (b) $\sigma_C = 0.1\%$.

2.3 Integrator Settling Error

In practice, dynamic limitations in SC integrators—basically due to the finite gain-bandwidth product *GB* and slew rate *SR* of the amplifiers—provoke errors in the charge transfer. The impact of the associated output voltage *settling error* on the modulator performance will be higher, the higher the sampling frequency. As the clock frequency increases in ΣΔ modulators to cope with wideband applications, integrator defective settling becomes one of the bottle necks in present SC designs. On the one hand, the time for the integrator operation is considerably shortened; on the other, the amplifier dynamic requirements must be minimized for reduced power consumption.

A proper knowledge of the mechanisms degrading the settling of SC integrators and quantitative analysis of the error becomes therefore mandatory to get efficient designs. Integrator settling errors have been previously studied by several authors, but although most SC integrator models take into account the amplifier finite *GB* and *SR* [Sans87] [Dias92] [Will94] [Wang97] [Mede99a], they do so only for the integration phase, whereas errors derived from the limited sampling dynamic are omitted. This may lead to an under-estimation of the defective settling error, specially important for high-speed applications. As an illustrative example, Fig. 2.11 shows the output voltage of one of the integrators in a high-speed ΣΔM. Note how the finite dynamic during sampling may also be a limiting factor.

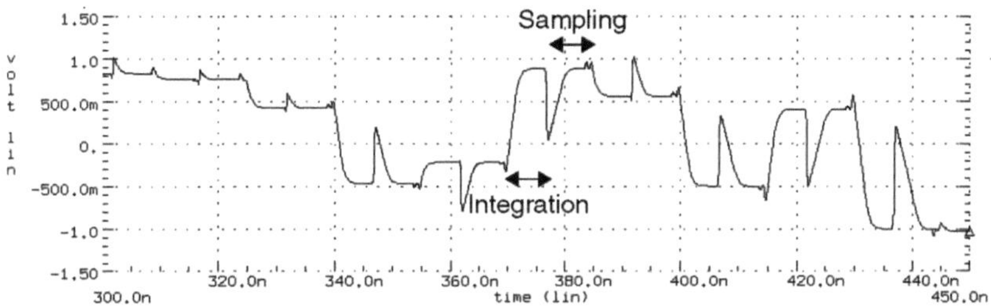

FIGURE 2.11 Influence of the opamp limited dynamic during both operation phases. Transient response of the 3rd-integrator output voltage in a 2-1-1 ΣΔM clocked at 64MHz [Rio01b].

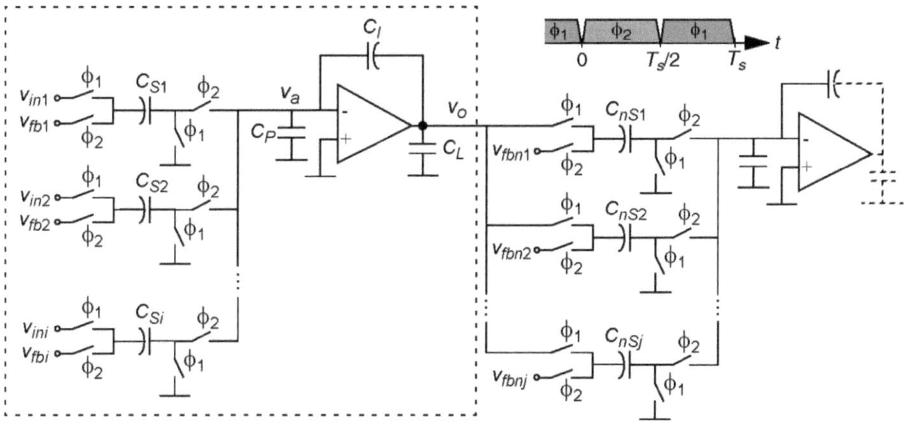

FIGURE 2.12 SC integrator (in dashed box), followed by a similar one for modeling purposes.

On the other hand, the SC integrator model considered in [Robe91] includes all former errors, but it focuses on filter design and the results cannot be easily extended to ΣΔ modulators.

The study developed here focuses on the analysis of the transient response of a generic SC integrator, during both integration and sampling. Results can be easily incorporated to CAD tools for the accurate simulation of high-speed ΣΔ modulators [Rio99]. Compact expressions are also derived for the estimation of the settling error power.

2.3.1 Model for the transient response of SC integrators

SC integrator model

In order to make a reliable analysis of the integrator transient response, the generic scheme in Fig. 2.12 is considered. This scheme includes:

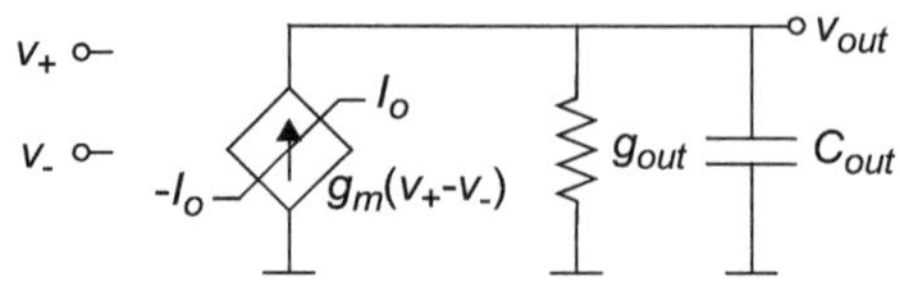

FIGURE 2.13 Amplifier model.

- A number i of input branches, each of them formed by a sampling capacitor C_{Si} and four switches—controlled by two non-overlapping phases, ϕ_1 and ϕ_2—which commute the sampling capacitor between voltages v_{ini} and v_{fbi}.
- A parasitic capacitor C_P associated to the summation node.
- A capacitive load C_L, which includes the amplifier output node parasitic and the one associated to the bottom plate of the integration capacitor (C_I) †5.
- A second integrator, whose j input branches are assumed to be connected to the first integrator output during the sampling phase. The j-th branch of the second integrator is connected to a voltage v_{fbnj} during the integration phase.

On the other hand, the model used for the amplifier is depicted in Fig. 2.13. This model includes:

- A single-pole dynamic.
- A non-linear characteristic, with maximum output current I_o.

This model of the SC integrator takes into account the amplifier GB and SR limitations, as well as the parasitic capacitors associated to its input and output nodes and switches. Moreover, the capacitive load at the integrator output is assumed to change from the integration to the sampling phase, reflecting the actual situation in most SC sections.

Transient during integration

We set out the time origin just at the beginning of an integration phase. Let $v_{a,es} = v_{a,n-1}$ and $v_{o,es} = v_{o,n-1}$ be the amplifier input and output voltages, respectively, at the end of the preceding sampling phase. At $t = 0$ (see Fig. 2.14), charge-conservation imposes a jump on these voltages to the following values

5. C_P and C_L can also include the respective switch parasitics, which can be a significant contribution, especially in high-speed high-resolution applications. This being the case, C_P and C_L are different during integration and sampling, because the state (on/off) of each switch varies from one phase to the other.

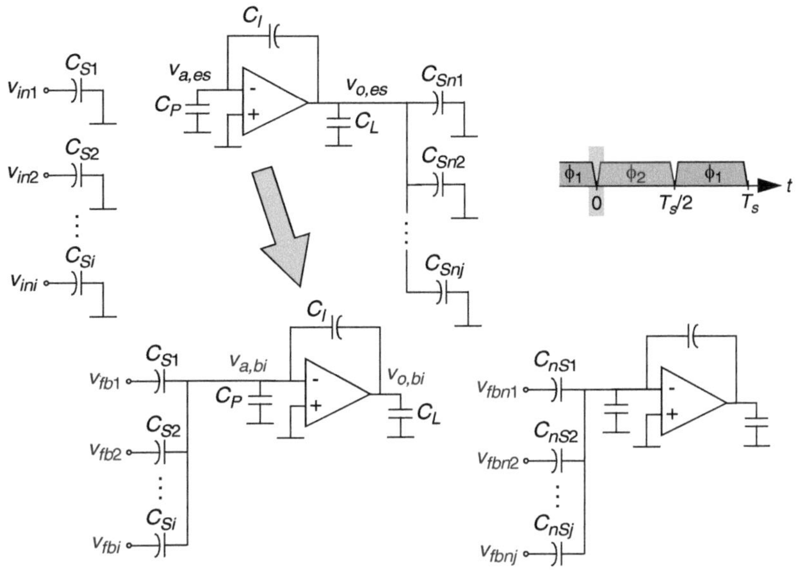

FIGURE 2.14 SC integrator at the beginning of the integration phase.

$$v_{a,bi} = \frac{\left(1+\frac{C_L}{C_I}\right)}{C_{eq,i}}[(v_{fb1,i}-v_{in1,es})C_{S1} + \dots + (v_{fbi,i}-v_{ini,es})C_{Si}] + \frac{C'}{C_{eq,i}}v_{a,es} \tag{2.33}$$

$$v_{o,bi} = v_{o,es} + \frac{C_I}{C_I + C_L}(v_{a,bi} - v_{a,es})$$

where $C' = C_P + C_L(1 + C_P/C_I)$ and $C_{eq,i}$ is the equivalent capacitive load at the amplifier output node during the integration phase, given by

$$C_{eq,i} = C_P + C_{S1} + \dots + C_{Si} + C_L\left(1 + \frac{C_P + C_{S1} + \dots + C_{Si}}{C_I}\right) \tag{2.34}$$

Note from eq(2.33) that v_a and v_o jump at $t = 0$ in the opposite direction to their final values. It must be also remarked that—unlike traditional models [Sans87] [Dias92] [Will94] [Wang97] [Mede99a]— $v_{a,es} \neq 0$, which reflects the possibility of inheriting some residual charge in C_P from the preceding sampling phase.

The integrator output voltage starts to evolve from the initial value in eq(2.33) towards its final value, according to the equivalent circuit model in Fig. 2.15. The evolution during integration depends on the initial value of the amplifier input voltage. Two cases can be distinguished:

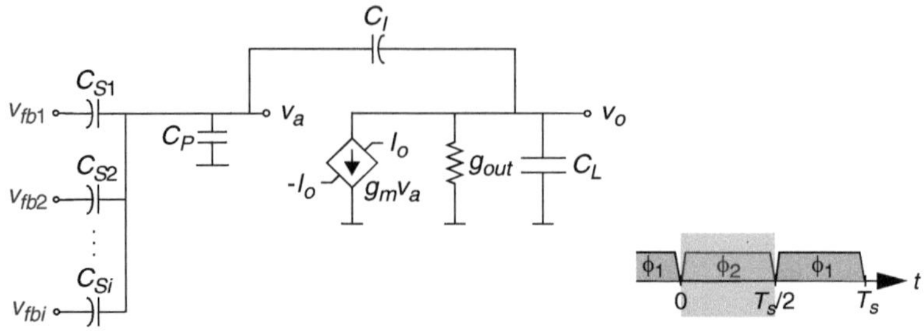

FIGURE 2.15 Equivalent circuit for the SC integrator during integration.

- $|v_{a,bi}| \le I_o/g_m$, with I_o being the amplifier maximum output current and g_m its transconductance. Then, the amplifier operates linearly and its input node discharges following the well-known exponential law

$$v_{a,i}(t) = v_{a,bi}\exp\left(-\frac{g_m}{C_{eq,i}}t\right) \tag{2.35}$$

where $g_m \gg g_{out}$ has been assumed.

- $|v_{a,bi}| > I_o/g_m$, so that the amplifier slews and its input node will then discharge linearly

$$v_{a,i}(t) = v_{a,bi} - \frac{I_o}{C_{eq,i}}\mathrm{sgn}(v_{a,bi})t \tag{2.36}$$

The slew mode will go on until $t = t_{o,i}$, when the condition for the amplifier to start operating linearly, $v_{a,i}(t_{o,i}) = I_o/g_m$, is satisfied. From this condition we get

$$t_{o,i} = \frac{C_{eq,i}}{I_o}|v_{a,bi}| - \frac{C_{eq,i}}{g_m} \tag{2.37}$$

Finally, for $t \ge t_{o,i}$, $v_{a,i}(t)$ relaxes exponentially following

$$v_{a,i}(t) = \frac{I_o}{g_m}\mathrm{sgn}(v_{a,bi})\exp\left[-\frac{g_m}{C_{eq,i}}(t - t_{o,i})\right] \tag{2.38}$$

During the integration phase $v_o(t)$ is given by

$$\begin{aligned} v_{o,i}(t) = {} & v_{o,es} - \left(1 + \frac{C_P}{C_I}\right)v_{a,es} + \frac{C_{S1}}{C_I}(v_{in1,es} - v_{fb1,i}) + \ldots \\ & \ldots + \frac{C_{Si}}{C_I}(v_{ini,es} - v_{fbi,i}) + \left(1 + \frac{C_P + C_{S1} + \ldots + C_{Si}}{C_I}\right)v_{a,i}(t) \end{aligned} \tag{2.39}$$

where $v_{a,i}(t)$ is calculated from equations (2.35), (2.38), or (2.36), depending on the integrator transient response during this phase: linear, partial-slew, or full-slew, respectively. These three evolutions are illustrated in Fig. 2.16.

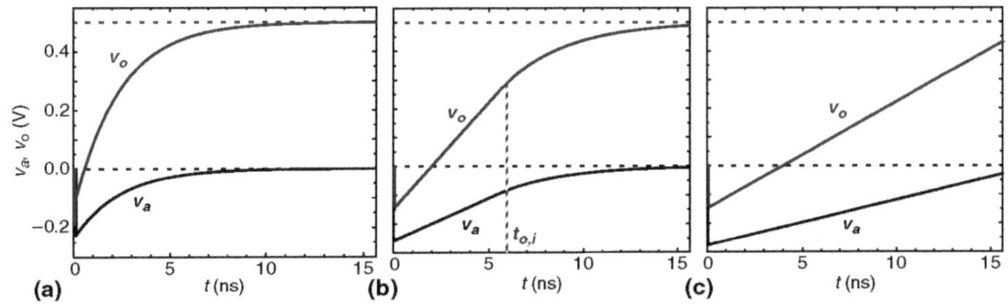

FIGURE 2.16 Transient response of the SC integrator during integration: (a) Linear, (b) Partial-slew, and (c) Full-slew operations.

At the end of integration, $t = T_s/2$, v_a and v_o will be given by

$$v_{a,ei} = \begin{cases} v_{a,bi}\exp\left(-\dfrac{g_m}{C_{eq,i}}\dfrac{T_s}{2}\right), & |v_{a,bi}| \le \dfrac{I_o}{g_m} \quad (2.40a) \\ v_{a,bi} - \dfrac{I_o}{C_{eq,i}}\mathrm{sgn}(v_{a,bi})\dfrac{T_s}{2}, & \dfrac{T_s}{2} \le t_{o,i}\ ,\ |v_{a,bi}| > \dfrac{I_o}{g_m} \quad (2.40b) \\ \dfrac{I_o}{g_m}\mathrm{sgn}(v_{a,bi})\exp\left[-\dfrac{g_m}{C_{eq,i}}\left(\dfrac{T_s}{2} - t_{o,i}\right)\right], & \dfrac{T_s}{2} > t_{o,i}\ ,\ |v_{a,bi}| > \dfrac{I_o}{g_m} \quad (2.40c) \end{cases}$$

$$\begin{aligned} v_{o,ei} = {} & v_{o,es} - \left(1 + \frac{C_P}{C_I}\right)v_{a,es} + \frac{C_{S1}}{C_I}(v_{in1,es} - v_{fb1,i}) + \ldots \\ & + \frac{C_{Si}}{C_I}(v_{ini,es} - v_{fbi,i}) + \left(1 + \frac{C_P + C_{S1} + \ldots + C_{Si}}{C_I}\right)v_{a,ei} \end{aligned} \quad (2.41)$$

Transient during sampling

Let $v_{a,ei}$ and $v_{o,ei}$ be the amplifier input and output voltage, respectively, at the end of the preceding integration phase. At $t = T_s/2$ (see Fig. 2.17) charge-conservation imposes a new jump on these voltages to †6

$$\begin{aligned} v_{a,bs} &= v_{a,ei} - \frac{C_{Sn1}}{C_{eq,s}}(v_{o,ei} - v_{fbn1,i}) - \ldots - \frac{C_{Snj}}{C_{eq,s}}(v_{o,ei} - v_{fbnj,i}) \\ v_{o,bs} &= v_{o,ei} + \left(1 + \frac{C_P}{C_I}\right)(v_{ai,s} - v_{a,ei}) \end{aligned} \quad (2.42)$$

6. To simplify the analytical expressions it is assumed that the summation node of the next integrator is totally relaxed at the end of the preceding integration phase. This is not rigorously true, especially if it has followed a full-slew operation (what must be avoided in a good design). The modification of the expressions above in order to account for this possibility is straight-forward and is incorporated to the behavioral simulation of ΣΔMs.

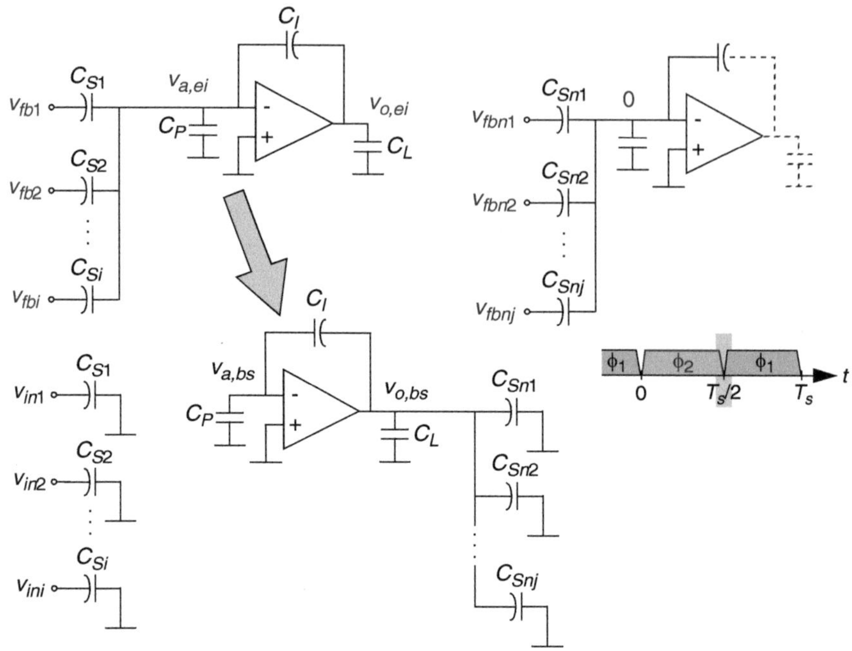

FIGURE 2.17 SC integrator at the beginning of the sampling phase.

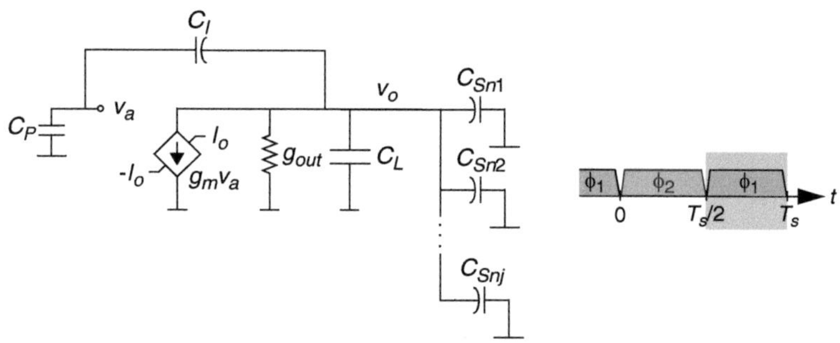

FIGURE 2.18 Equivalent circuit for the SC integrator during sampling.

with $C_{eq,s}$ being the equivalent capacitive load at the integrator output node during the sampling phase, given by

$$C_{eq,s} = C_P + (C_L + C_{Sn1} + \ldots + C_{Snj})\left(1 + \frac{C_P}{C_I}\right) \qquad (2.43)$$

According to eq(2.42), v_a and v_o once again jump in the opposite direction to their final values. Depending on the initial amplifier input voltage, two possibilities can be distinguished for the integrator evolution during sampling (see equivalent circuit in Fig. 2.18):

- $|v_{a,bs}| \le I_o/g_m$, and the amplifier operates linearly

$$v_{a,s}(t) = v_{a,bs}\exp\left[-\frac{g_m}{C_{eq,s}}\left(t-\frac{T_s}{2}\right)\right] \tag{2.44}$$

- $|v_{a,bs}| > I_o/g_m$, so that the amplifier slews and its input node discharges linearly following

$$v_{a,s}(t) = v_{a,bs} - \frac{I_o}{C_{eq,s}}\mathrm{sgn}(v_{a,bs})\left(t-\frac{T_s}{2}\right) \tag{2.45}$$

The slew will continue until $t = t_{o,s}$, when $v_{a,s}(t_{o,s}) = I_o/g_m$ and the amplifier enters linear operation. From this condition we get

$$t_{o,s} = \frac{T_s}{2} + \frac{C_{eq,s}}{I_o}|v_{a,bs}| - \frac{C_{eq,s}}{g_m} \tag{2.46}$$

For $t \ge t_{o,s}$, $v_{a,s}(t)$ will relax exponentially

$$v_{a,s}(t) = \frac{I_o}{g_m}\mathrm{sgn}(v_{a,bs})\exp\left[-\frac{g_m}{C_{eq,s}}(t-t_{o,s})\right] \tag{2.47}$$

During the sampling phase $v_o(t)$ is given by

$$v_{o,s}(t) = v_{o,ei} + \left(1+\frac{C_P}{C_I}\right)[v_{a,s}(t) - v_{a,ei}] \tag{2.48}$$

where $v_{a,s}(t)$ stands for equations (2.44), (2.47), or (2.45) depending on the integrator evolution during this phase: linear, partial-slew, or full-slew, respectively. These three possibilities are illustrated in Fig. 2.19.

At the end of the sampling phase, $t = T_s$, v_a and v_o are given by

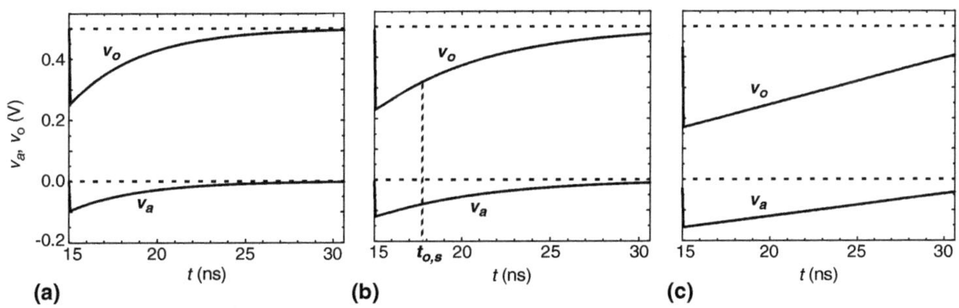

FIGURE 2.19 Transient response of the SC integrator during sampling: (a) Linear, (b) Partial-slew, and (c) Full-slew operations.

$$v_{a,n} \equiv v_{a,es} = \begin{cases} v_{a,bs}\exp\left(-\frac{g_m}{C_{eq,s}}\frac{T_s}{2}\right), & |v_{a,bs}| \le \frac{I_o}{g_m} \quad (2.49a) \\ v_{a,bs} - \frac{I_o}{C_{eq,s}}\operatorname{sgn}(v_{a,bs})\frac{T_s}{2}, & T_s \le t_{o,s},\ |v_{a,bs}| > \frac{I_o}{g_m} \quad (2.49b) \\ \frac{I_o}{g_m}\operatorname{sgn}(v_{a,bs})\exp\left[-\frac{g_m}{C_{eq,s}}(T_s - t_{o,s})\right], & T_s > t_{o,s},\ |v_{a,bs}| > \frac{I_o}{g_m} \quad (2.49c) \end{cases}$$

$$v_{o,n} \equiv v_{o,es} = v_{o,ei} + \left(1 + \frac{C_P}{C_I}\right)[v_{a,es} - v_{a,ei}] \tag{2.50}$$

Integration-sampling process

The analysis above of the SC integrator dynamic during integration and sampling can be easily concatenated, so that the transient evolution of the integrator output voltage is accurately described during the overall integration-sampling process. Nine different evolutions—summarized in Table 2.4—can be obtained for the complete process.

TABLE 2.4 Possible evolutions during the integration-sampling process.

Evolution	Integration	eq(2.41)	Sampling	eq(2.50)
1	Linear	eq(2.40a)	Linear	eq(2.49a)
2	Partial-slew	eq(2.40c)	Linear	eq(2.49a)
3	Full-slew	eq(2.40b)	Linear	eq(2.49a)
4	Linear	eq(2.40a)	Partial-slew	eq(2.49c)
5	Partial-slew	eq(2.40c)	Partial-slew	eq(2.49c)
6	Full-slew	eq(2.40b)	Partial-slew	eq(2.49c)
7	Linear	eq(2.40a)	Full-slew	eq(2.49b)
8	Partial-slew	eq(2.40c)	Full-slew	eq(2.49b)
9	Full-slew	eq(2.40b)	Full-slew	eq(2.49b)

Ideally, the integrator output voltage at the end of the process is

$$\begin{aligned} v_{o,n} = v_{o,n-1} &+ \frac{C_{S1}}{C_I}(v_{in1,n-1} - v_{fb1,n-1/2}) + \dots \\ &+ \frac{C_{Si}}{C_I}(v_{ini,n-1} - v_{fbi,n-1/2}) \end{aligned} \tag{2.51}$$

whereas the actual $v_{o,n}$ can be obtained for each possible evolution by linking the corresponding equations in Table 2.4. Every term obtained for the actual $v_{o,n}$, different from those in eq(2.51), derives from the integrator defective settling due to the finite amplifier GB and SR.

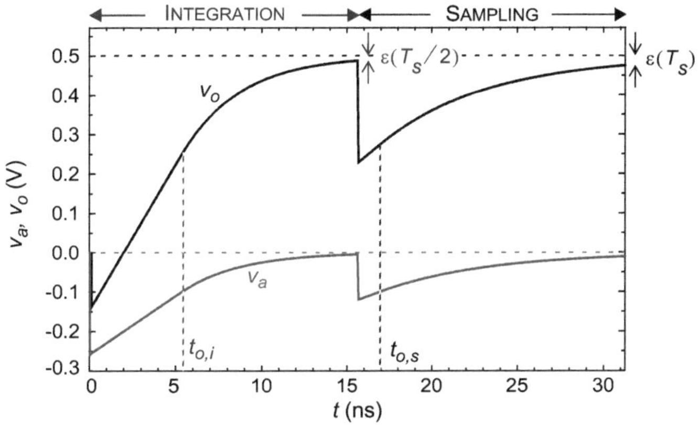

FIGURE 2.20 Transient evolution of an SC integrator, corresponding to case 5 in Table 2.4.

Fig. 2.20 shows an evolution with a partial-slew during both clock phases, where the influence of the sampling dynamic is evident. In this example, the deviation of the voltage at the end of the sampling phase is larger than that at the end of the integration phase. As the former contains the information being transmitted to the next integrator, considering only the integration dynamic would lead to an under-estimation of the defective settling error.

2.3.2 Validation of the proposed model

Comparison with experimental results

The former equations for the transient response of SC integrators were introduced in ASIDES [Mede99a], a behavioral simulation tool for ΣΔMs. A 2nd-order prototype in 0.7-μm CMOS [Mede97] was used for the validation of the proposed model. The ΣΔ modulator nominally operates at sampling frequency $f_s = 2.46\text{MHz}$ with reference voltages of $\pm 1.5\text{V}$ and provides 15-bit effective resolution at 19.2kS/s (oversampling ratio $OSR = 128$). Experimental measurements of the modulator in-band error power were taken increasing f_s from its nominal value, in order to make the defective settling error power the dominant error source. The dynamic of the two integrators in the modulator was also externally controlled, varying the amplifier bias currents.

Fig. 2.21 compares experimental and behavioral simulation results on the modulator in-band error power, for two biasing conditions of the amplifiers. Note how defective settling error power becomes dominant as f_s increases above 3MHz and 4MHz for the slow and fast case, respectively. Note also the good agreement between simulation and

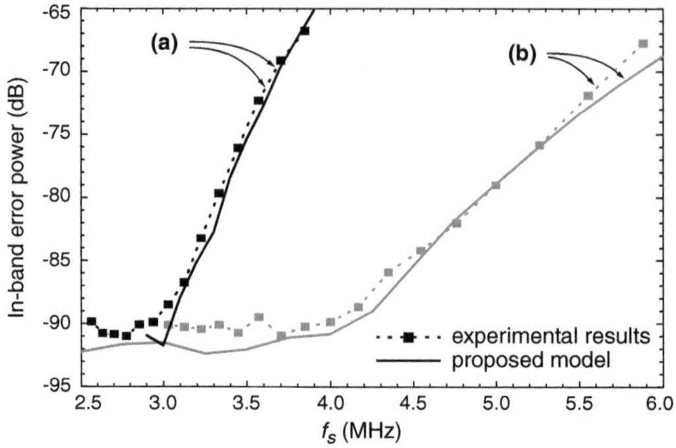

FIGURE 2.21 Comparison of the proposed model with experimental results on a 2nd-order ΣΔM: (a) Slow case ($g_m = 229\mu A/V$, $I_o = 16.7\mu A$), (b) Fast case ($g_m = 294\mu A/V$, $I_o = 28\mu A$).

experimental results. This agreement extends to the whole range of variation of the sampling frequency— $2.5MHz \le f_s \le 6.0MHz$—, in which the settling error power grows up to 25dB over other error contributions.

Comparison with traditional models

Traditional models for SC integrators take into account the amplifier *GB* and *SR* limitations during integration, whereas settling errors derived from the sampling dynamic are omitted. This is done assuming that the amplifier equivalent load during sampling is considerably smaller than that during the integration phase; that is $C_{eq,s} \ll C_{eq,i}$. Under this assumption, the integrator summation node is completely relaxed at the end of the sampling phase, so that $v_{a,es} = 0$ and eq(2.50) turns into

$$v_{o,es} = v_{o,ei} - \left(1 + \frac{C_P}{C_I}\right) v_{a,ei} \tag{2.52}$$

However, $C_{eq,s} \ll C_{eq,i}$ is not a general situation. In practice, for many ΣΔM designs, as soon as the sampling capacitors of the next integrator C_{Snj} are taken into account, $C_{eq,s}$ becomes comparable to the capacitive load during integration, and may be even larger than $C_{eq,i}$ in some cases. This leads to an incomplete discharge of the integrator summation node and therefore to an additional non-negligible error during sampling.

In order to illustrate this, behavioral simulations have been carried out for several high-speed ΣΔMs, using both the traditional and the proposed models for the SC integrator dynamic. One of the modulators considered is

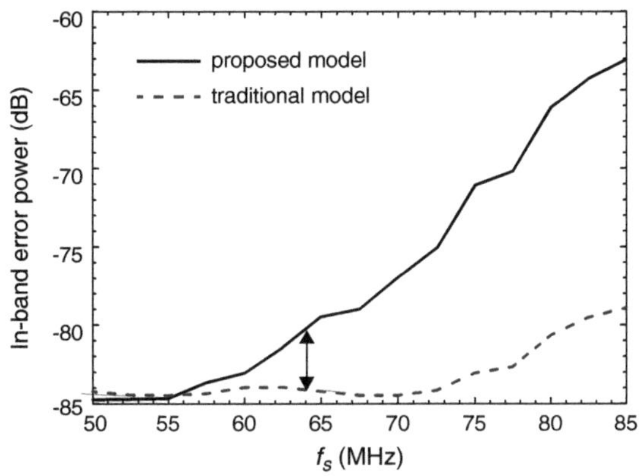

FIGURE 2.22 In-band error power vs. sampling frequency on a 2-1-1 ΣΔM.

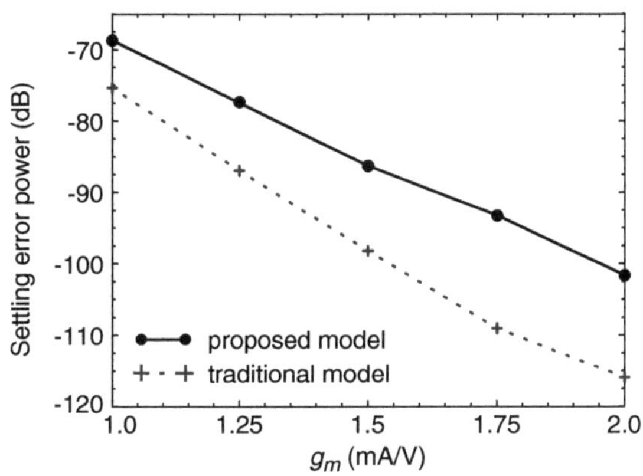

FIGURE 2.23 Settling error power vs. front-end amplifier g_m on a 2-1-1-1 ΣΔM.

a 2-1-1 multi-bit cascade that nominally operates at $f_s = 64\text{MHz}$, oversampling ratio $OSR = 16$, last-stage quantizer resolution $B = 4\text{bit}$, and reference levels $\pm V_{ref} = \pm 1\text{V}$. The ΣΔ modulator targets 14bit@4MS/s. Fig. 2.22 compares the simulation results obtained using both models, showing the in-band error power as a function of the sampling frequency. Note that the defective settling error power increases as f_s does, raising over the remaining noise contributions. Nevertheless, the rate of increase is considerably lower for the traditional model, what leads to unacceptably optimistic results because of the under-estimation of the defective settling error power.

Behavioral simulations have been also carried out on a 2-1-1-1 multi-bit cascade ΣΔM [Rio98] that targets 12bit@4MS/s. The modulator operates with a clock frequency of 32MHz, oversampling ratio of 8, 4-bit resolution in the last-stage quantizer, and reference levels of $\pm 1\,V$. Fig. 2.23 shows the settling error power as a function of the transconductance of the front-end amplifier, for both the traditional and the proposed model. Note that neglecting the settling error during the sampling phase —traditional model—leads in this case to under-estimations of the in-band error power as large as 13dB.

2.3.3 Effect of the amplifier finite gain-bandwidth product

In this section, first-order analytical expressions are obtained for the effect of the integrator settling error on the performance of ΣΔ modulators. For simplicity, first we consider only the amplifier finite GB †7. In the next section, the amplifier finite SR will be incorporated to the GB limitation and its effect will be examined.

With finite GB only, we are assuming that the integrator evolves linearly during both integration and sampling. This means that the amplifier is capable to deliver an output current I_o that is large enough to avoid any slewing; that is, the 'dominant' †8 integrator dynamic over the clock cycles corresponds to evolution 1 in Table 2.4. If we consider the front-end integrator in a ΣΔ modulator, the required I_o in order to fulfil this can be determined using a coarse, but efficient, approach: to suppose that the integrator dynamic is dominated by the situations in which its input is close to V_{ref}. Applying this approximation to eq(2.33), the dominant amplifier input voltage at the beginning of integration would be

$$v_{a,bi}{}^{*} = \left(1 + \frac{C_L}{C_I}\right)\frac{C_S}{C_{eq,i}}V_{ref} \tag{2.53}$$

so that, for a dominant linear integration—$|v_{a,bi}{}^{*}| \le I_o/g_m$—the required output current is

$$I_o \ge \left(1 + \frac{C_L}{C_I}\right)\frac{C_S}{C_{eq,i}}g_m V_{ref} \tag{2.54}$$

The required I_o for a linear sampling evolution can be determined likewise. In this case, a particular modulator architecture must be considered, because the connection of the front-end integrator to the next block must be taken into account. Let us consider a 2-1-1 cascade, with the first stage shown in Fig. 2.24. Comparing it with Fig. 2.12 and using eq(2.42), the dominant amplifier input voltage at the beginning of sampling would be

7. Obviously, considering that the slew rate is not limiting the integrator settling is a notable simplification of the model formerly illustrated in Section 2.3.1. Nevertheless, this first-order approximation to the problem will help to gain some insight of the settling effects on different modulator topologies, depending on the design parameters.

8. In the sense of 'most frequent' or 'most common'.

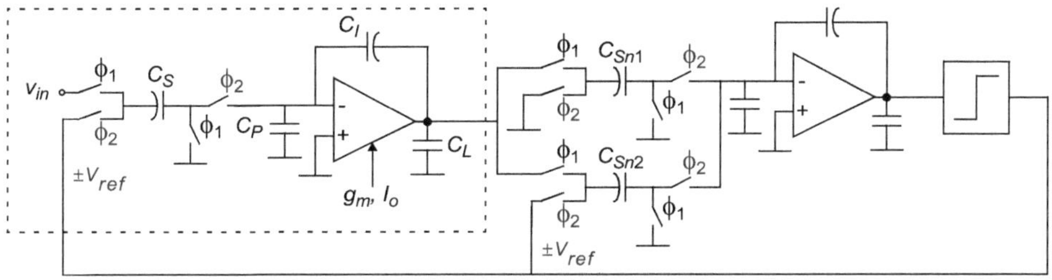

FIGURE 2.24 First stage of a 2-1-1 ΣΔ modulator; front-end integrator in dashed box.

$$\begin{aligned} v_{a,bs}^* &= \frac{C_{Sn1}}{C_{eq,s}}\left(\frac{C_S}{C_I}V_{ref} + 0\right) + \frac{C_{Sn2}}{C_{eq,s}}\left(\frac{C_S}{C_I}V_{ref} + V_{ref}\right) \\ &= \frac{1}{C_{eq,s}}\left(C_{Sntot}\frac{C_S}{C_I} + C_{Sn2}\right)V_{ref} \end{aligned} \tag{2.55}$$

where $C_{Sntot} = C_{Sn1} + C_{Sn2}$. Thus, for a dominant linear operation during sampling—$|v_{a,bs}^*| \le I_o/g_m$—the required I_o is

$$I_o \ge \frac{1}{C_{eq,s}}\left(C_{Sntot}\frac{C_S}{C_I} + C_{Sn2}\right)g_m V_{ref} \tag{2.56}$$

The dominant dynamic being linear, the incomplete settling causes a degradation of the integrator gain, so that in this simplified approach

$$v_{o,n} = v_{o,n-1} + \frac{C_S}{C_I}(1 - \varepsilon_{st})(v_{in,n-1} - v_{fb,n-1/2}) \tag{2.57}$$

where the settling error ε_{st} contains terms in $\exp(-g_m/C_{eq,i} \cdot T_s/2)$ and in $\exp(-g_m/C_{eq,s} \cdot T_s/2)$. If the settling error during integration dominates over that during sampling, ε_{st} can be approximated to

$$\varepsilon_{st} \approx \exp\left(-\frac{g_m}{C_{eq,i}}\frac{T_s}{2}\right) = \exp\left[-2\pi\frac{GB_i(\text{Hz})}{2f_s}\right] = \exp(-2\pi GB_i^{norm}) \tag{2.58}$$

where GB_i^{norm} stands for the amplifier GB during integration times the duration of this phase ($T_s/2$).

Under these assumptions, the output voltage of the front-end integrator turns into

$$v_o(z) = \frac{C_S}{C_I}(1 - \varepsilon_{st})\frac{z^{-1}v_{in}(z) - z^{-1/2}v_{fb}(z)}{1 - z^{-1}} \tag{2.59}$$

causing an error in the implemented weight. This effect—that can be extended to the rest of integrators in the ΣΔ modulator—will modify the relationships among integrator weights required in this topology for full performance, leading to a deviation of the noise transfer function from the ideal one †[9]. In this section we show that linear settling errors cause an increase of the in-band error power that depends on the modulator architecture and OSR.

Single-loop ΣΔ modulators

Let us consider the Lth-order ΣΔ modulator illustrated in Fig. 2.7*b*. Assuming a linear settling error ε_{st_i} in all the integrators, the modulator output in the z-domain can be calculated as

$$\begin{aligned} Y(z) &\cong z^{-L}X(z) + (1+\varepsilon_{st})(1-z^{-1})^L E(z) \\ \varepsilon_{st} &= \varepsilon_{st_1} + \varepsilon_{st_2} + \ldots + \varepsilon_{st_L} \end{aligned} \tag{2.60}$$

where terms in $\varepsilon_{st_i}^2$ have been neglected. Note from eq(2.60) that the linear settling errors increase the modulus of the noise transfer function, but the Lth-order shaping of the quantization error is preserved.

The output-referred in-band error power under linear settling error can be therefore calculated as

$$P_Q(\varepsilon_{st}) \cong \frac{\Delta^2}{12}\left[\frac{(1+\varepsilon_{st})^2 \pi^{2L}}{(2L+1)OSR^{(2L+1)}}\right] \tag{2.61}$$

and, by comparing eq(2.61) with the ideal expression in eq(1.43), the increase of the in-band error power is

$$\Delta P_Q(\varepsilon_{st})\big|_{\mathrm{dB}} = 20\log_{10}(1+\varepsilon_{st}) \tag{2.62}$$

Note that single-loop topologies are relatively insensitive to linear settling errors: the increase of the in-band error power is independent of the modulator OSR and exhibits a smooth increase for increasing orders of the loop, because more error terms will be added in ε_{st}.

Cascade ΣΔ modulators

Considering the Lth-order N-stage cascade in Fig. 2.3 and a linear settling of integrators, the ideal modulator output in eq(2.15) turns into

$$\begin{aligned} Y(z) \cong\ & z^{-L}X(z) + \varepsilon_1 z^{-(L-L_1)}(1-z^{-1})^{L_1} E_1(z) \\ & + d_1\varepsilon_2 z^{-(L-L_1-L_2)}(1-z^{-1})^{(L_1+L_2)} E_2(z) + \ldots \\ & + d_{2N-3}(1+\varepsilon_N)(1-z^{-1})^L E_N(z) \end{aligned} \tag{2.63}$$

with $\varepsilon_1, \varepsilon_2, \ldots, \varepsilon_N$ being the summation of the settling error of the different integrators in each stage

9. As for mismatch errors, for example.

$$\begin{aligned}\varepsilon_1 &= \varepsilon_{st_1} + \ldots + \varepsilon_{st_{L_1}} \qquad \varepsilon_2 = \varepsilon_{st_{L_1+1}} + \ldots + \varepsilon_{st_{L_1+L_2}} \\ \ldots \qquad \varepsilon_N &= \varepsilon_{st_{L_{N-1}+1}} + \ldots + \varepsilon_{st_L}\end{aligned} \tag{2.64}$$

Note from eq(2.63) that, besides an increase of the modulus of the last-stage quantization error—similar to the effect in single-loop topologies—, the settling error causes an incomplete cancellation of the quantization error of the remaining stages, that leak to the modulator output. Therefore, cascade topologies are more sensitive to settling errors than single-loop ΣΔMs, because $E_1(z)$ will appear at the output with a shaping of order L_1, $E_2(z)$ with a shaping of order $L_1 + L_2$, etc., always smaller than the overall modulator order.

The output-referred in-band error power under linear settling error $P_Q(\varepsilon_{st})$ can be obtained using eq(2.29). This equation can be particularized for common cascade topologies as shown in Table 2.2, using the expressions for $\varepsilon_1, \varepsilon_2, \ldots, \varepsilon_N$ given in eq(2.64).

In the case of cascade ΣΔMs using multi-bit quantization only in the last stage, the simplified expressions in eq(2.31) for $P_Q(\varepsilon_{st})$ and eq(2.32) for ΔP_Q are valid. Similar conclusions to those for the effect of capacitor mismatch can be therefore drawn in the case of a linear settling:

- The sensitivity of cascade architectures increases as OSR and/or the order of the modulator increases.
- For a given L, using a 1st-order front-end stage leads to a larger increase of the in-band error power than using a 2nd-order stage.
- For a given cascade topology, the impact increases as the resolution of the multi-bit quantizer does, because the last-stage quantization error is reduced, while the error leakage remains the same.

Fig. 2.25 shows the dynamic range of several 4th-order single-bit architectures versus the normalized amplifier GB, $GB^{norm} = GB/(2f_s)$, for several oversampling ratios. The curves labeled as '4th-order' correspond to an ideally stable single-loop, whereas the rest are cascade architectures. The ideal DR of the cascade topologies differs from that of the single-loop implementation due to the scaling factor d_{2N-3}—see eq(2.63).

Note that a normalized amplifier GB of $0.75 \sim 1$ is enough in the single-loop modulator to achieve full performance. This means an amplifier GB of $1.5 \sim 2$ times the clock frequency, regardless the oversampling ratio. The larger sensitivity of cascade implementations to settling errors is also manifest. Note that the linear settling requirements depend on the topology of the cascade, ranging from $GB \sim 1.5f_s$ to $3.5f_s$ as OSR increases.

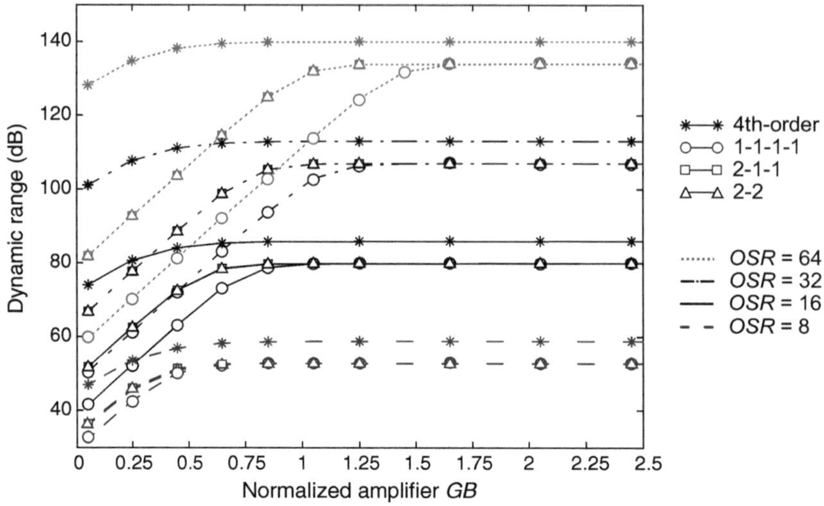

FIGURE 2.25 Modulator dynamic range versus the normalized amplifier GB during integration, for several 4th-order ΣΔ topologies.

2.3.4 Effect of the amplifier finite slew rate

As explained in Section 2.3.1, the linear integrator dynamic can be substantially altered as soon as the amplifier limited SR is considered. Due to the switching of the sampling capacitors in consecutive SC integrators (see Fig. 2.12), the capacitance at the summation and output nodes of a given integrator changes from integration to sampling. The total charge stored will be then shared at the beginning of each phase, inducing voltage jumps at the amplifier input and output. If the initial amplifier input voltage is outside its linear region, the amplifier will only be able to deliver its maximum output current I_o, so that the integrator transient response is slew-rate limited until the amplifier enters the linear region again. Whether this situation occurs or not during the clock phases—equations (2.37) and (2.46)—leads to partial-slew or to full-slew settling during the corresponding phase.

The slew rate of a single-stage amplifier during integration SR_i can be calculated using equations (2.40b) and (2.41), whereas the slew rate during sampling SR_s can be obtained from equations (2.49b) and (2.50), yielding

$$SR_i = \left(1 + \frac{C_P + C_S}{C_I}\right)\frac{I_o}{C_{eq,\,i}} \qquad SR_s = \left(1 + \frac{C_P}{C_I}\right)\frac{I_o}{C_{eq,\,s}} \tag{2.65}$$

On the other hand, the integrator linear operation is determined by the amplifier GB in each clock phase; i.e.,

$$GB_i = \frac{g_m}{C_{eq,\,i}} \qquad GB_s = \frac{g_m}{C_{eq,\,s}} \tag{2.66}$$

Obviously, if the dominant integrator dynamic along the clock cycles involves full-slew during integration or sampling, the performance of the modulator will be completely degraded. Nevertheless, that is not true for partial-slew, as long as the slew rate is large enough to let sufficient time for the linear transient to settle within the desired accuracy.

This effect is illustrated next through behavioral simulation of a 2-1-1 ΣΔM, using the model for the integrator settling described in Section 2.3.1. A normalized notation is used for GB and SR to gain generality, as follows

$$\begin{aligned} GB_i^{norm} &= \frac{GB_i(\text{Hz})}{2f_s} & SR_i^{norm} &= \frac{SR_i}{g_1'2V_{ref}}\frac{T_s}{2} = \frac{1}{g_1'2V_{ref}}\frac{SR_i}{2f_s} \\ GB_s^{norm} &= \frac{GB_s(\text{Hz})}{2f_s} & SR_s^{norm} &= \frac{SR_s}{g_1 2V_{ref}}\frac{T_s}{2} = \frac{1}{g_1 2V_{ref}}\frac{SR_s}{2f_s} \end{aligned} \tag{2.67}$$

where the expressions for SR^{norm} have been adapted from [Will94], whereas g_1 and g_1' stand for the signal and feedback weights, respectively.

Fig. 2.26*a* shows the *SNDR* obtained for an input sinewave of half-scale amplitude, as a function of the settling parameters of the front-end integrator. Capacitors used in the example lead to $GB_i = 1.3GB_s$ and $SR_i = 1.6SR_s$ in the first integrator. Note that, for a given amplifier GB, the modulator *SNDR* increases as the slew rate increases, reaching full performance if I_o is large enough. On the other hand, the slew rate required to achieve the optimum *SNDR* decreases as the GB increases. This highlights the existing trade-off between amplifier GB and SR specifications.

Fig. 2.26*b* shows the percentage of occurrence of each type of settling in the front-end integrator during both clock phases, as a function of I_o. The case $GB_s^{norm} = 1.5$ has been selected. Note that the *SNDR* in Fig. 2.26*a* starts increasing as soon as full-slew settling disappears. On the other hand, it is patent that full modulator performance can be achieved with dominant partial-slew settling, so that dominant linear evolution is not mandatory. Indeed, a dominant linear integration would require increasing the maximum output current to $I_o \geq 1.75\text{mA}$—eq(2.54)—, whereas $I_o \geq 1.34\text{mA}$ would be needed for a dominant linear sampling—eq(2.56)—, more than four times the maximum output current actually required for full performance ($\sim 400\mu\text{A}$).

Finally, note that unlike linear settlings, the integrator response is non-linear if it is slew-rate limited. Therefore, if the slew is partial or complete during the clock phases, distortion arises at the modulator output spectrum. This effect will be studied in detail in Section 2.6.4.

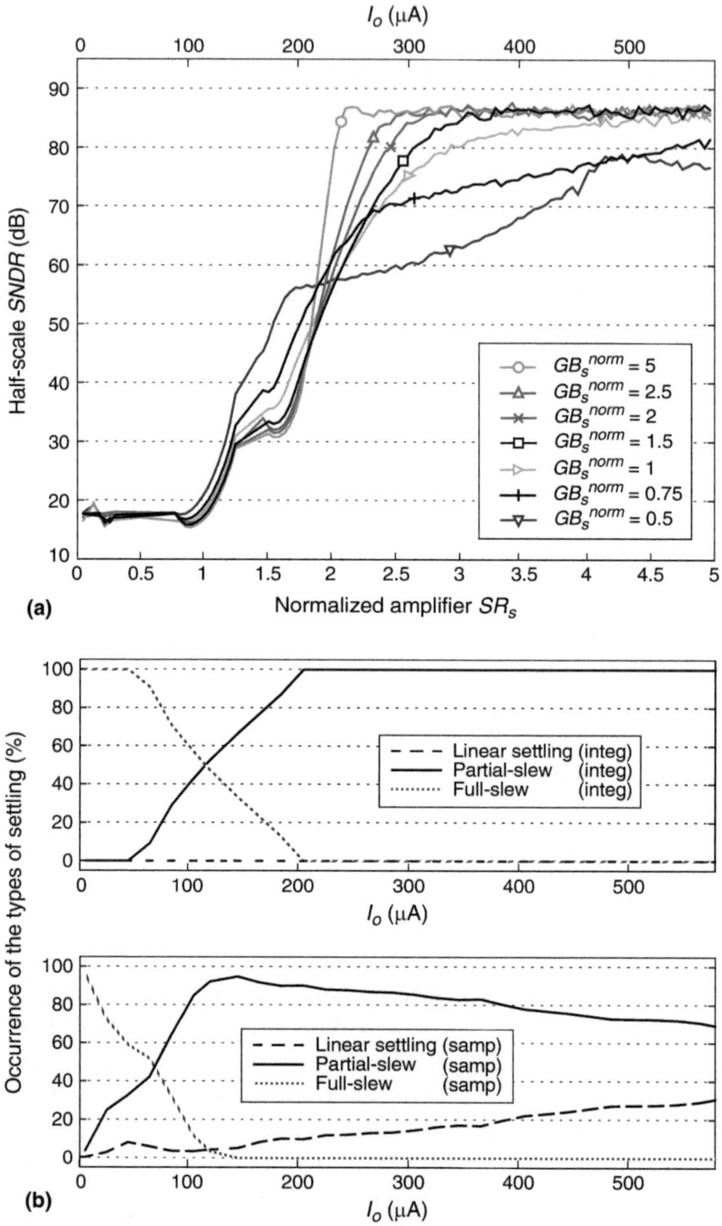

FIGURE 2.26 Influence of the SR of the front-end amplifier on a 2-1-1 ΣΔM: (a) Half-scale $SNDR$ vs. normalized SR for different values of the amplifier GB, (b) Percentage of occurrence of each type of settling during integration and sampling vs. I_o for $GB_s^{norm} = 1.5$. ($V_{ref} = \pm 1.5\text{V}$, $g_1 = g_1' = 0.25$, $C_S = 0.66\text{pF}$, $C_P = C_L = 0.1\text{pF}$, $C_{nS1} = C_{nS2} = 0.45\text{pF}$, $f_s = 70.4\text{MHz}$, and $OSR = 32$).

2.3.5 Effect of the switch finite on-resistance

Up until now the switches in SC integrators have been considered ideal. In practice, due to their implementation using MOSFETs, they exhibit non-zero on-resistance, capacitive parasitics, and leakage current †10. In this section, we look into the impact of the switch on-resistance (R_{on}) on the ΣΔM performance. First, we study the effect of R_{on} itself assuming an ideal SC integrator. Then, the combined effect of R_{on} and finite amplifier dynamic will be considered †11.

Effect on an ideal integrator

Let us consider the SC integrator in Fig. 2.27 and assume an ideal behavior, except for the on-resistance R_{on} of switches $S_1 - S_4$. This being the case, charge will be incompletely transferred to capacitor C_I during integration, due to the time constant $2R_{on}C_S$. A settling error $\varepsilon_{on,i}$ will then appear at the integrator output voltage at the end of the integration phase

$$v_{o,n} = v_{o,n-1} + \frac{C_S}{C_I}(1 - \varepsilon_{on,i})(v_{CS,n-1/2} - v_{fb,n-1/2}) \tag{2.68}$$

where $v_{CS,n-1/2}$ stands for the voltage across the sampling capacitor C_S at the beginning of integration and $\varepsilon_{on,i}$ is given by

$$\varepsilon_{on,i} = \exp\left(-\frac{1}{2R_{on}C_S}\frac{T_s}{2}\right) \tag{2.69}$$

Obviously, $v_{CS,n-1/2}$ will be also affected by a settling error due to the time constant formed by switches S_1 and S_2 with C_S during sampling, so that

$$v_{CS,n-1/2} = v_{in,n-1}(1 - \varepsilon_{on,s}) \tag{2.70}$$

where $\varepsilon_{on,s} = \varepsilon_{on,i}$, since the same duration is assumed for both phases.

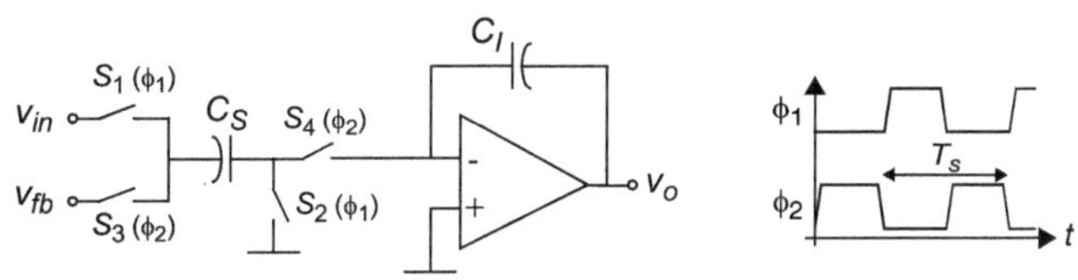

FIGURE 2.27 SC integrator and clock phases.

10. Out of these non-ideal features, on-resistance and capacitance are the ones compromising most the performance of high-speed ΣΔMs. The parasitic capacitances can be easily included in the appropriate capacitors in the settling model.
11. Through this section, a non-zero, but fixed, resistance is assumed. Section 2.6.3 shows that actual switches exhibit a voltage-dependent resistance and hence may generate distortion. Nevertheless, for the study of its impact on settling performance, R_{on} can be considered constant. In order to contemplate worst-case effects, R_{on} is assumed to be equal to the maximum switch resistance.

Thus, the integrator output voltage is given by

$$v_o(z) = \frac{C_S}{C_I}(1-\varepsilon_{on,i})\left[\frac{(1-\varepsilon_{on,s})z^{-1}v_{in}(z) - z^{-1/2}v_{fb}(z)}{1-z^{-1}}\right] \tag{2.71}$$

Note that R_{on} results in an integrator gain error and a systematic deviation of the signal and feedback contributions. Therefore, similar conclusions to those in Section 2.2 and Section 2.3.3 can be drawn for its effect on the performance of a ΣΔ modulator:

- The noise transfer function of single-loop topologies will increase in modulus, but the order of the quantization error shaping will be preserved.
- Cascade topologies will be more affected, since they rely in the cancellation of the quantization errors in all but the last stage through the fulfilment of relationships among integrator weights. Low-order quantization errors will then leak to the modulator output. The performance can be degraded if the time constant $2R_{on}C_S$ is not small enough.

Effect on the amplifier *GB*

A more realistic situation to estimate the effect of R_{on} is illustrated in Fig. 2.28*a*, considering the limited *GB* and the parasitic capacitances of the amplifier and switches. In this case, a two-pole system is created during integration (see Fig. 2.28*b*), where the poles can be obtained from

$$s^2 + s\left(\frac{C_{eq,i}}{C'}\frac{1}{2R_{on}C_S} + \frac{g_m}{C'}\right) + \frac{g_m}{C'}\frac{1}{2R_{on}C_S} = 0 \tag{2.72}$$

where $C' = C_P + C_L(1 + C_P/C_I)$ and $C_{eq,i}$ is given by eq(2.34).

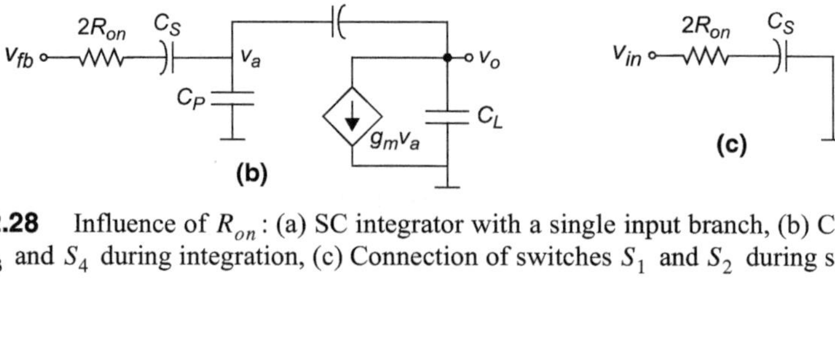

FIGURE 2.28 Influence of R_{on}: (a) SC integrator with a single input branch, (b) Connection of switches S_3 and S_4 during integration, (c) Connection of switches S_1 and S_2 during sampling.

Assuming the common situation in which the pole of the amplifier, $GB_i = g_m/C_{eq,i}$, is considerably smaller than the RC pole, $\omega_{on} = 1/(2R_{on}C_S)$, the poles p_1 and p_2 during integration can be approximated from eq(2.72) to

$$p_1 \cong \frac{g_m/C_{eq,i}}{1 + g_m/C_{eq,i} \cdot 2R_{on}C_S} = \frac{GB_i}{1 + GB_i/\omega_{on}}$$
$$p_2 \cong \frac{C_{eq,i}}{C'}\frac{1}{2R_{on}C_S} = \frac{C_{eq,i}}{C'}\omega_{on} \tag{2.73}$$

where $p_1 \ll p_2$, since $GB_i \ll \omega_{on}$ and $C_{eq,i} \geq C'$.

Thus, the amplifier GB during integration can be redefined, including the switch finite on-resistance

$$GB_{i,on} = \frac{GB_i}{1 + GB_i \cdot 2R_{on}C_S} \tag{2.74}$$

Note that the effect of R_{on} is to decrease the effective amplifier GB.

Taking into account equations (2.58) and (2.59)—for the effect of the amplifier GB—and equations (2.70) and (2.71)—for the effect of R_{on}—, the integrator output voltage can be re-written into first-order approximation as

$$v_o(z) = \frac{C_S}{C_I}(1 - \varepsilon_{st-on,i})\left[\frac{(1 - \varepsilon_{on,s})z^{-1}v_{in}(z) - z^{-1/2}v_{fb}(z)}{1 - z^{-1}}\right] \tag{2.75}$$

where the error during sampling $\varepsilon_{on,s}$ is introduced by R_{on} (see Fig. 2.28*c*)

$$\varepsilon_{on,s} = \exp\left(-\frac{1}{2R_{on}C_S}\frac{T_s}{2}\right) = \exp\left(-2\pi\frac{f_{on}}{2f_s}\right) = \exp(-2\pi f_{on}^{norm}) \tag{2.76}$$

and the settling error during integration $\varepsilon_{st-on,i}$ is introduced by the combined effect of the amplifier limited linear dynamic and the $2R_{on}C_S$ time constant

$$\varepsilon_{st-on,i} = \exp\left(-GB_{i,on}\frac{T_s}{2}\right) = \exp\left(-2\pi\frac{GB_i^{norm}}{1 + GB_i^{norm}/f_{on}^{norm}}\right) \tag{2.77}$$

A normalized notation has been adopted in equations (2.76) and (2.77) by multiplying GB (in Hz) and f_{on} by the duration of the clock phases.

Note that eq(2.75) is similar to eq(2.71), but the integrator gain error is now expected to be larger than the systematic deviation of the signal and feedback contributions, since $GB_i \ll \omega_{on}$ is assumed.

Fig. 2.29 illustrates the effect of the switch on-resistance on a 2-1-1 cascade, showing the attainable dynamic range as a function the amplifier GB,

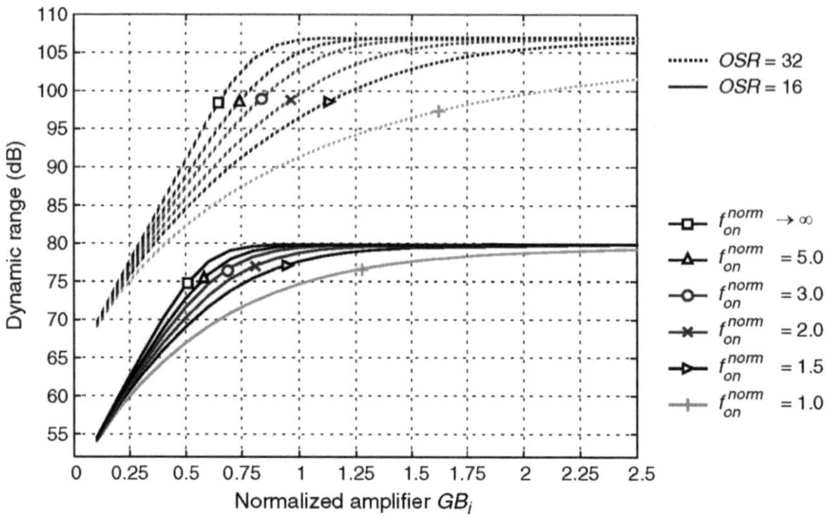

FIGURE 2.29 Dynamic range of a 2-1-1 cascade versus the normalized amplifier GB during integration, for several values of the normalized RC pole $f_{on}^{norm} = 1/(2\pi 2R_{on}C_S) \cdot 1/(2f_s)$.

for different values of f_{on}. Note that, as stated in eq(2.74), the effect of R_{on} is to decrease the effective amplifier GB. If $R_{on} = 0$ ($f_{on} \to \infty$), $GB_i^{norm} \approx 1$ for $OSR = 16$ and $GB_i^{norm} \approx 1.25$ for $OSR = 32$ are enough to achieve full performance. As the switch resistance increases, the amplifier must be faster to compensate for the error introduced. Note also that the performance degradation introduced by R_{on} is larger as the oversampling ratio of the cascade modulator increases.

Effect on the amplifier *SR*

Unlike previously assumed in Section 2.3.1, charge is not immediately redistributed at the beginning of the clock phases when the on-resistance of the switches is considered. This and other related effects are next illustrated using electrical simulation results. Fig. 2.30 shows the transient evolution of an SC integrator during one clock cycle, for different values of R_{on}. The amplifier finite GB and slew rate are taken into account, and the border between linear and slew-rate limited operation is shown. Note that the jumps at the beginning of integration are clearly affected by the RC time constant, so that the spikes in $v_a(t)$ are softened as R_{on} increases. Since the magnitude of the initial jumps decreases, the duration of the slew-rate limited evolution also decreases. This means that the integrator will have more time to evolve linearly, but with a slowed-down dynamic, since the effective GB is decreased—as stated in eq(2.74). Therefore, the amplifier GB should be increased in order to let the linear evolution settle within the desired accuracy.

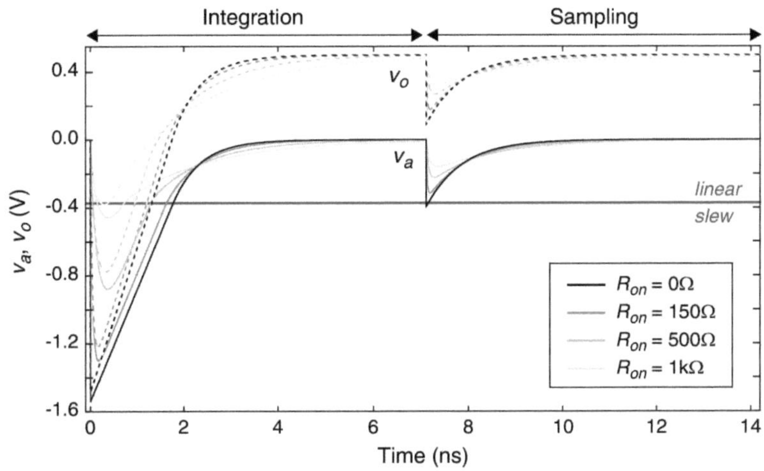

FIGURE 2.30 Electrical simulation of the transient evolution of an SC integrator during a clock cycle, for different values of the switch on-resistance.

The same applies during sampling, but the RC constant now affecting the integrator dynamic will be given by the resistance R_{non} of the switches and the sampling capacitor C_{nS} of the next integrator. As for the integration phase in eq(2.74), the amplifier GB during sampling can be approximated to

$$GB_{s,\,on} = \frac{g_m / C_{eq,\,s}}{1 + g_m / C_{eq,\,s} \cdot 2R_{non}C_{nS}} = \frac{GB_s}{1 + GB_s \cdot 2R_{non}C_{nS}} \tag{2.78}$$

Note that the formerly described effects of the switch on-resistance can not be easily extended to the model for the transient response of a generic SC integrator described in Section 2.3.1. If we consider the integrator in Fig. 2.12, a system with $j+1$ poles will result during sampling—due to the j input branches of the following integrator and the single-pole of the amplifier—, whereas the system will have $i+1$ poles during integration. If we restrict to SC integrators in ΣΔ modulators, where the maximum number of branches required is usually 3 (see the cascade modulators in Fig. 1.28 to Fig. 1.31), the complexity of the problem can be reduced to 2 poles during sampling and 3 poles during integration, but is still analytically prohibitive.

Therefore, an approximate method has been adopted in order to include the switch on-resistance in the model proposed in Section 2.3.1. Only its effect on the amplifier GB during integration and sampling is contemplated, but charge is still considered to instantaneously redistribute at the beginning of the phases. The approach is illustrated in Fig. 2.31, showing the transient evolution implemented in behavioral simulations when the switch on-resistance is taken

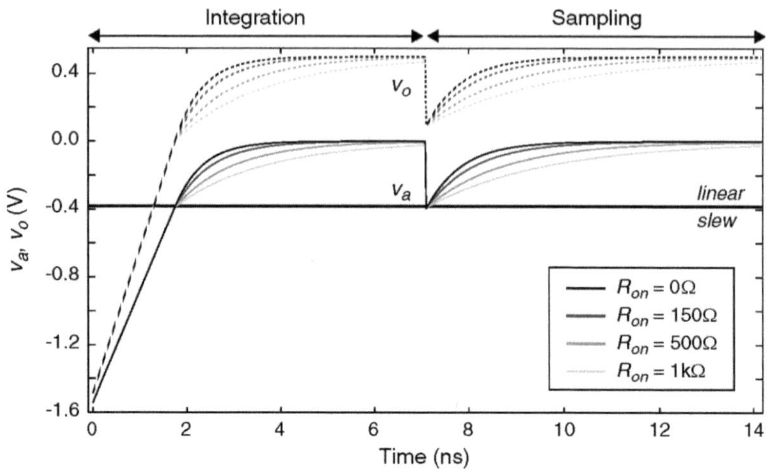

FIGURE 2.31 Behavioral simulation of the transient evolution of an SC integrator during a clock cycle, for different values of the switch on-resistance.

into account. For comparison purposes, the same integrator and values of R_{on} as in Fig. 2.30 are considered. Note that only the slow-down effect of the switch resistance on the linear evolutions can be observed and their duration is independent of R_{on}, since charge is considered to redistribute and the slew-rate limited dynamic remains unchanged. This makes this approach conservative in comparison with the dynamic exhibited in practice, where slewing is less frequent and the duration of the linear dynamic is larger as R_{on} increases.

Fig. 2.32 shows behavioral simulation results on the 2-1-1 cascade modulator in Section 2.3.4, applying the approximation formerly described to include the effect of the switch on-resistance. The half-scale $SNDR$ is depicted versus the slew rate of the front-end amplifier. The case $GB_s^{norm} = 1.5$ in Fig. 2.26*a* has been selected and different values of R_{on} are considered. Note that as R_{on} increases, the SR requirements to achieve full performance increase, since more time is needed for the slowed-down linear dynamic to settle within the desired accuracy. For instance, for the case $R_{on} = 150\Omega$, the pole formed by the $2R_{on}C_S$ time constant is located at $11.4f_s$, whereas the pole given by $2R_{on}C_{nS}$ is at $8.4f_s$. Given that $GB_s = 3f_s$ and $GB_i = 1.3GB_s = 3.9f_s$, the effective amplifier GB is reduced to $GB_{s,\,on} = 2.2f_s$ and $GB_{i,\,on} = 2.9f_s$. Thus, the maximum amplifier output current must be increased from $I_o \geq 400\mu\text{A}$ to $I_o \geq 450\mu\text{A}$ in order to achieve the expected $SNDR$.

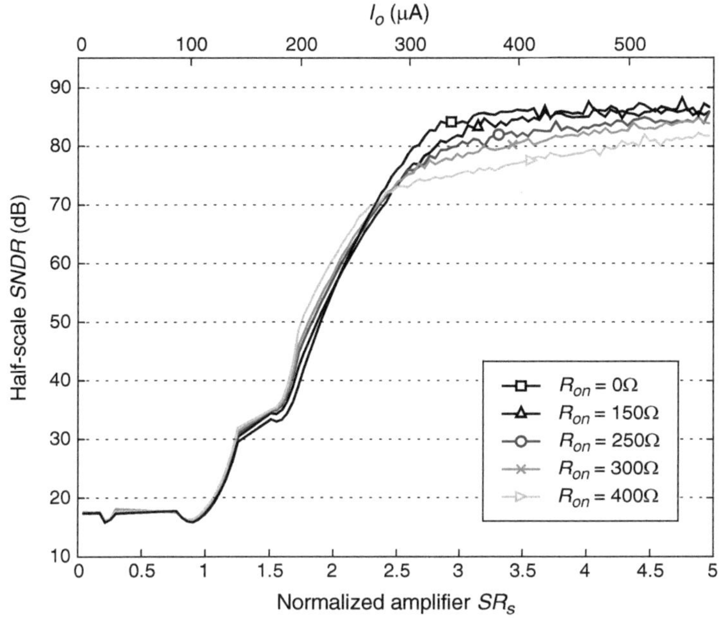

FIGURE 2.32 Half-scale $SNDR$ of a 2-1-1 ΣΔM vs. normalized SR of the front-end amplifier and different values of R_{on}. ($GB_s^{norm} = 1.5$, $V_{ref} = \pm 1.5\text{V}$, $g_1 = g_1' = 0.25$, $C_S = 0.66\text{pF}$, $C_P = C_L = 0.1\text{pF}$, $C_{nS} = 0.9\text{pF}$, $f_s = 70.4\text{MHz}$, and $OSR = 32$).

2.4 Circuit Noise

Circuit (or electronic) noise generated in transistors and resistors is present in all circuit implementations. Nevertheless, its importance is even higher in switched-capacitor implementations, because circuit noise is also sampled at the clock frequency. Given that many circuit noise sources have a wide-band spectrum—such as thermal noise generated in the switches and in the amplifier of the SC integrators—, they are undersampled by the SC network. This derives in aliasing, so that circuit noise is several times folded-back over the baseband, considerably increasing the in-band noise power.

This section studies in detail the sources and effects of electronic noise in SC networks. The derived results—applicable in the noise analysis of general switched-capacitor circuits—are particularized for SC ΣΔ modulators at the end of the section.

2.4.1 Noise in track-and-holds

Let us consider the track-and-hold depicted in Fig. 2.33*a*, which consists of a simple switched-capacitor †[12]. We make use of this basic circuit as a vehicle to the analysis of circuit noise, prior to considering SC integrators and ΣΔMs.

Assume that the input signal v_{in} is zero and that the switch transistor M has an on-resistance R_{on} when $\phi = 1$. Thus, the transistor can be modeled by a resistor in series with a voltage source v_S to account for its thermal noise, leading to the equivalent *static noise* model in Fig. 2.33*b*. The noise waveform across the capacitor v_C will then have the shape shown in Fig. 2.33*d*: during $\phi = 1$, v_C tracks the noise v_S with time constant $R_{on}C$; when ϕ goes low, the last noise value remains sampled in v_C. The waveform in Fig. 2.33*d* can be therefore partitioned into a *track component* v_C^T—also known as *direct* component—(Fig. 2.33*e*) and a *sampled-and-held component* v_C^{SH} (Fig. 2.33*f*) [Gobet81] [Fisc82] [Greg86]. Since the track and sample-and-hold operations take place during non-overlapping time intervals and thermal noise is white by nature †[13], their power spectral densities can be added up to obtain the noise *PSD* at the track-and-hold circuit

$$S_C(f) = S_C^T(f) + S_C^{SH}(f) \tag{2.79}$$

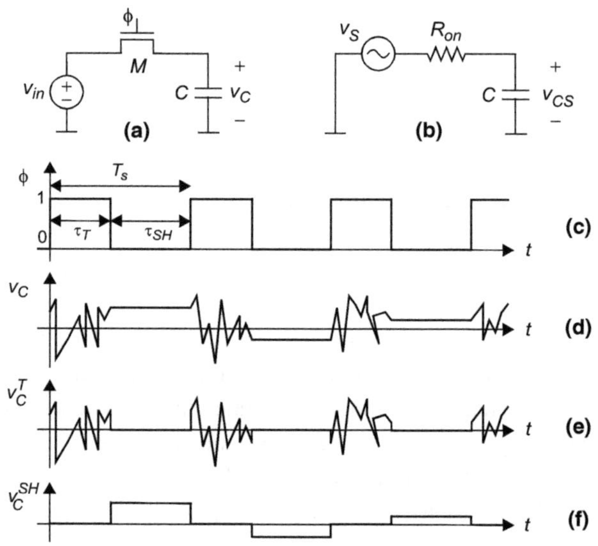

FIGURE 2.33 Noise in a track-and-hold: (a) Switched-capacitor, (b) Equivalent circuit for the static noise, (c) Clock signal, (d) Output noise waveform, (e) Track noise component, (f) Sampled-and-held noise component.

---

12. Actually, a buffer is needed to get the hold operation, but it is not depicted for clarity.
13. Thermal noise is uncorrelated from sample to sample.

Track component

The track component of the capacitor voltage can be expressed as

$$v_C^T(t) = v_{CS}(t) g_{\tau_T}(t) \tag{2.80}$$

where $v_{CS}(t)$ stands for the capacitor voltage in the static noise model and $g_{\tau_T}(t)$ is the gating function, defined as

$$g_{\tau_T}(t) = \begin{cases} 1, & nT_s < t < nT_s + \tau_T \\ 0, & nT_s + \tau_T < t < (n+1)T_s \end{cases} \tag{2.81}$$

with τ_T being the duration of the track operation.

Applying the Fourier transform to eq(2.80), the spectral density of $v_C^T(t)$ can be written as the convolution of the respective PSDs

$$\begin{aligned} S_C^T(f) &= S_{CS}(f) \otimes Sg_{\tau_T}(f) \\ &= S_{CS}(f) \otimes \left[\left(\frac{\tau_T}{T_s}\right)^2 \sum_{n=-\infty}^{+\infty} \mathrm{sinc}^2(n\pi\tau_T f_s)\delta(f - nf_s) \right] \\ &= \left(\frac{\tau_T}{T_s}\right)^2 \sum_{n=-\infty}^{+\infty} S_{CS}(f - nf_s) \cdot \mathrm{sinc}^2(n\pi\tau_T f_s) \end{aligned} \tag{2.82}$$

where $\otimes$ is the convolution product operator and $f_s = 1/T_s$ is the clock frequency. The power spectral density of the thermal noise generated in the switch transistor (in a double-sided representation in frequency) is $S_S = 2kTR_{on}$ and, hence, the PSD of the static noise in the capacitor yields

$$S_{CS}(f) = S_S |H(f)|^2 = \frac{2kTR_{on}}{1 + (2\pi f R_{on} C)^2} \tag{2.83}$$

where $H(s)$ is the transfer function from the noise source to the capacitor voltage

$$H(s) = \frac{1}{1 + sR_{on}C} \tag{2.84}$$

Sampled-and-held component

The sampled-and-held component of $v_C(t)$ can be written as

$$v_C^{SH}(t) = \left[\sum_{n=-\infty}^{+\infty} v_{CS}(t)\delta(f - nf_s) \right] \otimes g_{\tau_{SH}}(t) \tag{2.85}$$

where $g_{\tau_{SH}}(t)$ is the gating function defined for the duration of the hold phase τ_{SH}.

Taking the Fourier transform of eq(2.85) yields

$$S_C^{SH}(f) = \left(\frac{\tau_{SH}}{T_s}\right)^2 \operatorname{sinc}^2(\pi\tau_{SH}f) \sum_{n=-\infty}^{+\infty} S_{CS}(f - nf_s) \qquad (2.86)$$

The summations in $S_C^T(f)$ and $S_C^{SH}(f)$ can be simplified to obtain $S_C(f)$ by using the concept of equivalent noise bandwidth. Eq(2.83) shows that $S_{CS}(f)$ is a narrow-band noise resulting from low-pass filtering S_S with $H(f)$, as illustrated in Fig. 2.34. The total static noise power in the capacitor can be calculated as

$$\overline{v_{CS}^2} = \int_{-\infty}^{+\infty} S_{CS}(f)df = \int_{-\infty}^{+\infty} S_S|H(f)|^2 df = \frac{kT}{C} \qquad (2.87)$$

leading to the well-known kT/C expression.

The equivalent noise bandwidth BW_n can be defined as the bandwidth over which a constant spectral density S_S—i.e., the white noise—has to be integrated in order to obtain the same total static noise power. Note from Fig. 2.34 that eq(2.87) can be re-written as

$$\overline{v_{CS}^2} = S_S \cdot 2BW_n = 2S_S \int_0^{+\infty} |H(f)|^2 df \qquad (2.88)$$

so that the equivalent noise bandwidth (in Hz) for the track-and-hold circuit is

$$BW_n = \int_0^{+\infty} |H(f)|^2 df = \frac{1}{4R_{on}C} \qquad (2.89)$$

Folding-back effect

By using the concept of equivalent noise bandwidth, the particular shape of $H(f)$, and consequently that of $S_{CS}(f)$, can be obviated in the following calculations. Considering equations (2.82) and (2.86), two possibilities can be distinguished:

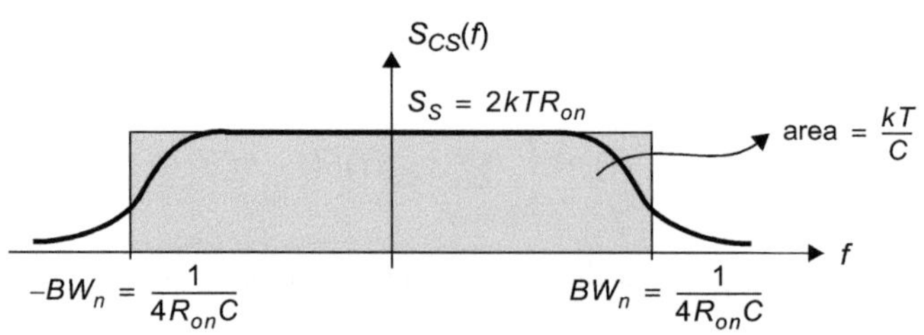

FIGURE 2.34 *PSD* for the track-and-hold circuit and equivalent noise bandwidth.

(a) $BW_n \leq f_s/2$

As illustrated in Fig. 2.35*a*, no aliasing occurs in this case and the two components of $S_C(f)$—equations (2.82) and (2.86)—coincide within the baseband with their contributions for $n = 0$. So

$$S_C(f) = \left(\frac{\tau_T}{T_s}\right)^2 S_S + \left(\frac{\tau_{SH}}{T_s}\right)^2 S_S \cdot \mathrm{sinc}^2(\pi\tau_{SH}f) \tag{2.90}$$

(b) $BW_n > f_s/2$

In this case, the noise is undersampled and aliasing occurs, increasing the noise power in the baseband. This situation is illustrated in Fig. 2.35*b* for the case $BW_n = 6f_s/2$. Since the number of bands that overlap in the interval $[-f_s/2, +f_s/2]$ is $2BW_n/f_s$, the *PSD* of the track-and-hold circuit can be written as

$$S_C(f) = \left(\frac{\tau_T}{T_s}\right)^2 S_S \sum_{n=-BW_n/f_s}^{+BW_n/f_s} \mathrm{sinc}^2(n\pi\tau_T f_s) + \left(\frac{\tau_{SH}}{T_s}\right)^2 \cdot \frac{2BW_n}{f_s} \cdot S_S \cdot \mathrm{sinc}^2(\pi\tau_{SH}f) \tag{2.91}$$

The following approximation can be applied if $BW_n \geq 10f_s$ [Fisc82]

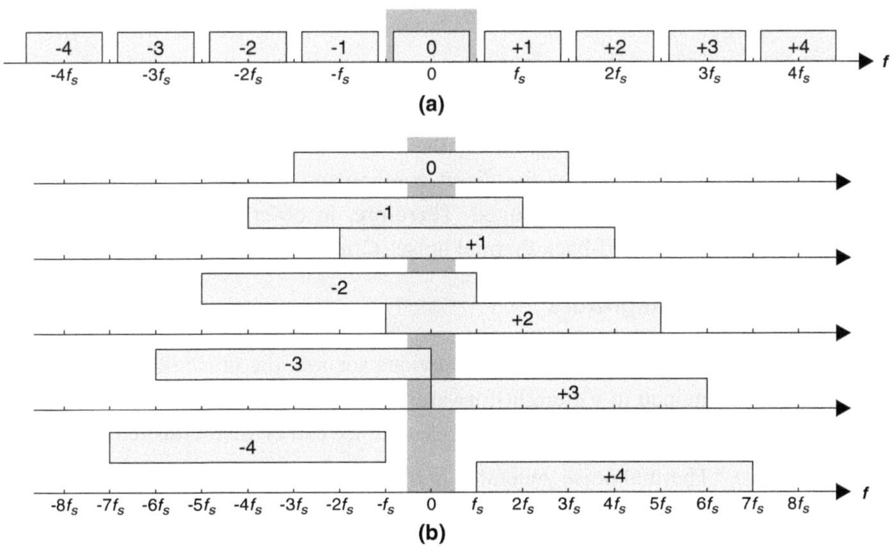

FIGURE 2.35 Replication of the equivalent noise bandwidth: (a) No aliasing occurs ($BW_n \leq f_s/2$), (b) Aliasing occurs ($BW_n > f_s/2$). The corresponding index n is shown inside each band.

$$\sum_{n=-BW_n/f_s}^{+BW_n/f_s} \mathrm{sinc}^2(n\pi\tau_T f_s) \cong \frac{T_s}{\tau_T} \tag{2.92}$$

yielding

$$S_C(f) = \left(\frac{\tau_T}{T_s}\right) \cdot S_S + \left(\frac{\tau_{SH}}{T_s}\right)^2 \cdot \frac{2BW_n}{f_s} \cdot S_S \cdot \mathrm{sinc}^2(\pi\tau_{SH} f) \tag{2.93}$$

From equations (2.90) and (2.93) we obtain that the noise PSD for the track-and-hold circuit in the interval $[-f_s/2, +f_s/2]$ is

$$S_C(f) \cong \begin{cases} \left(\frac{\tau_T}{T_s}\right)^2 S_S + \left(\frac{\tau_{SH}}{T_s}\right)^2 S_S \cdot \mathrm{sinc}^2(\pi\tau_{SH} f), & BW_n \le \frac{f_s}{2} \\ \left(\frac{\tau_T}{T_s}\right) S_S + \left(\frac{\tau_{SH}}{T_s}\right)^2 \cdot \frac{2BW_n}{f_s} \cdot S_S \cdot \mathrm{sinc}^2(\pi\tau_{SH} f), & BW_n > \frac{f_s}{2} \end{cases} \tag{2.94}$$

where $BW_n = 1/(4R_{on}C)$.

Given that the switched-capacitor time constant must be much smaller than the clock period for an accurate settling during the charge transfer

$$R_{on}C \ll T_s \quad \Rightarrow \quad BW_n \gg f_s/2 \tag{2.95}$$

aliasing occurs, so that the noise PSD yields

$$S_C(f) = \left(\frac{\tau_T}{T_s}\right) \cdot 2kTR_{on} + \left(\frac{\tau_{SH}}{T_s}\right)^2 \frac{kT}{Cf_s} \mathrm{sinc}^2(\pi\tau_{SH} f) \tag{2.96}$$

The second term in eq(2.96)—the sampled-and-held component—is dominant due to the undersampling of the wide-band white noise. Note that this makes that reducing R_{on} below the value required for settling conditions is useless, since it reduces the direct component of the PSD, but the dominant component remains unchanged. Therefore, in order to effectively reduce the power of the folded-back thermal noise, C must be increased.

2.4.2 Noise in SC integrators

Using the results derived in the previous section, the noise in an SC integrator can be obtained in a straight-forward way. Let us consider the SC integrator in Fig. 2.36. Three main sources of circuit noise can be distinguished:

- Thermal noise generated in the switches
- Thermal and flicker noise generated in the amplifier [†14]

---

14. Flicker (or $1/f$) noise is caused by the fluctuation in the number of carriers in the transistor channel, so that a device with no current flowing through it has no $1/f$ noise.

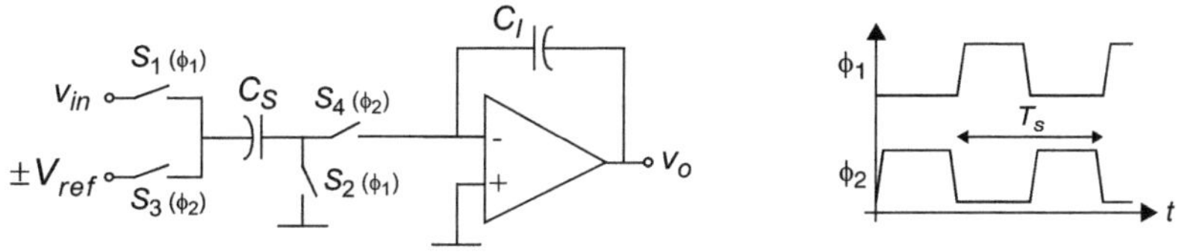

FIGURE 2.36 SC integrator with a single input branch and clock phases.

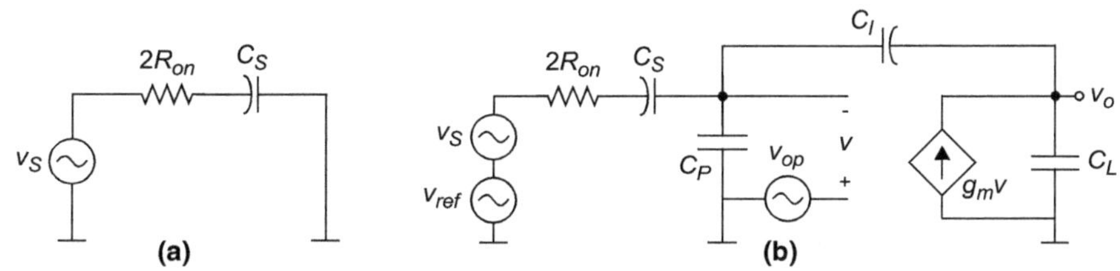

FIGURE 2.37 Static equivalent circuits for noise evaluation in an SC integrator: (a) Noise generated in switches S_1 and S_2, (b) Noise generated in switches S_3 and S_4, and in the opamp.

- Thermal and flicker noise accompanying the voltage references

Fig. 2.37 illustrates the static models used for the evaluation of each noise source. Fig. 2.37*a* shows the model for the thermal noise introduced by switches S_1 and S_2, which are controlled by clock phase ϕ_1. Their on-resistances are supposed to be equal with value R_{on}, which are in series with a noise voltage source v_S. Fig. 2.37*b* shows the model for the thermal noise introduced by switches S_3 and S_4, controlled by clock phase ϕ_2, and also the noise in the amplifier and in the voltage references. A single-pole model with infinite DC gain is supposed for the amplifier and its equivalent input noise is modeled by a voltage source v_{op} at the positive input terminal. The opamp parasitic input and output capacitances—C_P and C_L, respectively—are explicitly shown.

Switches controlled by ϕ_1

Considering Fig. 2.37*a*, the transfer function from the noise source to the voltage of capacitor C_S is

$$H_{S\phi_1}(s) = \frac{1}{1 + s \cdot 2R_{on}C_S} \tag{2.97}$$

so that the equivalent noise bandwidth yields

$$BW_{n,\,S\phi_1} = \int_0^{+\infty} \left|H_{S\phi_1}(f)\right|^2 df = \frac{1}{4 \cdot 2R_{on}C_S} \tag{2.98}$$

This implies that switches that are fully 'on' or 'off' contribute no $1/f$ noise. That is why flicker noise is only of concern in the amplifier [Bran97b].

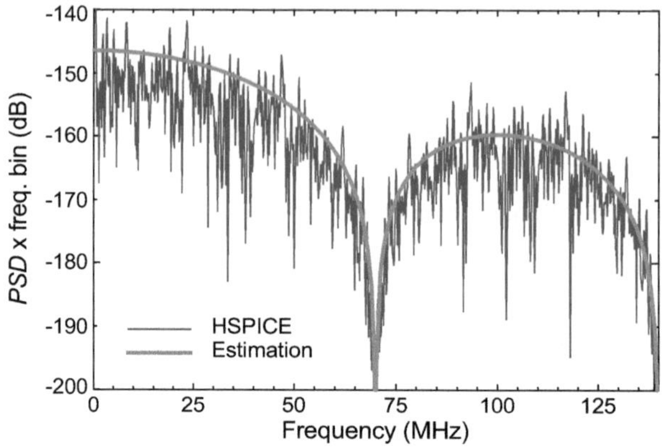

FIGURE 2.38 *PSD* of the input-referred noise generated by switches controlled by ϕ_1.

Given that the time constant $2R_{on}C_S$ is normally much smaller than the clock period for settling reasons, aliasing occurs and the input-referred noise *PSD* in the frequency range $[-f_s/2, +f_s/2]$ is given by

$$S_{in, S\phi_1}(f) = \frac{2BW_{n, S\phi_1}}{f_s} \cdot S_S \cdot \mathrm{sinc}^2(\pi f/f_s) = \frac{kT}{C_S f_s}\mathrm{sinc}^2(\pi f/f_s) \qquad (2.99)$$

where $S_S = 2kT \cdot 2R_{on}$ is the *PSD* of the thermal noise generated in switches S_1 and S_2, added together because they are assumed uncorrelated.

Note that, contrary to eq(2.96), no track term is present in eq(2.99), since there is no direct path between the noise source and the integrator output [Gobet81]. On the other hand, the sampled noise voltage is actually held during a complete clock cycle (the held component is resampled during ϕ_2), so that $\tau_{SH} \cong T_s$ [Fisc82] [Olia00].

Fig. 2.38 compares the input-referred noise *PSD* obtained using eq(2.99) to electrical simulation results from HSPICE on an SC integrator clocked at $f_s = 70\mathrm{MHz}$. Since HSPICE does not include noise sources in the transient analysis, a white noise was externally generated and then included in the electrical simulation using the DATA command.

Switches controlled by ϕ_2

Considering Fig. 2.37*b*, the transfer function from the noise source v_S to the voltage of capacitor C_S yields

$$H_{S\phi_2}(s) = \frac{1 + s/z_1}{(1 + s/p_1)(1 + s/p_2)}$$

$$\text{with} \quad z_1 = \frac{g_m}{C'} \ , \ p_1 \cong \frac{g_m}{C_{eq,\,i}} \ , \ p_2 \cong \frac{C_{eq,\,i}}{C'} \cdot \frac{1}{2R_{on}C_S} \qquad (2.100)$$

where $C' = C_P + C_L(1 + C_P/C_I)$ and $C_{eq,i}$ is the equivalent capacitive load at the amplifier output node during integration, given by

$$C_{eq,i} = C_P + C_S + C_L\left(1 + \frac{C_P + C_S}{C_I}\right) \tag{2.101}$$

SC integrators are designed so that their settling is constrained by the opamp bandwidth and not by the $2R_{on}C_S$ time constant. Therefore, the zero and the poles of $H_{S\phi_2}(s)$ fulfil $p_1 \le z_1 \ll p_2$ and the equivalent noise bandwidth can be calculated as

$$BW_{n,S\phi_2} = \int_0^{+\infty} \left|H_{S\phi_2}(f)\right|^2 df = \frac{p_1 p_2 (p_1 p_2 + z_1^2)}{4 z_1^2 (p_1 + p_2)} \cong \frac{p_2}{4} \approx \frac{1}{4 \cdot 2R_{on}C_S} \tag{2.102}$$

where it has been assumed that $C_{eq,i} \approx C'$. Note that the noise bandwidth is limited by the high-frequency pole, so that $BW_{n,S\phi_2} > f_s/2$ and the white noise folds back over the baseband. Since the sampled-and-held component is commonly dominant, if the track term is neglected, the input-referred noise PSD can be written as

$$S_{in,S\phi_2}(f) \cong \frac{2BW_{n,S\phi_2}}{f_s} \cdot S_S \cdot \mathrm{sinc}^2(\pi f/f_s) = \frac{kT}{C_S f_s} \mathrm{sinc}^2(\pi f/f_s) \tag{2.103}$$

where $S_S = 2kT \cdot 2R_{on}$ is the spectral density of the thermal noise generated in switches S_3 and S_4, assumed uncorrelated.

Opamp noise

The noise introduced by the amplifier—modeled in Fig . 2.37*b* by means of the voltage source v_{op}—will have a thermal and a flicker component.

The input-referred thermal noise of a MOSFET in saturation is approximately

$$S_{MOS}^t = \frac{4kT}{3g_m}, \qquad -\infty < f < +\infty \tag{2.104}$$

whereas the flicker (or $1/f$) noise, which is nearly independent of the bias condition, is approximately given by

$$S_{MOS}^f(f) = \frac{K}{C_{ox}WL} \cdot \frac{1}{|f|}, \qquad -\infty < f < +\infty \tag{2.105}$$

where K is a process- and temperature-dependent parameter. Both noises are uncorrelated in the same device, so their PSDs can be directly added.

Since the opamp noise is originated in its MOSFETs—mostly usually in its input devices—, its PSD will have the generic shape shown in Fig.2.39.

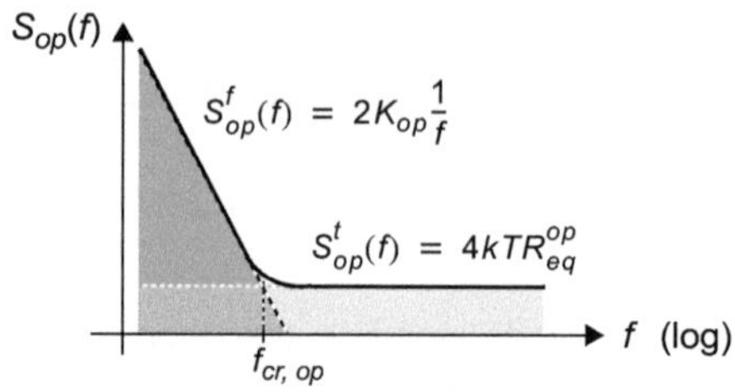

FIGURE 2.39 PSD of the opamp noise showing the contributions of $1/f$ and thermal noise. (Single-sided representation).

Flicker noise dominates at low frequencies and decreases with frequency. The frequency at which $1/f$ noise is equal to the white thermal noise is called *corner frequency* f_{cr}. Beyond f_{cr}, thermal noise dominates. Taking this into account, the PSD of the opamp noise, referred to its input, can be written as

$$S_{op}(f) = S^t_{op}(f) + S^f_{op}(f) = S^t_{op}\left(1 + \frac{f_{cr,\,op}}{|f|}\right) \tag{2.106}$$

where S^t_{op} stands for the input-referred PSD of the opamp thermal noise.

In order to calculate the contribution of the opamp thermal noise at the integrator input, its equivalent bandwidth can be used. The transfer function from v_{op} to the input voltage can be calculated as

$$\begin{aligned} H_{op}(s) &= \frac{1}{(1 + s/p_1)(1 + s/p_2)} \\ \text{with} \quad p_1 &\cong \frac{g_m}{C_{eq,\,i}} \quad , \quad p_2 \cong \frac{C_{eq,\,i}}{C'} \cdot \frac{1}{2R_{on}C_S} \end{aligned} \tag{2.107}$$

where $p_1 \ll p_2$, because the integrator settling is normally dominated by the opamp bandwidth. Thus, the thermal noise equivalent bandwidth yields

$$BW_{n,\,op} = \int_0^{+\infty} |H_{op}(f)|^2 df = \frac{p_1 p_2}{4(p_1 + p_2)} \cong \frac{p_1}{4} \cong \frac{g_m}{4C_{eq,\,i}} \tag{2.108}$$

so that it is limited by the gain-bandwidth product of the opamp during integration, $GB_i = g_m/C_{eq,\,i}$ (in rad/s). Therefore, the sampled-and-held thermal noise aliases into the baseband ($BW_{n,\,op} > f_s/2$), leading to

$$S^t_{in,\,op}(f) \cong \frac{2BW_{n,\,op}}{f_s} \cdot S^t_{op} \cdot \mathrm{sinc}^2(\pi f/f_s) = \frac{GB_i}{2f_s} \cdot S^t_{op} \cdot \mathrm{sinc}^2(\pi f/f_s) \tag{2.109}$$

For a single-stage opamp, its thermal noise PSD can be written as

$$S^t_{op} = 2 \cdot \frac{4kT}{3g_m} \cdot (1 + n_t) \tag{2.110}$$

where g_m is the transconductance of the input transistor and n_t is the noise contribution factor of the remaining transistors —i.e., the summation of the ratio of their transconductances to g_m. The factor 2 accounts for the fully-differential structure assumed for the opamp. Replacing eq(2.110) in eq(2.109), the *PSD* of the opamp thermal noise at the integrator input yields

$$S_{in,\,op}^{t}(f) \cong \frac{4kT(1+n_t)}{3C_{eq,\,i}f_s} \cdot \mathrm{sinc}^2(\pi f/f_s) \tag{2.111}$$

In order to calculate the contribution of the opamp flicker noise, it must be first noted that the opamp is usually designed to have a corner frequency considerably smaller than the sampling rate, $f_{cr,\,op} \ll f_s$ —using large sizes for the transistors contributing most [eq(2.105)]. On the other hand, the opamp bandwidth must be several times larger than f_s for an appropriate settling of the integrator. Thus, the low-pass filtered version of the opamp flicker noise at the integrator input is normally about the original flicker component at the opamp input—i.e., $S_{op}^{f}(f)|H_{op}(f)|^2 \cong S_{op}^{t}f_{cr,\,op}/|f|$, where $H_{op}(f)$ stands for eq(2.107) in the frequency domain.

Even so, the sample-and-held flicker noise will also contain a folded-back component due to the aliasing of the high-frequency $1/f$ noise. This effect is illustrated in Fig. 2.40, where the tails of the $1/f$ noise fold back to the baseband and originate a white component that adds to the original flicker *PSD*. For the case considered ($f_{cr} = f_s/10$), this extra white contribution is about $1.1S_{op}^{t}$. However, note that the aliasing of opamp thermal noise is considerably larger. For instance, assuming that the opamp bandwidth is 4 times the sampling frequency, the opamp thermal noise at the integrator input is approximately $\pi GB_i(\mathrm{Hz})/f_s = 4\pi = 12.6$ times S_{op}^{t} [see eq(2.109)]. The flicker noise that folds back is then less than 10% of the thermal component, so

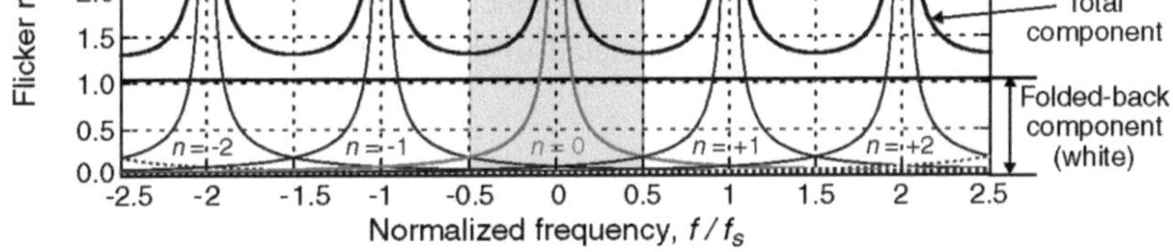

FIGURE 2.40 Illustration of the folding-back effect on flicker noise for $f_{cr} = f_s/10$. Fundamental, sidebands, and total component of a sampled flicker noise.

that omitting it will only represent an underestimation of the white noise power due to the opamp of 0.4dB within a given band †15.

Thus, the folded-back component of the $1/f$ noise will be generally 'submerged' by the aliased thermal noise [Fisc82] [Gobet83] [Enz96], so that it can be neglected for noise computation. Thus, the input-referred PSD of the sample-and-held opamp flicker noise can be approximated to

$$S_{in,\,op}^{f}(f) \cong S_{op}^{t}\frac{f_{cr,\,op}}{|f|} \cdot \mathrm{sinc}^2(\pi f/f_s) \tag{2.112}$$

Noise in the references

The voltage references of the converter are normally generated from band-gap circuits or resistor ladders and are then buffered to the input branches of the integrators in the ΣΔ modulator. They can not be absolutely 'clean', and besides the corresponding DC voltage, they usually contain a thermal and a $1/f$ component, apart from other spurious signals.

Let us assume that the noise PSD at the buffer output contains thermal and flicker noise, which are low-pass filtered by the reference buffer; i.e.,

$$S_{ref}(f) = S_{ref}^{t}(f) + S_{ref}^{f}(f) = S_{ref}^{t}\left(1 + \frac{f_{cr,\,ref}}{|f|}\right)|H(f)|^2 \tag{2.113}$$

where $f_{cr,\,ref}$ stands for the corner frequency of the noise in the references and $H(s)$ for the transfer function of the reference buffer, which we will approximate to the following single-pole function

$$H(s) = \frac{1}{1 + s/GB_{ref}} \tag{2.114}$$

with GB_{ref} being the bandwidth of the reference buffer (in rad/s).

Thus, the noise in the references (source v_{ref} in Fig. 2.37*b*) contains a thermal component with PSD equal to S_{ref}^{t} within an equivalent bandwidth $BW_{n,\,ref} = GB_{ref}/4$, as well as a $1/f$ component with filtered tails. On the other hand, note from Fig. 2.37*b* that the noise source v_{ref} is in series with the switch noise source v_S, so that $H_{S\phi_2}(s)$ [eq(2.100)] is the transfer function from v_{ref} to the integrator input. Then, the static noise associated to the references at the integrator input can be obtained as

$$S_{ref}^{t}\left(1 + \frac{f_{cr,\,ref}}{|f|}\right)|H(f)|^2 |H_{S\phi_2}(f)|^2 \tag{2.115}$$

15. The underestimation is indeed lower, because the tails (high-frequency components) of the $1/f$ noise are in practice low-pass filtered by $H_{op}(f)$.

Assuming that the bandwidth of the reference buffer is similar to that of the opamp †16, and thus much smaller than $1/(2R_{on}C_S)$, the filtering effect of $H_{S\phi_2}(s)$ on the noise PSD can be neglected [see eq(2.102)]; i.e.,

$$S^t_{ref}\left(1+\frac{f_{cr,\,ref}}{|f|}\right)|H(f)|^2\left|H_{S\phi_2}(f)\right|^2 \cong S^t_{ref}\left(1+\frac{f_{cr,\,ref}}{|f|}\right)|H(f)|^2 \tag{2.116}$$

Therefore, the sampled-and-held noise associated to the references contains a folded-back component due to the aliasing of its thermal noise, a flicker component, and a white component due to the aliasing of the high-frequency $1/f$ noise. Assuming $f_{cr,\,ref} \ll f_s$, the latter can be neglected and the input-referred PSD of the noise in the references can be approximated to

$$S_{in,\,ref}(f) \cong S^t_{ref}\left(\frac{GB_{ref}}{2f_s}+\frac{f_{cr,\,ref}}{|f|}\right)\cdot \mathrm{sinc}^2(\pi f/f_s) \tag{2.117}$$

In some cases, the voltage reference buffer is connected to large bypass capacitors—usually off-chip, what requires an extra pad/pin. Whenever this is possible, the current peaks needed for charging and discharging the reference SC branches are provided by the external capacitor. Hence, the mission of the on-chip reference buffer is to re-charge the external capacitor and to keep the output impedance low within the signal band [Gust00] [Maul00]. This can be achieved with very little power dissipation (see Section 4.4.7). Furthermore, when this strategy can be applied, the noise in the references is limited to a very narrow bandwidth and does not alias in the baseband. In telecom applications, in which the low-frequency region is usually out of the signal band, its contribution can habitually be neglected.

Total noise

Adding the thermal noise introduced by the switches [eq(2.99) and eq(2.103)], the opamp noise contributions [eq(2.109) and eq(2.112)], and the noise in the references [eq(2.117)], the total input-referred noise PSD in the interval $[-f_s/2, +f_s/2]$ for the SC integrator in Fig. 2.36 yields

$$\begin{aligned} S_{eq,\,in}(f) \cong{}& \frac{2kT}{C_S f_s}\cdot \mathrm{sinc}^2(\pi f/f_s) + S^t_{op}\left(\frac{GB_i}{2f_s}+\frac{f_{cr,\,op}}{|f|}\right)\cdot \mathrm{sinc}^2(\pi f/f_s) \\ &+ S^t_{ref}\left(\frac{GB_{ref}}{2f_s}+\frac{f_{cr,\,ref}}{|f|}\right)\cdot \mathrm{sinc}^2(\pi f/f_s) \end{aligned} \tag{2.118}$$

If the SC integrator has two input branches, like that in Fig. 2.41, a similar procedure can be used to calculate the noise PSD in capacitor C_{S2}, leading to

16. This is a normal choice, because otherwise the dynamic of the first integrator would be limited by that of the reference buffer.

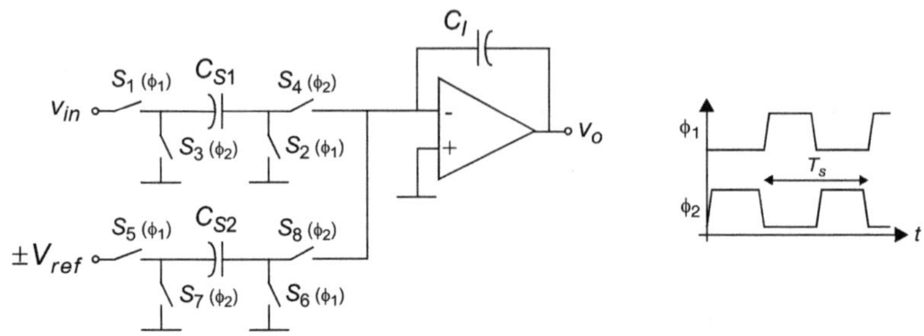

FIGURE 2.41 Two-branch SC integrator and clock phases.

$$S_{C_{S2}}(f) \cong \frac{2kT}{C_{S2}f_s} \cdot \mathrm{sinc}^2(\pi f/f_s) + S^t_{op}\left(\frac{GB_i}{2f_s} + \frac{f_{cr,\,op}}{|f|}\right) \cdot \mathrm{sinc}^2(\pi f/f_s) + S^t_{ref}\left(\frac{GB_{ref}}{2f_s} + \frac{f_{cr,\,ref}}{|f|}\right) \cdot \mathrm{sinc}^2(\pi f/f_s) \tag{2.119}$$

where the first term stands for the thermal noise introduced by switches S_5 to S_8, the second term corresponds to the opamp noise contribution, and the last term to the noise introduced by the references. Note that, in this case, $C_{eq,\,i}$ is given by

$$C_{eq,\,i} = C_P + C_{S1} + C_{S2} + C_L\left(1 + \frac{C_P + C_{S1} + C_{S2}}{C_I}\right) \tag{2.120}$$

Eq(2.119) can be referred to the integrator input, yielding

$$S_{in,\,C_{S2}}(f) \cong S_{C_{S2}}(f)\left(\frac{C_{S2}}{C_{S1}}\right)^2 \cong \frac{2kT}{C_{S1}f_s}\left(\frac{C_{S2}}{C_{S1}}\right) \cdot \mathrm{sinc}^2(\pi f/f_s) + S^t_{op}\left(\frac{GB_i}{2f_s} + \frac{f_{cr,\,op}}{|f|}\right)\left(\frac{C_{S2}}{C_{S1}}\right)^2 \mathrm{sinc}^2(\pi f/f_s) + S^t_{ref}\left(\frac{GB_{ref}}{2f_s} + \frac{f_{cr,\,ref}}{|f|}\right)\left(\frac{C_{S2}}{C_{S1}}\right)^2 \mathrm{sinc}^2(\pi f/f_s) \tag{2.121}$$

The noise *PSD* of the two branches [eq(2.118) and eq(2.121)] can be added, keeping in mind that the contributions from the switches are uncorrelated, but the terms derived from the opamp noise are correlated †17. Note that in this case the noise in the references is associated to the second input

17. Two noise sources v_1, v_2, with *PSD*s $S_1(f)$, $S_2(f)$ respectively, add yielding [Motc93]

$$\overline{v_1^2} + \overline{v_2^2} + 2\rho\sqrt{\overline{v_1^2}\,\overline{v_2^2}} \quad \Rightarrow \quad S_1(f) + S_2(f) + 2\rho\sqrt{S_1(f)S_2(f)}$$

where ρ is the correlation coefficient of the two random processes ($-1 \le \rho \le 1$).
If the noise sources are uncorrelated, $\rho = 0$ and the noise *PSD*s directly add up. If the noise sources are fully correlated—as in this case—$\rho = 1$.

branch and must be omitted from eq(2.118). Taking this into account, the total input-referred noise PSD of the two-branch SC integrator in Fig. 2.41 yields

$$\begin{aligned} S_{eq,\,in}(f) \cong {} & \frac{2kT}{C_{S1}f_s}\left(1+\frac{C_{S2}}{C_{S1}}\right)\cdot \mathrm{sinc}^2(\pi f/f_s) \\ & + S^t_{op}\left(\frac{GB_i}{2f_s}+\frac{f_{cr,\,op}}{|f|}\right)\left(1+\frac{C_{S2}}{C_{S1}}\right)^2 \mathrm{sinc}^2(\pi f/f_s) \\ & + S^t_{ref}\left(\frac{GB_{ref}}{2f_s}+\frac{f_{cr,\,ref}}{|f|}\right)\left(\frac{C_{S2}}{C_{S1}}\right)^2 \mathrm{sinc}^2(\pi f/f_s) \end{aligned} \qquad (2.122)$$

Note that:

- Thermal noise introduced by the switches can be reduced by increasing the integrator input capacitor C_{S1}.
- In order to reduce the contribution of the opamp thermal noise, its GB should be as low as the settling requirements allow. The same applies for the thermal noise associated to the references.
- The $1/f$ noise components can be reduced by decreasing the corner frequency. Also, cancellation techniques such as correlated double sampling, chopper stabilization, and autozeroing can be employed for further reduction of the flicker component [Enz96].

2.4.3 Circuit noise in ΣΔ modulators

In a ΣΔ modulator, all integrators contribute to the input-referred circuit noise of the modulator. Nevertheless, the noise contributions of the integrators other than the first one is (at least) divided by the gain of the first integrator when referred to the modulator input. Since the gain of an integrator within the baseband is high—although not infinite (see Section 2.1)—, the noise contribution of the remaining integrators can be considered negligible in comparison with that of the front-end integrator. The input-referred circuit noise of a ΣΔ modulator is therefore pretty much the same as that of the first integrator.

For the calculation of the in-band error power, two cases can be distinguished, depending on the values of the signal weight ($g_1 = C_{S1}/C_I$) and the feedback weight ($g_1' = C_{S2}/C_I$):

(a) $g_1 = g_1'$, so that the modulator gain ($G = g_1/g_1'$) equals unity.

In this case, the front-end integrator can have a single input branch (see Fig. 2.36). Assuming that a single-stage opamp is used, the input-referred circuit noise PSD is given by eq(2.118) and the input-referred error power due to circuit noise $P_{CN,\,in}$ in the baseband $[-f_b, +f_b]$ yields

$$P_{CN,\,in} = \int_{-f_b}^{+f_b} S_{eq,\,in}(f)df \cong \int_{-f_b}^{+f_b} \frac{1}{f_s}\left(\frac{2kT}{C_S} + S^t_{op}\frac{GB_i}{2} + S^t_{ref}\frac{GB_{ref}}{2}\right)df \\ + 2\int_{+f_o}^{+f_b}\left[S^t_{op}\frac{f_{cr,\,op}}{f} + S^t_{ref}\frac{f_{cr,\,ref}}{f}\right]df \qquad (2.123)$$

where the approximation $\mathrm{sinc}(\pi f/f_s) \cong 1$, $f \ll f_s$ has been used. Note that the $1/f$ noise component is integrated from a frequency $f_o > 0$ in order to exclude DC and make the integral converge, yielding

$$P_{CN,\,in} \cong \left[\frac{2kT}{C_S} + S^t_{op}\frac{GB_i}{2} + S^t_{ref}\frac{GB_{ref}}{2}\right]\frac{1}{OSR} \\ + 2\ln\left(\frac{f_b}{f_o}\right)(S^t_{op}f_{cr,\,op} + S^t_{ref}f_{cr,\,ref}) \qquad (2.124)$$

where $OSR = f_s/(2f_b)$ stands for the modulator oversampling ratio.

(b) $g_1 \neq g_1'$, so that the modulator has a gain $G \neq 1$.

The front-end integrator needs to be implemented using two input branches (see Fig. 2.41). From eq(2.122), the input-referred in-band error power due to circuit noise can be calculated as †18

$$P_{CN,\,in} \cong \frac{2kT}{C_{S1}}\left(1 + \frac{C_{S2}}{C_{S1}}\right)\frac{1}{OSR} \\ + \left[S^t_{op}\frac{GB_i}{2}\left(1 + \frac{C_{S2}}{C_{S1}}\right)^2 + S^t_{ref}\frac{GB_{ref}}{2}\left(\frac{C_{S2}}{C_{S1}}\right)^2\right]\frac{1}{OSR} \qquad (2.125) \\ + 2\ln\left(\frac{f_b}{f_o}\right)\left[S^t_{op}f_{cr,\,op}\left(1 + \frac{C_{S2}}{C_{S1}}\right)^2 + S^t_{ref}f_{cr,\,ref}\left(\frac{C_{S2}}{C_{S1}}\right)^2\right]$$

Note from equations (2.124) and (2.125) that the terms due to thermal noise are inversely proportional to the oversampling ratio, whereas the low-frequency flicker components are independent of OSR.

Fully-differential circuitry

In case the SC integrator is fully-differential, the number of switches will be doubled and the first term in equations (2.124) and (2.125) will be multiplied by a factor 2. Although the in-band power due to the switches thermal noise

18. Note that the output-referred in-band error power due to circuit noise is calculated as

$$P_{CN} = \int_{-f_b}^{+f_b} S_{eq,\,in}(f)|STF(f)|^2 df$$

If $g_1 = g_1'$, $|STF(f)| = 1$ and $P_{CN} = P_{CN,\,in}$.
If $g_1 \neq g_1'$, $|STF(f)| = G$ and $P_{CN} = P_{CN,\,in} \cdot (C_{S1}/C_{S2})^2$.

will increase by 3dB, the signal power will increase by 6dB, so that the signal-to-(switch noise) ratio will improve by 3dB in a fully-differential implementation. On the other hand, the noise power introduced in the baseband by the opamp and by the references will be the same, so that the signal-to-opamp + reference noise) ratio improves by 6dB.

2.5 Clock Jitter

Jitter can be defined as a short-term, non-cumulative variation of the switching instant of a digital signal from its ideal position in time [Lee01] (see Fig. 2.42*a*). This inaccuracy is inherent to every clock generation circuitry—i.e., crystal oscillators, PLL-based oscillators, etc.—and is mainly caused by thermal noise, phase noise, and spurious components.

The effect of clock jitter on an SC ΣΔ modulator can be calculated by analyzing its effect on the sampling of the input signal. Clock jitter during the integration phase will only cause a higher-order error to be added to the integrator settling error, and can therefore be neglected. This implies that the effect is independent of the structure or the order of the modulator [Bran97b].

Clock jitter in the sampling phase—also called *sampling time uncertainty*—results in non-uniform sampling of the converter input signal, increasing the total error power. The magnitude of this increase is a function of the statistical properties of the jitter and the input signal, but simple and effective estimations of the error induced by jitter can be derived [Boser88].

For an input sinewave $x(t)$ with amplitude A_x and frequency f_x , the value of the error in the sampled signal at each clock cycle is given by (see Fig. 2.42*b*)

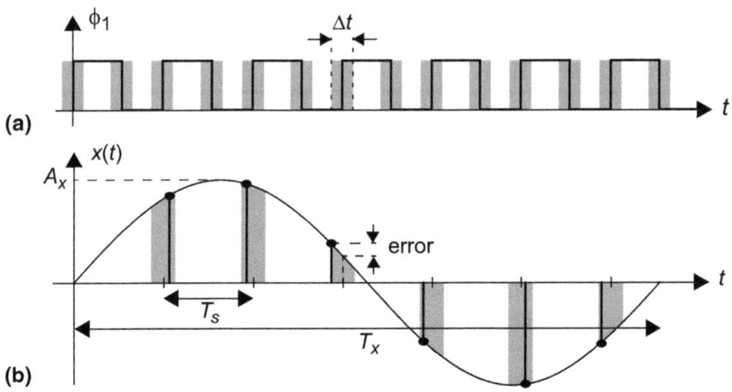

FIGURE 2.42 (a) Jitter in a digital signal; (b) Effect of clock jitter in the sampling.

$$x(nT_s+\Delta t)-x(nT_s)\cong \left.\frac{dx(t)}{dt}\right|_{nT_s}\Delta t = 2\pi f_x A_x \cos(2\pi f_x nT_s)\Delta t \tag{2.126}$$

where Δt is the uncertainty in the sampling instant.

Assuming that the sampling uncertainty Δt is an uncorrelated Gaussian random process with standard deviation σ_J, the power of the error signal will be uniformly distributed in the interval $[-f_s/2, +f_s/2]$, and the *PSD* is

$$S_J = \frac{A_x^2}{2}\frac{(2\pi f_x \sigma_J)^2}{f_s} \tag{2.127}$$

The in-band jitter noise power is then given by †19

$$P_J = \int_{-f_b}^{+f_b} S_J df = \frac{A_x^2}{2}\frac{(2\pi f_x \sigma_J)^2}{OSR} \tag{2.128}$$

Note that the error caused by clock jitter is inversely proportional to the oversampling ratio and increases as the amplitude and/or frequency of the input signal increase.

Since $A_x \le V_{ref}$ and $f_x \le f_b$, an upper bound (worst-case value) can be calculated for P_J, yielding [Yin94a]

$$P_{J,wc} = \frac{V_{ref}^2}{2}\frac{(2\pi f_b \sigma_J)^2}{OSR} = \frac{V_{ref}^2}{2}\frac{(\pi f_s \sigma_J)^2}{OSR^3} \tag{2.129}$$

Assuming that σ_J has a fixed value, independent of the clock frequency, eq(2.129) shows that a ΣΔ modulator is *OSR* times less sensitive to jitter than a Nyquist converter. If the jitter is proportional to the clock frequency, $f_s\sigma_J$ will be constant, leading to a reduction of OSR^3 in the in-band noise power †20.

2.6 Sources of Distortion

The non-idealities studied so far are linear and, as such, they can seriously affect the modulator performance. In this section, the main non-linear effects present in SC ΣΔMs will be treated. These non-linearities will generate distortion, limiting the peak *SNDR* achievable at large input levels. In general, non-

19. If $g_1 \neq g_1'$, $|STF(f)| = g_1/g_1' = G$ with G being the modulator gain. In this case the output-referred in-band jitter noise power will be multiplied by a factor G^2.

20. If the clock jitter has a $1/f$ characteristic—*close-in-noise*—, the oversampling will not reduce the in-band noise and the ΣΔ converter will be as sensitive to jitter as a Nyquist converter [Bran97b].

linearities are much more difficult to handle analytically that linear effects and, although the effect depends on the structure of the modulator, in general there is no clear advantage to either single-loop or cascade topologies [Bran97b].

Several simplifications have to be made before treating them. The most extended approach is to consider each individual non-linearity as a small perturbation of the ideal modulator response—i.e., as a *weak non-linearity.* On the other hand, the non-linearities in the front-end integrator are considered the ones directly affecting the overall converter linearity. When referred to the input, non-linear effects in the remaining integrators will be attenuated by the gain of the integrator(s) before them and their contributions are consider negligible. It is also assumed in the analysis that the power of the harmonics decreases when its order increases and, therefore, only 2nd- and 3rd-order non-linear terms are usually considered.

Distortion introduced by an SC integrator is mainly due to voltage-dependent behavior of capacitors and switches, and non-linearities associated to the gain of the amplifier as well as its settling behavior. The non-linearity due to the charge injection from the switches can be neglected employing clock phases with delayed falling edges [Haigh83]. We will return to this in Chapters 3 and 4.

2.6.1 Distortion due to the non-linear capacitors

In the previous sections the capacitance of a capacitor has been assumed constant, whereas in practice it exhibits a dependence on the stored voltage. This relationship is commonly expressed with a Taylor expansion as

$$C(v) = C(1 + a_1 v + a_2 v^2 + \dots) \tag{2.130}$$

where C is the capacitance when the capacitor is discharged and $a_1, a_2, \dots$ are the non-linear coefficients, usually given in ppm/V, ppm/V^2, ... [†21].

Let us consider the SC integrator in Fig. 2.43, where the integration and sampling capacitors are assumed to be non-linear. The charge transferred to the integration capacitor after one clock cycle can be calculated as

$$\int_{v_{o,n-1}}^{v_{o,n}} C_I(v)dv = \int_0^{v_{1,n-1}} C_{S1}(v)dv + \int_0^{v_{2,n-1}} C_{S2}(v)dv \tag{2.131}$$

yielding to the following difference equation for the integrator output voltage

21. Note that the representation in eq(2.130) does not mean that the total charge in a non-linear capacitor is obtained as the product of its voltage and its capacitance evaluated at that voltage. On the contrary, it means that an increment dv in the capacitor voltage v requires an increment of charge dq; i.e., $dq = C(v)dv$ [Raza95].

$$C_I v_{o,n}\left(1+\frac{a_1}{2}v_{o,n}+\frac{a_2}{3}v_{o,n}^2\right) = C_I v_{o,n-1}\left(1+\frac{a_1}{2}v_{o,n-1}+\frac{a_2}{3}v_{o,n-1}^2\right)$$
$$+C_{S1}v_{1,n-1}\left(1+\frac{a_1}{2}v_{1,n-1}+\frac{a_2}{3}v_{1,n-1}^2\right) \quad (2.132)$$
$$+C_{S2}v_{2,n-1}\left(1+\frac{a_1}{2}v_{2,n}+\frac{a_2}{3}v_{2,n}^2\right)$$

The distortion introduced by a non-linear capacitor is strongly dependent on the placement of that capacitor within the modulator [Bran97b]. If the considered integrator is at the front-end of a ΣΔ modulator, as in Fig. 2.41, one of the branches is connected to the converter input ($v_{1,n} = v_{in,n}$), whereas the other feeds the converted output back ($v_{2,n} = v_{fb,n} = -y_n V_{ref}$). Let us consider, into first-order approximation, that the dominant source of distortion is associated to capacitor C_{S1}, which samples the modulator input. In this case eq(2.132) can be approximated to

$$v_{o,n} \cong v_{o,n-1} + g_1 v_{in,n-1}\left(1+\frac{a_1}{2}v_{in,n-1}+\frac{a_2}{3}v_{in,n-1}^2\right) - g_1' y_{n-1} V_{ref} \quad (2.133)$$

with $g_1 = C_{S1}/C_I$ and $g_1' = C_{S2}/C_I$ being the signal and feedback weights, respectively. Considering that the converter input is a sinewave with amplitude A_x, the input-referred amplitude and distortion of the 2nd- and 3rd-order harmonics is given by

$$A_2 \cong \frac{1}{2}\left(\frac{a_1}{2}A_x^2\right) \quad \Rightarrow \quad HD_2 \cong 20\log_{10}\left(\frac{a_1}{4}A_x\right)$$
$$A_3 \cong \frac{1}{4}\left(\frac{a_2}{3}A_x^3\right) \quad \Rightarrow \quad HD_3 \cong 20\log_{10}\left(\frac{a_2}{12}A_x^2\right) \quad (2.134)$$

If the ΣΔ modulator employs a fully-differential topology, a similar expression to eq(2.133) is obtained, yielding

$$v_{o,n} = v_{o,n}^+ - v_{o,n}^- \cong v_{o,n-1} + g_1 v_{in,n-1}\left(1+\frac{a_2}{12}v_{in,n-1}^2\right) - 2g_1' y_{n-1} V_{ref} \quad (2.135)$$

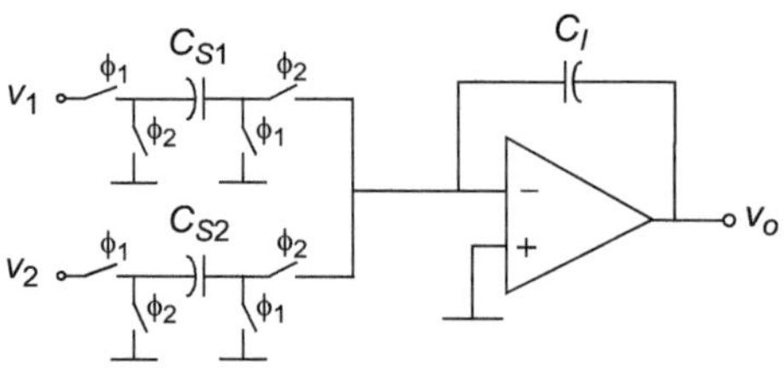

FIGURE 2.43 Two-branch SC integrator.

Note that, as known, the 2nd-order non-linearity is translated into a common-mode voltage, which cancels out to first-order in a fully-differential implementation. On the other hand, the input signal amplitude can be doubled ($A_{xd} = 2A_x$) maintaining the same distortion level of a single-ended version:

$$A_{3,\,FD} \cong \frac{1}{4}\left(\frac{a_2}{12}A_{xd}^3\right) \quad \Rightarrow \quad \begin{aligned} HD_{3,\,FD} &\cong 20\log_{10}\left[\frac{1}{4}\left(\frac{a_2}{12}A_{xd}^2\right)\right] \\ &= 20\log_{10}\left(\frac{a_2}{12}A_x^2\right) \end{aligned} \tag{2.136}$$

The validity of the approximated results above can be demonstrated through behavioral simulation, using eq(2.132) to account for the non-linearity of the integrator capacitors. Note that in each clock cycle $v_{o,\,n}$ must be numerically solved, but a fast convergence is obtained using a simple iterative loop.

Fig. 2.44*a* compares the spectra of the single-ended and the fully-differential versions of a 2-1-1 modulator. The front-end integrator is considered to have a non-linear sampling capacitor with $a_1 = 500\text{ppm/V}$ and $a_2 = 500\text{ppm/V}^2$, whereas the input amplitude is 0.7V in the single-ended modulator and twice that value in the differential modulator. The measured 2nd- and 3rd-order distortion components from the single-ended spectrum are -81.2dB and -94.0dB, respectively. The corresponding values calculated according to eq(2.134) are in very good agreement (-81.2dB and -93.8dB). On the other hand, Fig. 2.44*a* also illustrates how the 2nd-order non-linearity is cancelled in the fully-differential implementation, while preserving the 3rd-order harmonic distortion for doubled input amplitude.

Fig. 2.44*b* compares the spectra obtained for the fully-differential 2-1-1 modulator in Fig. 2.44*a* with that obtained when the non-linearity in the integration capacitor is also considered. Note that the effect is a rise in the noise floor plus an increase of the 3rd-order harmonic distortion, which is measured to be -89.9dB †22. Finally, Fig. 2.44*b* also includes the output spectra when the non-linearity in all the integrators is considered. Note that the contribution of the rest of integrators is negligible in comparison with that at the front-end, leading to a small increase of the noise floor and the harmonic distortion (-89.8dB in this case).

22. This effect is not included in eq(2.134) because is more difficult to handle analytically and leads to results dependent on the structure and the order of the modulator [Bran97b].

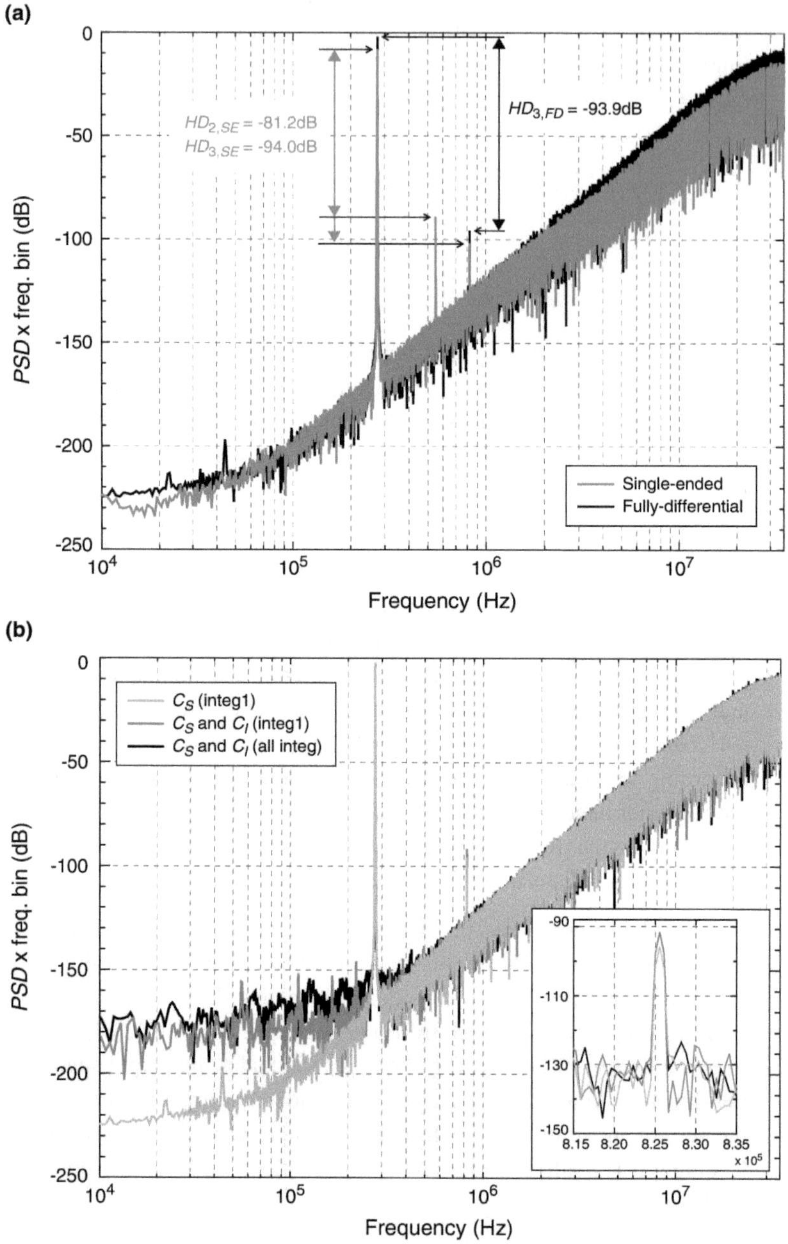

FIGURE 2.44 Simulated spectra of the 2-1-1 ΣΔM with non-linear constants $a_1 = 500\text{ppm/V}$ and $a_2 = 500\text{ppm/V}^2$ in capacitors: (a) Comparison between the single-ended and the fully-differential implementations, (b) Effect of the different non-linear capacitors on the fully-differential implementation. ($A_x = 0.7\text{V}$ for single-ended, $A_{xd} = 2 \cdot 0.7\text{V}$ for fully-differential).

2.6.2 Distortion due to the amplifier non-linear gain

In Section 2.1 the effect of a fixed finite DC gain of the amplifiers was analyzed. This gain is not constant in practice, but voltage-dependent: when the amplifier output voltage increases, the drain-to-source voltage of the output transistors decreases, causing a degradation of the amplifier output impedance. This variation can be expressed with a Taylor expansion as

$$A_{DC}(v_o) = A_{DC}(1 + \gamma_1 v_o + \gamma_2 v_o^2 + \ldots) \tag{2.137}$$

where v_o stands for the amplifier output voltage, A_{DC} is the DC gain in the quiescent point, and $\gamma_1, \gamma_2, \ldots$ are the non-linear coefficients, usually given in %/V, %/V^2, ….

The analysis of the leaky integrator presented in Section 2.1 can be easily adapted to take into account the non-linear amplifier DC gain. Considering that the ΣΔ modulator has a front-end integrator like that in Fig. 2.41, eq(2.2) can be re-written as the following difference equation

$$v_{o,n} = \frac{\left[1 + \dfrac{1 + C_P/C_I}{A_{DC}(v_{o,n-1})}\right] \cdot v_{o,n-1}}{1 + \dfrac{k_{fbi}}{A_{DC}(v_{o,n})}} + \frac{g_1 v_{in,n-1} - g_1' y_{n-1} V_{ref}}{1 + \dfrac{k_{fbi}}{A_{DC}(v_{o,n})}} \tag{2.138}$$

where $v_{1,n} = v_{in,n}$, $v_{2,n} = v_{fb,n} = -y_n V_{ref}$, and $k_{fbi} = 1 + g_1 + g_1' + C_P/C_I$ is the inverse of the capacitive feedback factor during integration. Note that the integrator output voltage in each clock cycle depends on the finite DC gain evaluated for the previously stored voltage—$A_{DC}(v_{o,n-1})$—, as well as on the actual voltage—$A_{DC}(v_{o,n})$.

Let us consider, into first-order estimation, that the distortion is mainly introduced by the non-linear terms directly affecting the converter input signal [Yin94a]. In this case, eq(2.138) can be approximated to

$$v_{o,n} \cong v_{o,n-1} + g_1 v_{in,n-1}\left[1 - \frac{k_{fbi}}{A_{DC}}(1 - \gamma_1 v_{o,n} - \gamma_2 v_{o,n}^2)\right] - g_1' y_{n-1} V_{ref} \tag{2.139}$$

assuming that $A_{DC} \gg 1$ and the non-linear coefficients γ_1, γ_2 are small. Since the output voltage $v_{o,n}$ obviously depends on the converter input, the second term in the above equation leads to the following non-linear terms

$$\begin{aligned} & g_1 v_{in,n-1}\left[1 - \frac{k_{fbi}}{A_{DC}}(1 - \gamma_1 v_{o,n} - \gamma_2 v_{o,n}^2)\right] \\ & \propto g_1 v_{in,n-1}\left[1 + \frac{k_{fbi}}{A_{DC}}(\gamma_1 g_1 v_{in,n-1} + \gamma_2 g_1^2 v_{in,n-1}^2)\right] \end{aligned} \tag{2.140}$$

If the converter input is a sinewave with amplitude A_x, the input-referred amplitude and distortion of the 2nd- and 3rd-order harmonics is then given by

$$
\begin{aligned}
A_2 &\cong \frac{1}{2}\left(\frac{k_{fbi}}{A_{DC}}\gamma_1 g_1 A_x^2\right) \quad \Rightarrow \quad HD_2 \cong 20\log_{10}\left(\frac{\gamma_1}{2}\frac{k_{fbi}}{A_{DC}} g_1 A_x\right) \\
A_3 &\cong \frac{1}{4}\left(\frac{k_{fbi}}{A_{DC}}\gamma_2 g_1^2 A_x^3\right) \quad \Rightarrow \quad HD_3 \cong 20\log_{10}\left(\frac{\gamma_2}{4}\frac{k_{fbi}}{A_{DC}} g_1^2 A_x^2\right)
\end{aligned}
\qquad (2.141)
$$

Note that the distortion is proportional to k_{fbi}; i.e., decreasing the integrator weights will decrease the amplitude of the harmonics. Also, if the relative non-linearity does not change, increasing the amplifier DC gain will help to reduce distortion [Bran97b]. Besides this, if the amplifier is fully-differential, the 3rd-order component is expected to dominate. Using a fully-differential structure in the modulator will further reduce A_2, while preserving HD_3.

These results can be validated through behavioral simulation, implementing eq(2.138) to describe the integrator behavior under non-linear amplifier DC gain. Note that again $v_{o,n}$ must be numerically solved in each clock cycle. An iterative loop has been used, showing a fast convergence.

Fig. 2.45*a* compares the spectra of a single-ended 2-1 modulator for a 0.8-V input amplitude and different non-linear coefficients. The front-end integrator has weights $g_1 = g_1' = 0.25$ and a finite DC gain $A_{DC} = 500$ with $\gamma_1 = 10\%/\mathrm{V}$, whereas γ_2 is 20, 30, and $50\%/\mathrm{V}^2$. The measured 2nd-order distortion is $-90.8\mathrm{dB}$ in all cases, that matches well the value of $-90.5\mathrm{dB}$ calculated according to eq(2.141).

The 3rd-order distortion components are measured to be $-97.4\mathrm{dB}$, $-94.6\mathrm{dB}$, and $-90.8\mathrm{dB}$, respectively. The corresponding calculated values are $-104.4\mathrm{dB}$, $-100.9\mathrm{dB}$, and $-96.5\mathrm{dB}$, respectively. Note that there is a significant deviation on the calculated values, but it can be explained considering Fig. 2.45*b*, that compares the spectrum of the ideal modulator to that obtained with a linear DC gain of 500. We can see that a 3rd-order component exists even with $\gamma_1, \gamma_2 = 0$; it is not due to distortion but to the uncancelled pattern noise generated in the modulator first stage [Yin94a]. If the power of this noise pattern is added to the values calculated with eq(2.141), the estimations for HD_3 become $-97.9\mathrm{dB}$, $-96.0\mathrm{dB}$, and $-93.3\mathrm{dB}$, respectively. Note these new values are in better agreement to the ones measured. Nevertheless, an under-estimation of HD_3 is still noticed for $\gamma_2 = 50\%/\mathrm{V}^2$. This deviation will increase for higher non-linear coefficients, since the non-linearity can be no longer treated as a weak perturbation to the ideal modulator response.

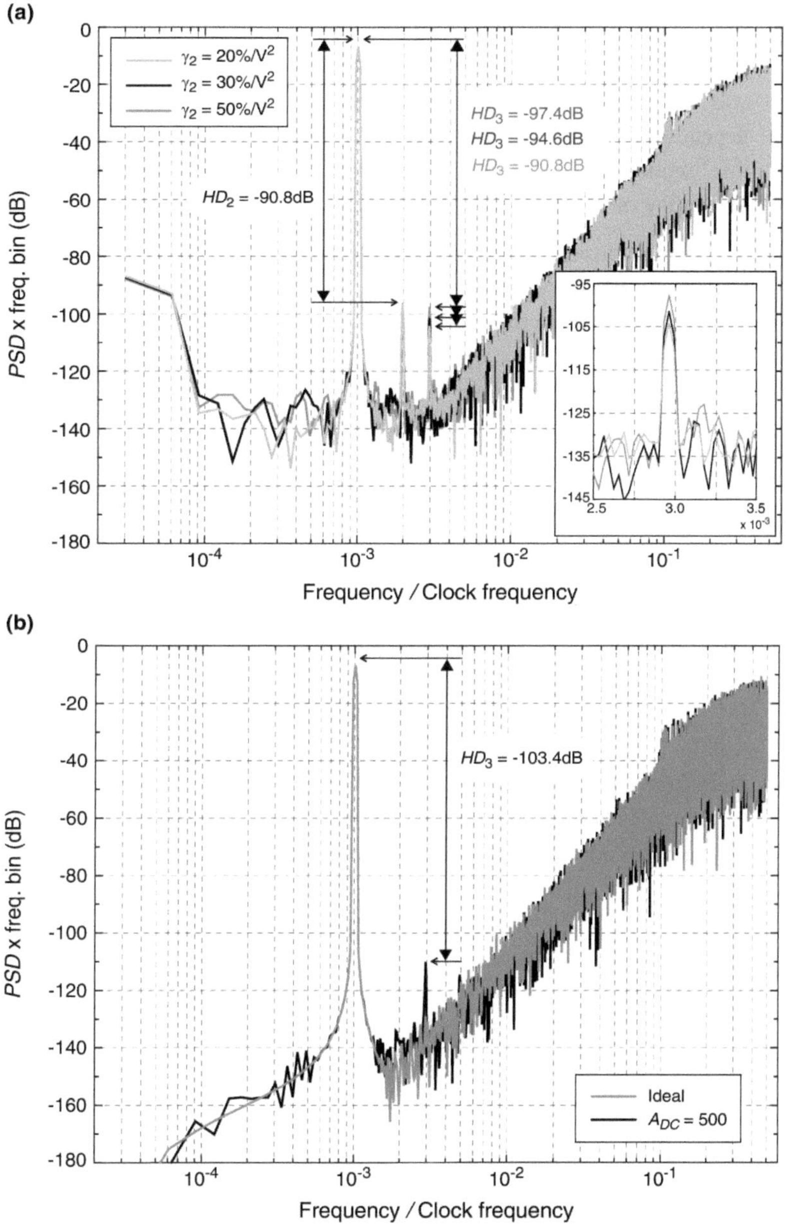

FIGURE 2.45 Simulated spectra of the 2-1 modulator with $A_{DC} = 500$ in the first opamp: (a) Comparison of the non-linearity effects with $\gamma_1 = 10\%/\mathrm{V}$ and varying γ_2, (b) Illustration of the generated noise pattern with $A_{DC} = 500$ and $\gamma_1 = \gamma_2 = 0$. ($A_x = 0.8\mathrm{V}$ in all cases).

2.6.3 Distortion due to the switch non-linear on-resistance

The effect of the non-zero switch on-resistance has already been studied in Section 2.3.5 when analyzing the integrator settling. During the analysis R_{on} was considered constant, but in practice the switch on-resistance heavily depends on the voltage across the nodes of the MOS transistors. As we show in this section, this voltage dependency can generate noticeable distortion depending on the location of the switch inside the integrator.

Let us consider the CMOS switch shown in Fig. 2.46*a*. Considering that the nMOS and pMOS transistors operate in the linear region, their on-resistances can be approximated to

$$
\begin{aligned}
R_N &= \frac{1}{K_N \frac{W_N}{L_N}\left[(v_G - V_{TN}) - \frac{v_D + v_S}{2}\right]} = \frac{1}{K_N \frac{W_N}{L_N}[(V_{DD} - V_{TN}) - v_{CM}]} \\
R_P &= \frac{1}{K_P \frac{W_P}{L_P}\left[\frac{v_D + v_S}{2} - (v_G - V_{TP})\right]} = \frac{1}{K_P \frac{W_P}{L_P}[v_{CM} - (V_{SS} - V_{TP})]}
\end{aligned}
\tag{2.142}
$$

where $v_{CM} = (v_D + v_S)/2$ is the common-mode voltage—i.e., the switch input voltage. The total resistance of the transmission gate is obtained as

$$\frac{1}{R_{on}} = \frac{1}{R_N} + \frac{1}{R_P} \tag{2.143}$$

Note from eq(2.142) that reducing the supply voltage when migrating to smaller technologies causes an increase of the switch on-resistance, since threshold voltages are not scaled down by the same amount [Bult00]. Also, R_{on} heavily

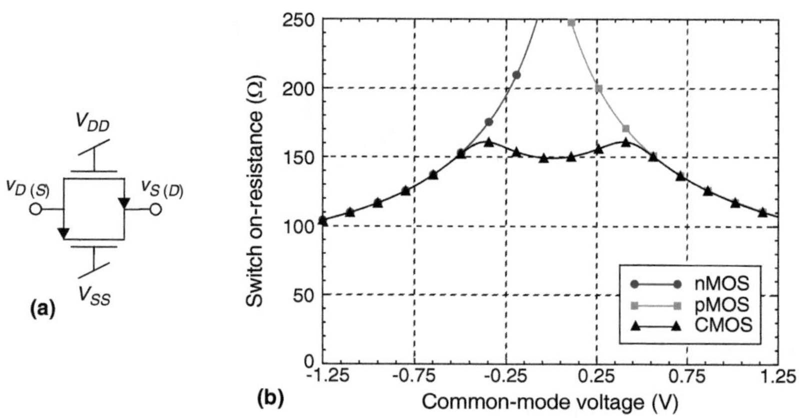

FIGURE 2.46 CMOS switch: (a) Schematic, (b) On-resistance versus common-mode voltage. (Electrical simulation: 2.5-V supply, $W_N/L_N = 8.5/0.25$, and $W_P/L_P = 36.5/0.25$).

depends on the voltage being transmitted. This is illustrated in Fig. 2.46*b*, that shows electrical simulation results for the switch on-resistance as a function of the input voltage. This switch is used in the implementation of a 2-1-1 ΣΔ converter in 2.5-V 0.25-μm CMOS technology (see Chapter 4).

In order to analyze the distortion introduced by the non-linear switches at the front-end integrator, let us consider the integrator schematic shown in Fig. 2.36. During the sampling phase the converter input signal is sampled on capacitor C_S through switches S_1 and S_2. Since S_1 is directly connected to the input, the non-linear on-resistance of this switch can introduce considerable distortion. On the other hand, switch S_2 has one terminal connected to a fixed voltage—the analog ground—, so that at the end of the sampling phase the voltage across S_2 remains approximately constant over clock periods. This makes the distortion introduced by the switch S_2 considerably lower than that generated in S_1, and it can be neglected in practice. The same applies for switches S_3 and S_4 during the integration phase: S_3 has one terminal connected to the reference voltage and S_4 is connected to the virtual ground of the amplifier.

Therefore, in an approximated analysis of the induced distortion, only the sampling operation through switches S_1 and S_2 needs to be considered. The model used is shown in Fig. 2.47 and the procedure followed is similar to that in [Geer02]. The operation during sampling can be described by the following differential equations

$$C_S\frac{dv_{C_S}(t)}{dt} = \frac{v_{in}(t) - v_b(t)}{R_{on,1}[v_{in}(t), v_b(t)]} = \frac{v_b(t) - v_{C_S}(t) - V_{gnd}}{R_{on,2}[V_{gnd}]} \tag{2.144}$$

where the on-resistance of S_1 is considered to be a function of the voltage across its terminals, whereas $R_{on,2}$ is just a function of the analog ground. For the reasons above, this simplification introduces a negligible error in comparison with considering $R_{on,2} = f[v_b(t) - v_{C_S}(t), V_{gnd}]$.

First, as in Fig. 2.46*b*, the characteristic of the switch is obtained by

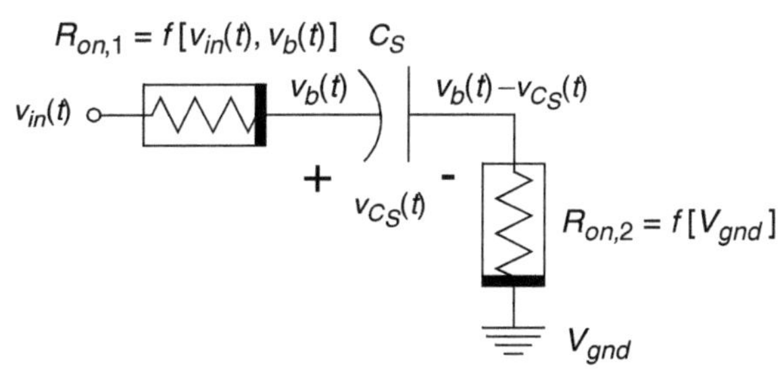

FIGURE 2.47 Model used to study the distortion induced by the switches.

electrical simulation, for given process and transistor sizes. Then, the set of equations in eq(2.144) is numerically solved for an input sinewave of fixed frequency and amplitude, including the non-linear characteristic of switches S_1 and S_2 with a table look-up method. Finally, the transient signal is sampled at the clock frequency in order to process the data.

Fig. 2.48*a* shows the 2nd- and 3rd-order harmonic components of the sampled signal as functions of the frequency f_x of the input signal, for switches as those in Fig. 2.46. The amplitude of the sinewave is $A_x = 0.8\text{V}$ and $f_s = 70.4\text{MHz}$. Given the dynamic nature of the distortion, both HD_2 and HD_3 increase proportionally to f_x, at a rate of 20dB/dec. Assuming that the amplitude of the harmonics decreases for increasing orders, the worst-case distortion in a $\Sigma\Delta$ converter will then be obtained for an input frequency equal to $1/3$ of the signal bandwidth (2.2MHz in the case considered).

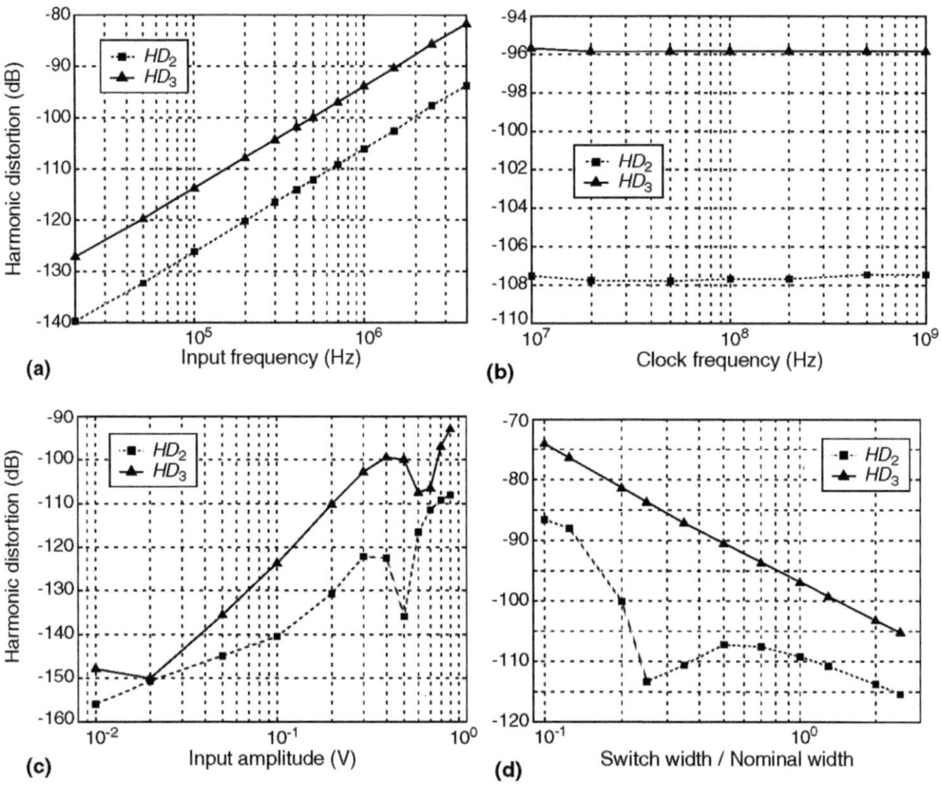

FIGURE 2.48 Harmonic distortion versus: (a) Input frequency, (b) Clock frequency, (c) Input amplitude, (d) Width of the switch transistors. Nominal switches are those in Fig. 2.46 ($W_N/L_N = 8.5/0.25$, $W_P/L_P = 36.5/0.25$) and $C_S = 0.66\text{pF}$. When fixed, $f_s = 70.4\text{MHz}$, $f_x = 700\text{kHz}$, and $A_x = 0.8\text{V}$.

Fig. 2.48*b* shows HD_2 and HD_3 as a function of the clock frequency for a 0.8V @700kHz input tone. Since the model in Fig. 2.47 considers only the sampling operation, the constant behavior of the harmonic distortion is expected. However, note that, in practice, high clock frequencies will cause an increase of the settling error, degrading the converter performance.

Due to the high non-linearity of the switch characteristic (see Fig. 2.46*b*), the dependence of distortion on the input amplitude is not as evident as for the former parameters. Fig. 2.48*c* shows the harmonic components versus A_x. As expected, both HD_2 and HD_3 basically increase for increasing amplitudes. Also, the 3rd-order harmonic distortion rises at 40dB/dec in the amplitude range $[0.02\text{V}, 0.3\text{V}]$. However, the rate of increase for HD_2 differs from 20dB/dec for medium amplitudes. Moreover, both components are non-monotonic for high input levels, where the power of higher-order harmonics starts becoming comparable with that of the 2nd- and 3rd-order components.

The dependence of distortion on the switch dimensions is investigated in Fig. 2.48*d*. The widths of the nMOS and the pMOS transistors were varied from their nominal values—$W_N = 8.5\mu\text{m}$ and $W_P = 36.5\mu\text{m}$—, scaling them by the same amount in order to preserve the shape of the characteristic shown in Fig.2.46*b*. Note that the harmonic components decrease as the width of the switch increases. For HD_3 this dependence is very linear, showing an improvement of 25dB/dec approximately. However, a trade-off exists when augmenting the switch dimensions. On the one hand, the average value of the on-resistance decreases, reducing the harmonic distortion and the nominal $R_{on}C_S$ constant. On the other, the parasitic capacitive load introduced by the switches can increase considerably, affecting the driving of the clock phases, the integrator settling error, etc.

Finally, the shape of the switch characteristic is modified in order to examine its effect on harmonic distortion. To that purpose, the size of the pMOS transistor of the switch in Fig. 2.46*b* is maintained, but the width of the nMOS transistor increases from its nominal value ($8.5\mu\text{m}$) to equal W_P ($36.5\mu\text{m}$). Fig. 2.49 shows the characteristic of each switch sizing as a function of the input voltage. Note that, as W_N tends to equal W_P, the non-linearity of the switch on-resistance increases. This is illustrated in Fig. 2.50, that plots HD_2 and HD_3 as a function of W_N, for input amplitudes of 0.8V and 0.2V. Note that the distortion obtained when $W_N = W_P$ is higher than that for the nominal W_N, especially in the 2nd-order component. Nevertheless, even order harmonics can be strongly reduced employing fully-differential circuitry.

Note also from Fig. 2.49 that the average value of the on-resistance decreases when W_N tends to equal W_P—137Ω for $W_N = 8.5\mu\text{m}$ and 74Ω for $W_N = 36.5\mu\text{m}$. Again a trade-off can be established when selecting the

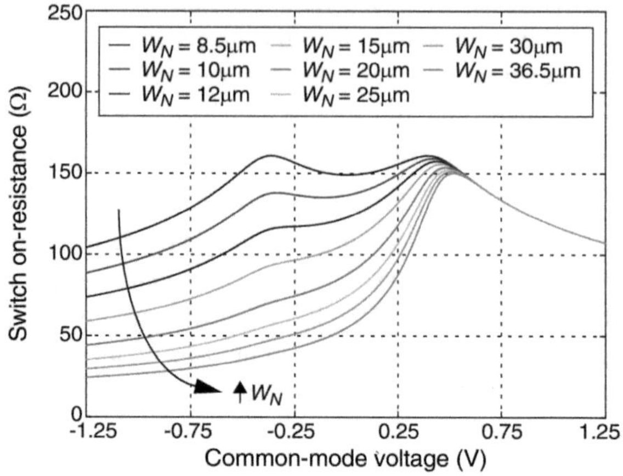

FIGURE 2.49 Switch resistance versus common-mode voltage, for varying width of the nMOS transistor. (Electrical simulation: 2.5-V supply, $W_P/L_P = 36.5/0.25$, and $L_N = 0.25\mu m$).

shape of the characteristic. If the sizing compensates the difference in the transconductance parameter of the nMOS and pMOS transistors—i.e., $K_N W_N \approx K_P W_P$, as in the nominal switch—the non-linearity is low, but the average on-resistance is larger than for $W_N = W_P$ (used in [Yin94a] [Geer02]). In the latter case, the area occupied and the switch capacitive parasitics increase, but the effect of the finite resistance on the settling performance decreases. Therefore, special attention must be paid to the selection of the switch during the design phase of the converter, since the former trade-off must be solved depending on the requirements of a given application.

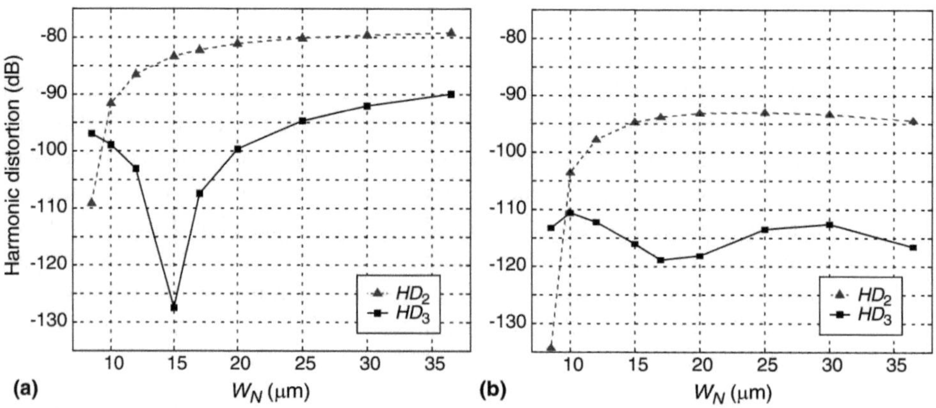

FIGURE 2.50 Harmonic distortion versus width of the nMOS transistor, for: (a) $A_x = 0.8\text{V}$, (b) $A_x = 0.2\text{V}$. ($W_P/L_P = 36.5/0.25$, $L_N = 0.25\mu m$, $C_S = 0.66\text{pF}$, $f_s = 70.4\text{MHz}$, and $f_x = 700\text{kHz}$).

2.6.4 Distortion due to the non-linear settling

As stated in Section 2.3.4, if the most common dynamic of an integrator is slew-rate limited during integration or sampling over the clock cycles, the integrator response is not linear and distortion will appear at the modulator output spectrum. This effect is illustrated next through behavioral simulation performed on the same 2-1-1 cascade modulator used through Section 2.3.4 — the reader can refer to Fig. 2.26 for the sake of completeness.

Fig. 2.51 shows the modulator output spectrum for different values of the normalized slew rate during sampling. These are $SR_s^{norm} = 3, 4, 5$, which correspond to maximum output currents $I_o = 350, 465, 580\mu\text{A}$ in the front-end integrator, respectively. As shown in Fig. 2.26, these values of I_o led to a dominant situation in which the dynamic is partial-slew during both integration and sampling. Note from Fig. 2.51 that a 3rd-order harmonic component arises in all the spectra, due to the non-linearity of the settling and the fully-differential implementation of the modulator. The corresponding values for the harmonic distortion are $HD_3 = -90.9, -109.9, -115.8\text{dB}$, respectively. Thus, the expected $SNDR$ in Fig. 2.26 is not achieved for $SR_s^{norm} = 3$ due to the power of the harmonic, but not to an excessive error in the final linear settling. Besides this, the 3rd-order harmonic component rapidly decreases when increasing I_o, showing that full performance can be achieved although the dominant integrator dynamic is partially slew-rate limited.

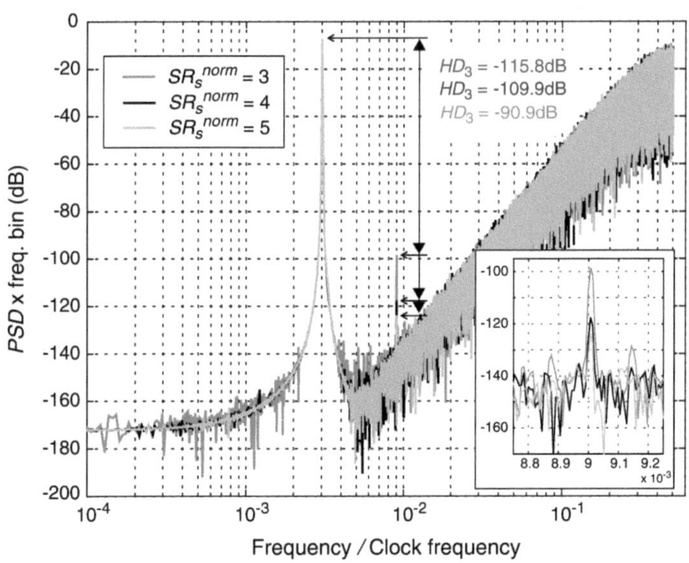

FIGURE 2.51 Simulated spectra of a 2-1-1 modulator considering the slew-rate limitation of the front-end integrator ($GB_s^{norm} = 1.5$ and $A_{xd} = 0.75\text{V}$).

2.7 Summary

In this chapter the main error mechanisms degrading the performance of ΣΔ modulators have been studied. These errors are caused by different non-idealities affecting the analog building blocks and produce extra errors that add to the in-band quantization error power and/or distortion. For demanding specifications on the ΣΔ converter, they can severely limit the modulator performance. Therefore, non-idealities have been studied in detail and behavioral models and, whenever possible, closed-form expressions to estimate their effects have been presented.

First, errors modifying the noise transfer function of the modulator have been analyzed, such as the finite DC gain of amplifiers, the limited dynamic of integrators, and the mismatch in capacitor ratios. As shown, single-loop topologies are less sensitive to these errors than cascade topologies, which suffer from noise leakage; i.e., the degradation of the shaping due to the leakage of lower-order quantization errors to the modulator output.

Second, errors that can be associated to the modulator input have been studied. The main sources of electronic noise have been identified and their effects have been derived. The effect of the clock jitter on high-speed converters has been also taken into account. Finally, non-linear non-idealities have been considered and the generation of distortion has been studied.

The models and guidelines derived in this chapter have been extensively used during the design phases of the two high-speed cascade ΣΔ modulators presented in Chapters 3 and 4.

CHAPTER 3

A Wideband ΣΔ Modulator in 3.3-V 0.35-μm CMOS

NOWADAYS, THE NEED FOR EVER FASTER DATA RATES in broadband communication systems has boosted industrial interest in high-performance data converters capable of achieving 12-to-16-bit accuracy for signal bandwidths well in excess of 1MHz. In addition, reducing production costs pressures mixed-signal designers to pursue these challenging specifications in mainstream digital CMOS processes, where analog primitives are not fully optimized.

In this scenario, ΣΔ converters are gaining ground in a field traditionally dominated by Nyquist converters—especially pipeline. This change is being endorsed by the smaller analog content of ΣΔ modulators and their larger robustness against poorly matched devices—a natural benefit of oversampling and noise-shaping techniques. Although initially employed for high-quality digital audio, these pros have motivated designers to explore the use of ΣΔ converters for signal bands up to telecom and video. The inherent increase of complexity and speed of the digital post-processing required for these high-speed ΣΔ modulators fit into the very application scenario of modern digital CMOS technologies.

This chapter presents a high-speed ΣΔ modulator that targets 14-bit effective resolution for a 2-MHz signal bandwidth. These specifications pose a significant design challenge, especially considering the implementation in a deep-submicron digital CMOS process. The architecture selected is a 4th-order cascade with low oversampling ratio. It employs a robust dual-quantization strategy for increasing resolution, which needs no correction/calibration techniques, thus simplifying the circuitry and reducing power dissipation.

The modulator design follows a top-down approach. First, the topology selection is discussed in Sections 3.2 and 3.3. The requirements of the building blocks in its switched-capacitor implementation are then derived in Section 3.4, whereas their design at the transistor level is presented in Section 3.5. Finally, experimental results of the fabricated prototype are presented.

3.1 Design Methodology

Design methods are of prime importance for completing an efficient implementation in an acceptably short design cycle. As an answer to this necessity, a top-down methodology was developed in [Mede99a] for assisting the design of ΣΔ modulators. Here we will make use of this methodology for designing the targeted ΣΔ modulator. Its flow diagram is illustrated in Fig. 3.1 and considers three linked levels of the design flow:

- modulator level
- building block level
- transistor level

Each level receives specifications as inputs and serves design parameters on the next level down in the hierarchy. Thus, e.g., design parameters obtained from the modulator level are amplifier DC gain, bandwidth, etc., which are specifications for the building block level.

At each level, specifications are translated into design parameters using either equations, behavioral, or electrical models for evaluation purposes and statistical optimization routines for providing 'good-enough' solutions. As shown in Fig. 3.1, this process is supported by the following dedicated CAD tools [Mede99a]:

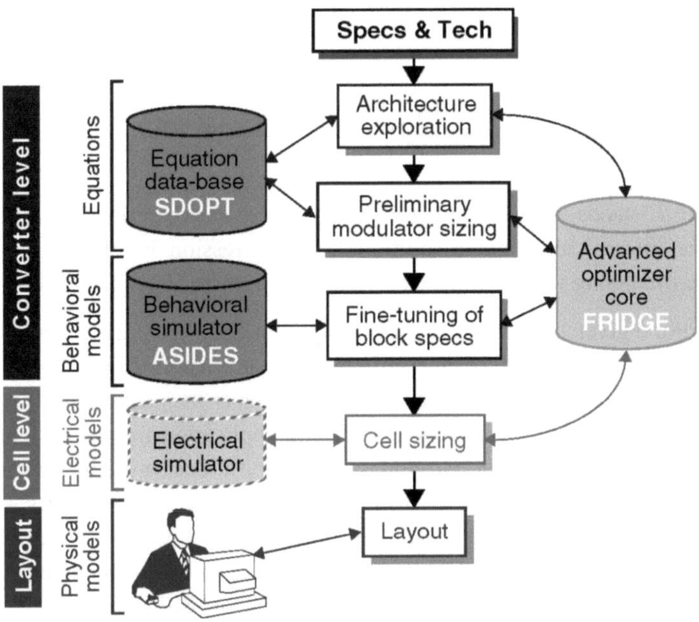

FIGURE 3.1 Flow diagram of the methodology for the design of ΣΔ modulators [Mede99a].

- SDOPT, a synthesis tool for ΣΔ modulators that combines an equation database and a statistical optimizer. It allows architecture exploration and transmission of the modulator specifications down to the building block level.
- ASIDES, a time-domain simulator for ΣΔ modulators based on detailed behavioral models of the building block non-idealities. It is used as a design validation tool and for the fine-tuning of the building block requirements.
- FRIDGE, a general-purpose tool for automatic sizing of basic IC cells, based on electrical simulation and a statistical optimizer. It allows to obtain power-efficient designs of the modulator building blocks at the transistor level.

These CAD tools have been updated with the results from the analysis developed in Chapter 2 for the influence of the main block non-idealities on the operation of ΣΔ modulators. Closed-form expressions have been compiled into the database of design equations associated to SDOPT, whereas the more complex behavioral models have been incorporated to ASIDES.

3.2 Topology Selection

Let us consider the following expression that estimates the dynamic range of a ΣΔ modulator at the architectural level, considering only quantization noise,

$$DR|_{\text{dB}} \approx 10\log_{10}\left[\frac{3}{2}(2^B - 1)^2 \cdot \frac{(2L+1)OSR^{2L+1}}{\pi^{2L}}\right] \tag{3.1}$$

where L stands for the modulator order, OSR is the oversampling ratio, and B is the internal quantizer resolution. This expression highlights the existing trade-off among the three primary design parameters: L, OSR, and B.

For a given bandwidth, the sampling frequency f_s of the ΣΔ modulator—and, thus, the amount of oversampling—is ultimately limited by the speed capabilities of the intended technology. Whenever oversampling ratios as high as 128 and above are compatible with the application frequency—such as in energy metering, speech, and audio—, the 2nd-order single-bit ΣΔ modulator has become a popular choice because of its simplicity and robustness. However, high-frequency applications with medium/high resolutions—typically 12 bits or more are needed for telecom—, are only feasible with moderate values of OSR, thus forcing us to either increasing L or B in order to achieve the required dynamic range.

Whereas increasing *OSR* does not impose an additional complexity on the ΣΔM architecture (except for the increased dynamic requirements of the building blocks), in practice, augmenting *L* and/or *B* considerably involves the modulator design, principally because of two important drawbacks already explained in Chapter 1:

- Unlike 1st- and 2nd-order loops, higher-order loops are not unconditionally stable.
- The linearity of the multi-bit DAC in the feedback path of a multi-bit ΣΔM compromises that of the overall converter.

As stated through Chapter 1, both problems can be partially overcome. On the one hand, high-order ΣΔMs can be stabilized through different techniques; e.g., by using optimized noise transfer functions with reduced out-of-band gain [Schr93] [Adams97a], by properly choosing the integrator scaling factors [OptE90] [Marq98b], or by resetting the integrators if unstable operation is detected [OptE91] [Mous94]. However, as shown in Section 1.3.3, stable high-order single-loop ΣΔMs suffer from a considerable reduction of the achievable *DR* in comparison with the ideal one, eq(3.1)—see for instance Fig. 1.24.

On the other hand, correction and calibration methods can be included in multi-bit ΣΔMs, in either digital or in the analog domain—see Section 1.5. However, the performance of these techniques depends on the resolution of the DAC being calibrated. DACs cannot be efficiently linearized within an arbitrarily large resolution, so that the use of low-order multi-bit modulation may not be enough to obtain a given dynamic range.

A direct solution to this problems is to increase both the modulator order and the internal quantizer resolution, giving rise to moderate-order ($3 \sim 5$) multi-bit architectures. In fact, the use of multi-bit quantization in high-order single-loop modulators inherently improves their stability [Carl97] [Geer00] [Broo02], so that they can be good candidates to obtain high-resolution high-frequency operation, provided that the problem of the DAC non-linearity is solved. Dynamic element matching techniques are normally included to attenuate its impact, but often at the cost of higher circuit complexity and larger occupation area.

With the same objective, the combination of high-order cascade (MASH) architectures with multi-bit quantization was proposed. Cascade ΣΔMs employing dual-quantization techniques gather the unconditional stability of low-order loops (whenever only 1st- and 2nd-order stages are used) and the advantages of multi-bit quantization, with relaxed linearity requirements for the latter. The feasibility and efficiency of this approach, because it needs no correction/calibration mechanisms, has been proved in many reported designs [Bran91b] [Feld98] [Mede99b] [Mori00].

The performance targeted for the ΣΔ modulator here is 14-bit effective resolution (86-dB dynamic range) in a signal bandwidth of 2MHz. Given that the signal bandwidth is high and a 0.35-μm mainstream digital CMOS process is to be used, oversampling must be restricted to low/moderate values ($OSR \leq 24$) in order to run the modulator at a feasible clock rate ($f_s \leq 100\text{MHz}$). As shown in Fig. 1.24, such a maximum value of the oversampling ratio discards the use of low-order single-loop single-bit ΣΔ topologies. Instead, cascade ΣΔ modulators can be considered, since the required high-order shaping can be obtained without the loss of performance derived from stabilizing high-order loops.

Several cascade architectures have been considered; namely:

- a 3rd-order 2-stage cascade (2-1 ΣΔM)
- a 4th-order 2-stage cascade (2-2 ΣΔM)
- a 4th-order 3-stage cascade (2-1-1 ΣΔM)
- a 5th-order 4-stage cascade (2-1-1-1 ΣΔM)
- a 6th-order 3-stage cascade (2-2-2 ΣΔM)

These topologies are depicted in Fig. 1.28 to Fig. 1.31 and the relationships required for correct functioning are shown in Table 1.3 to Table 1.6.

As shown in Section 1.4, under ideal conditions, the z-domain output of an Lth-order N-stage cascade ΣΔM can be generalized as

$$Y(z) = z^{-L}X(z) + d_{2N-3}(1 - z^{-1})^L E_N(z) \tag{3.2}$$

where L is the summation of the stages orders, and $X(z)$ and $E_N(z)$ stand for the z-domain signal and last-stage quantization error, respectively. The scalar d_{2N-3} accounts for the signal scaling required to avoid premature overload of the stages and equals the inverse of the product of the inter-stage coupling factors. In general, this results in a value of d_{2N-3} larger than unity, which means an amplification of the last-stage quantization error, thus generating a systematic loss of resolution in comparison with the ideal case in eq(3.1).

With proper selection of the inter-stage couplings, the scaling factor can be reduced to only 2 for all the cascades considered [Marq98b] [Rio00], except for the 2-2-2 ΣΔM, in which the scalar equals 8 [Feld98]. The former implies a 6-dB (1-bit) reduction in DR, whereas the latter leads to a 18-dB (3-bit) reduction. In any case, these systematic losses are considerably lower than those found in practical realizations of their single-loop counterparts.

Fig. 3.2 compares the ideal performance of the considered cascades as a function of the oversampling ratio. Note that the use of a 2-1 architecture can be discarded at this point, since 3rd-order shaping is not enough to achieve the required dynamic range with $OSR \leq 24$. Ideally, modulator resolutions above

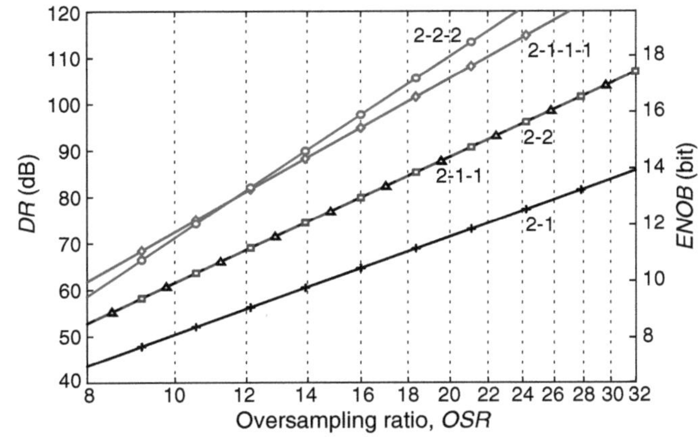

FIGURE 3.2 Ideal performance of cascade ΣΔMs versus oversampling ratio.

14 bits can be obtained with the 2-2 and the 2-1-1 ΣΔM operating with $20 \leq OSR \leq 24$, or the 2-1-1-1 and the 2-2-2 ΣΔM with $14 \leq OSR \leq 24$.

Using multi-bit quantization can help to reduce the oversampling required by these cascade ΣΔMs. Moreover, if multi-bit quantization is exclusively used in the last stage, the linearity requirements for the multi-bit DAC are relaxed, since DAC error is high-pass filtered and most of its power falls out of the band. Linear analysis shows that the z-domain output of an Lth-order N-stage cascade ΣΔM using multi-bit quantization only in the last stage becomes

$$Y(z) = z^{-L}X(z) + d_{2N-3}(1 - z^{-1})^{L}E_N(z) + d_{2N-3}(1 - z^{-1})^{(L - L_N)}E_D(z) \quad (3.3)$$

where $E_D(z)$ is the z-transform of the non-linearity error in the last-stage DAC. Such error presents a shaping of order $L - L_N$; i.e., the order of the overall modulator minus that of the last stage. This means that the influence of the DAC non-linearity is attenuated inside the band, and hence some non-linearity can be tolerated without correction/calibration. From eq(3.3), the in-band error power for each of the cascades considered can be expressed as follows,

$$\begin{aligned}
P_Q\big|_{2\text{-}2\text{mb}} &= d_1^2\left(\sigma_Q^2 \cdot \frac{\pi^8}{9OSR^9} + \sigma_D^2 \cdot \frac{\pi^4}{5OSR^5}\right) \\
P_Q\big|_{2\text{-}1\text{-}1\text{mb}} &= d_3^2\left(\sigma_Q^2 \cdot \frac{\pi^8}{9OSR^9} + \sigma_D^2 \cdot \frac{\pi^6}{7OSR^7}\right) \\
P_Q\big|_{2\text{-}1\text{-}1\text{-}1\text{mb}} &= d_5^2\left(\sigma_Q^2 \cdot \frac{\pi^{10}}{11OSR^{11}} + \sigma_D^2 \cdot \frac{\pi^8}{9OSR^9}\right) \\
P_Q\big|_{2\text{-}2\text{-}2\text{mb}} &= d_3^2\left(\sigma_Q^2 \cdot \frac{\pi^{12}}{13OSR^{13}} + \sigma_D^2 \cdot \frac{\pi^8}{9OSR^9}\right)
\end{aligned} \quad (3.4)$$

where $\sigma_Q^2 = [\Delta/(2^B - 1)]^2/12$ is the power of the last-stage quantization error (Δ stands for the quantizer full scale and B for its resolution) and $\sigma_D^2 \cong (\Delta \cdot INL/100)^2/2$ represents the DAC-induced error power [Mede99a], with INL being the DAC integral non-linearity expressed in percentage of the full scale (%FS). Note that this error power is attenuated by OSR^{-7} in the 2-1-1 modulator, whereas it is attenuated by OSR^{-9} in the 2-1-1-1 and the 2-2-2 cascades. They outperform the 2-2 modulator, in which DAC errors are only 2nd-order shaped (OSR^{-5}). The larger immunity of the former modulators to DAC imperfections is a valuable characteristic taking into account the poor performance of devices in digital CMOS processes and it can be exploited to implement a calibration-free multi-bit ΣΔM.

However, the sensitivity of cascade modulators to noise leakage can hide some of the benefits of high-order filtering and multi-bit quantization. As shown in Chapter 2, finite amplifier DC gain and capacitor mismatch cause leakage of low-order shaped quantization error to the modulator output, which degrades the dynamic range. This problem must be carefully tackled in digital submicron technologies, where degradation of both MOSFET output conductance and capacitor matching is more than foreseeable.

For the cascades considered here, finite amplifier DC gain causes a leakage dominated by 1st-order shaped quantization error (Table 2.1), whereas capacitor mismatch leads to 2nd-order shaping dominant leakage terms (Table 2.2). These extra error contributions can indeed mask those in eq(3.4), because they are attenuated by only OSR^3 and OSR^5, respectively. Hence, the feasibility of these cascades depends on how demanding the requirements for the DC gain and capacitor matching are, regarding the process capabilities.

These issues have been taken into account to identify possible selections of $\{OSR, B\}$ pairs for the cascades above in the presence of reasonable DC gain, mismatch, and DAC errors. Results are summarized in Fig. 3.3, which shows the effective resolution achieved by the cascade ΣΔMs as a function of the last-stage quantizer resolution B, with the oversampling acting as a parameter. The value of OSR is depicted next to each curve. A DC gain of 2500, a capacitor mismatch of 0.12%, and a DAC INL of 0.4%FS have been assumed in all cases. Note that curves saturate in the presence of non-idealities, leading to a practical useful limit for multi-bit quantization. For a given OSR, increasing B over this limit will not further improve the modulator dynamic range. Nevertheless, quantizer resolutions below this limit can be enough to significantly relax the oversampling ratio, and hence the dynamic requirements of the analog circuitry, with respect to single-bit approaches.

There are different alternatives to achieve 14-bit resolution depending on the topology considered (encircled in Fig. 3.3*a* to Fig. 3.3*d*):

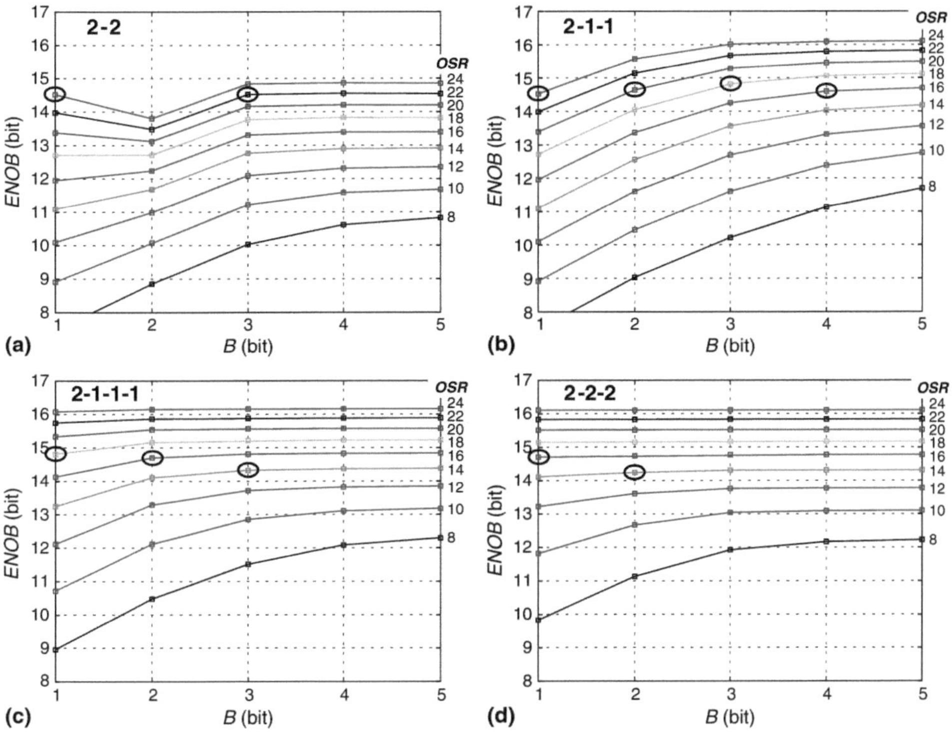

FIGURE 3.3 *ENOB* of cascade ΣΔMs versus the resolution in the last-stage quantizer, for oversampling ratios from 8 to 24: (a) 2-2 ΣΔM, (b) 2-1-1 ΣΔM, (c) 2-1-1-1 ΣΔM, (d) 2-2-2 ΣΔM. (A_{DC} = 2500, σ_C = 0.12%, and DAC *INL* = 0.4%FS).

- For the 2-2 ΣΔM, $\{OSR, B\} = \{24, 1\}$ and $\{22, 3\}$ can be selected. Because of the larger sensitivity of this topology to DAC non-linearities, an intermediate case using 2-bit quantization in the last stage is discarded, because DAC errors mask the benefits of multi-bit quantization.
- For the 2-1-1 ΣΔM, the pairs $\{OSR, B\} = \{24, 1\}$, $\{20, 2\}$, $\{18, 3\}$, and $\{16, 4\}$ can be selected.
- For the 2-1-1-1 ΣΔM, an effective resolution larger than 14 bits is obtained for the pairs $\{OSR, B\} = \{18, 1\}$, $\{16, 2\}$, and $\{14, 3\}$.
- For the 2-2-2 ΣΔM, possible pairs are $\{OSR, B\} = \{16, 1\}$ and $\{14, 2\}$.

After comparing the former figures, the 2-2 and the 2-2-2 ΣΔM were discarded for the targeted application, since they are outperformed by the 2-1-1 and the 2-1-1-1 cascade, respectively. On the one hand, the oversampling—and thus the sampling frequency—required in the 2-2 modulator is high and the use of dual-quantization does not relax it, because of the sensitivity to DAC errors. For the same number of integrators, the 2-1-1 topology is more flexible and benefits from a considerable OSR reduction if multi-bit quantization is used. On the other hand, the 2-2-2 ΣΔM employs one more integrator than the 2-1-1-1, but the benefits from a higher noise shaping are masked by its 3-bit systematic loss in dynamic range and its larger sensitivity to mismatch.

Among the possibilities left, the best choice was considered to be the 2-1-1 ΣΔM working with 4-bit quantization in the last stage and an oversampling ratio of only 16. The reduction from 16 to 14 in oversampling ratio attainable with the 2-1-1-1 cascade did not justify the presence and additional power/area consumption of the extra stage.

The set of integrator weights of the 2-1-1 cascade were optimized attending to the following criteria:

- Fulfillment of the relationships in Table 1.4 for error cancellation.
- Minimization of the scaling factor d_3 that determines systematic losses.
- Maximization of the modulator overload level (X_{OL}).
- Minimization of the output swing (OS) required for the integrators.
- Easy implementation of capacitor ratios using unit elements.
- Minimization of the total number of unit capacitors to save silicon area.
- Easy implementation of the digital multipliers in the cancellation logic.

Table 3.1 shows the integrator weights selected for the 2-1-1 ΣΔM after optimization (column [Rio00]), together with other sets reported in open literature. For comparison purposes, resulting features—like systematic loss, overload level, output swing, and number of unit capacitors—are also included.

Note that all sets of coefficients lead to $d_3 = 2$ and, therefore, to a systematic reduction of the dynamic range of only 6dB (1bit). The overload

TABLE 3.1 Reported coefficients for the 2-1-1 cascade ΣΔM.

Weights	[Rio00]	[Yin94b] [Mede99b]	[Marq98a] [Geer99]
g_1, g_1'	1/4, 1/4	1/4, 1/4	1/3, 1/3
g_2, g_2'	1, 1/2	2/4, 1/4	3/5, 2/5
g_3, g_3', g_3''	1, 1/2, 1/2	1, 3/8, 2/8	5/6, 3/6, 2/6
g_4, g_4', g_4''	1, 1/2, 1/2	1, 1/4, 1/4	1, 1/3, 1/3
d_0, d_1, d_2, d_3	1, 2, 0, 2	2, 2, 0, 2	2, 2, 0, 2
ΔDR due to scaling	–6dB (–1bit)	–6dB (–1bit)	–6dB (–1bit)
$X_{OL}/(\Delta/2)$	–3dBFS	–2.5dBFS	–2dBFS
$OS/(\Delta/2)$	0.75, 1, 1, 1	0.75, 0.7, 0.6, 0.6	1, 1, 0.9, 0.8
Unit capacitors	17 (5 + 4 + 4 + 4)	35 (5 + 6 + 16 + 8)	29 (4 + 8 + 11 + 6)

level of the modulators is also similar (around –2.5dBFS). The main differences are related to the required amplifier output swing and the number of unit capacitors needed for the implementation of the weights. Regarding the output swing of the amplifiers, coefficients in [Yin94b] and [Mede99b] impose the more relaxed requirements—normalized OS of 0.75 for the amplifier in the first integrator and 0.6 for the third and fourth integrators—, whereas those in [Marq98a] and [Geer99] require a normalized OS close to 1 in all integrators. Coefficients proposed here are halfway between these two sets in regard to output swing demands, but also offer relaxed requirements for the front-end integrator. As we will show, a large DC gain is usually required in this integrator to attenuate distortion, so that a relaxed output swing is a desirable feature for a low-voltage implementation. On the other hand, the set of coefficients in [Rio00] requires fewer unit capacitors for its implementation. This leads to a considerable reduction of the modulator occupation area, especially considering the digital process to be used, in which capacitors are implemented as large multi-metal structures with thick oxide. The proposed coefficients exhibit the additional advantage of requiring only two-branch integrators, because the largest weight in the three-weight integrators (third and fourth) can be obtained as a combination of the other two.

Fig. 3.4 shows the block diagram of the selected 2-1-1 ΣΔM. Coefficients in the multi-bit last stage are scaled by a factor 2 in comparison with those in Table 3.1. This is done to make $g_3'' = 1$ and have a loop gain of 1 when the gain of the multi-bit ADC and DAC is 1. This way, their input and output full scales coincide and their design is simplified, as shown in Section 3.5.5.

As shown in Fig. 3.3*b*, the 2-1-1 architecture can achieve the targeted resolution (14 bits) for $\{OSR, B\} = \{16, 4\}$ and still provides some margin for the contribution of other error mechanisms. However, another viable choice

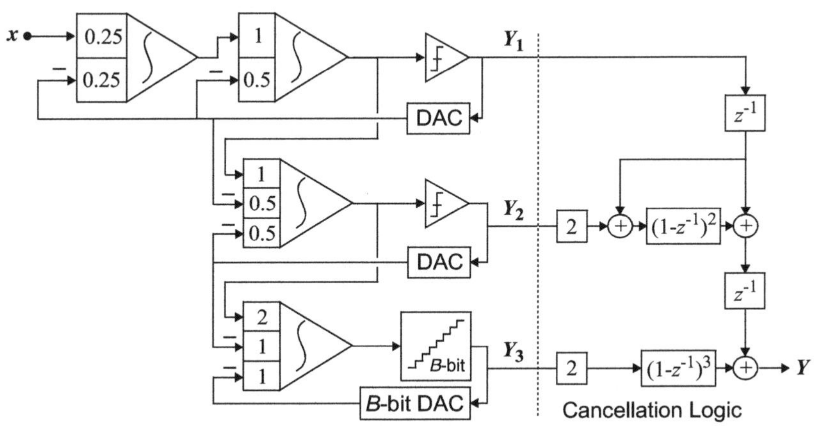

FIGURE 3.4 Block diagram of the dual-quantization 2-1-1 cascade ΣΔM.

can result from $\{OSR, B\} = \{16, 3\}$. In order to gain flexibility, the 2-1-1 ΣΔM in Fig. 3.4 has been implemented with programmable A/D/A converter in the last stage, capable to operate with either 2-, 3-, or 4-bit resolution. Thus, combining this programming capability with the possibility of modifying the oversampling ratio (decimation will be performed off-chip) will result into a quite valuable test vehicle to validate the architectural study.

3.3 Switched-Capacitor Implementation

The block diagram of the 2-1-1 ΣΔ modulator in Fig. 3.4 maps onto the fully-differential SC schematic shown in Fig. 3.5. The first stage of the cascade consists of two SC integrators and a comparator. The front-end integrator contains a single input branch, since its signal and feedback weights are equal. The second integrator has two input branches. At the back-end of the stage the comparator performs the 1-bit quantization of the signal. The analog version of that signal is fed back to the integrators using differential voltage references (V_r^+, V_r^-) and switches controlled by the comparator output. The second stage also uses an integrator with two input branches. Although three weights must be implemented for this integrator ($g_3 = 1$, $g_3' = 0.5$, and $g_3'' = 0.5$), the first one is distributed between the two branches in order to save area. The same applies for g_4 in the fourth integrator, which only requires two input branches as well. This integrator drives a programmable ADC ($B = 4, 3, 2$) and the 3rd-stage loop is closed through a programmable DAC. The 1-of-16 output code of the ADC for $B = 4$ (1-of-8 if $B = 3$ and 1-of-4 if $B = 2$) is converted into binary using a ROM that provides the 4-bit outputs $Y_{3,\,0\text{-}3}$ ($Y_{3,\,0\text{-}2}$ if $B = 3$ and $Y_{3,\,0\text{-}1}$ if $B = 2$).

The timing scheme of ΣΔ modulator is illustrated in Table 3.2. The modulator operation is controlled by two non-overlapped clock phases, ϕ_1 and ϕ_2. The integrator input signals are sampled during phase ϕ_1 and then integrated together with the corresponding feedback signals during phase ϕ_2. The comparators and the ADC are activated at the end of ϕ_2 (using $\bar{\phi}_2$ as strobe) to avoid any possible interference of the integrator transient response at the beginning of sampling. The operation over time of each block for the conversion of an input sample is summarized in Table 3.2. Note that the timing guarantees a single delay per clock cycle.

Delayed versions of the two main phases are provided (ϕ_{1d} and ϕ_{2d}) in order to attenuate signal-dependent charge injection [Lee85]. Complementary versions of all the phases are also used for the control of the switches, which are implemented as CMOS transmission gates.

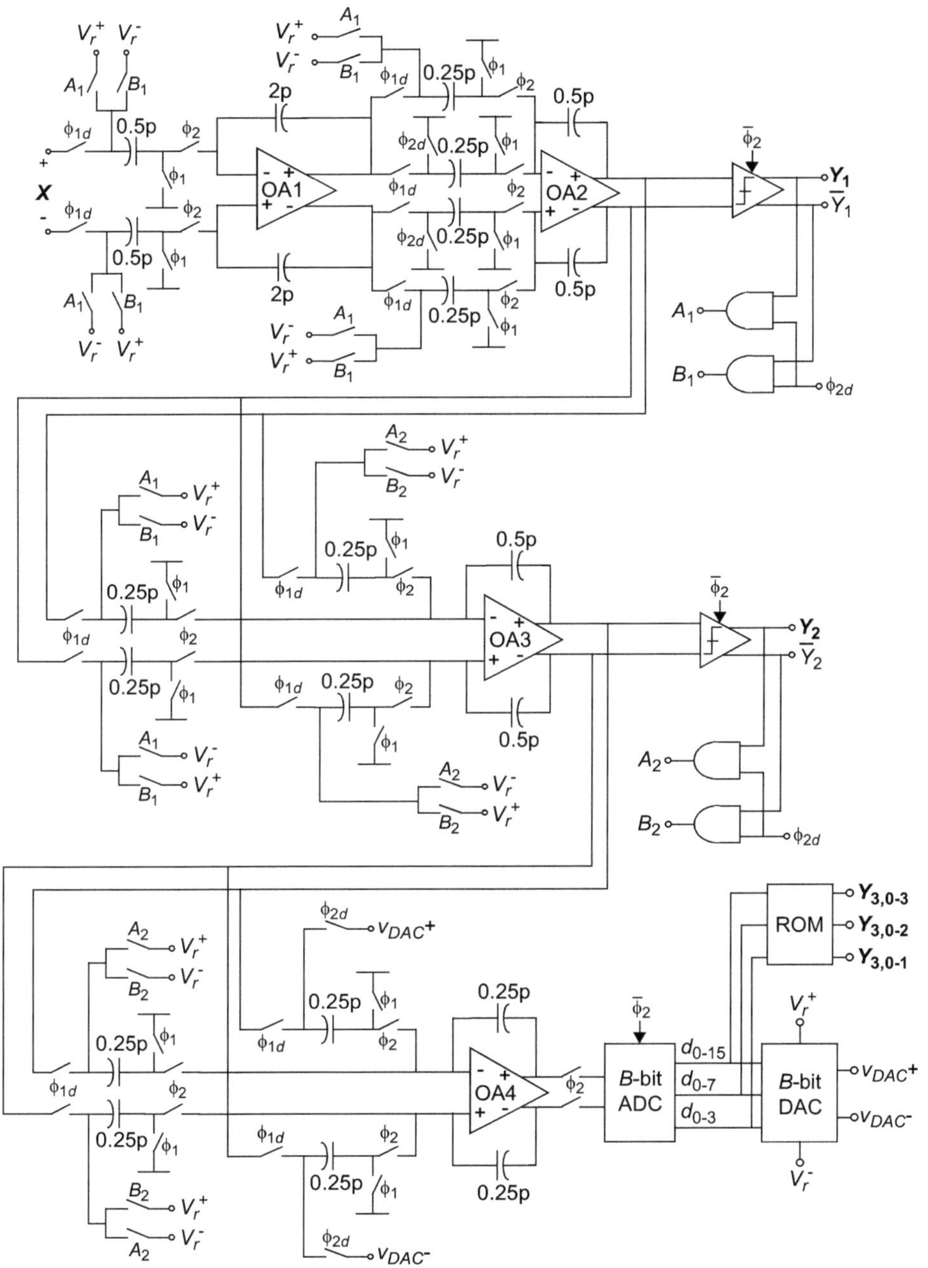

FIGURE 3.5 SC implementation of the 2-1-1 ΣΔM with programmable dual-quantization.

TABLE 3.2 Timing of the ΣΔ modulator.

	ϕ_1	ϕ_2	ϕ_1	ϕ_2
INTEGRATORS	**sample**	**integrate**	sample	integrate
COMPARATORS	regenerate	**sample**	**regenerate**	sample
	refresh output	NC (no change)	**refresh output**	**NC**
ADC	**store references**	--	store references	--
	regeneration	**sample**	**regeneration**	sample
	refresh output	NC	**refresh output**	**NC**
DAC	refresh output	NC	**refresh output**	**NC**

3.4 Specifications for the Building Blocks

This section derives the most important requirements for the modulator building blocks †1. First, closed-form equations for the main errors that determine performance are used to derive an initial modulator sizing fulfilling the specifications. The architecture is then extensively simulated using more complex behavioral models in order to fine-tune the initial sizing.

3.4.1 Modulator sizing

As illustrated in Fig. 3.1, the second step of the modulator design flow consists of a fast estimate of the building block requirements; i.e., the specifications for the amplifiers, capacitors, switches, comparators, and multi-bit quantizers in the SC implementation. For this task, the closed-form expressions derived in Chapter 2 for the influence of the main block non-idealities on the modulator operation are used. These expressions have been compiled into a database of design equations associated to the tool SDOPT [Mede95] [Mede99a]—see Fig. 3.1. Covered issues include thermal noise, integrator leakage and settling errors, capacitor mismatch, analog switch errors, jitter noise, etc. Quantization error is linearly modeled using the additive white noise approximation (see Section 1.1.2).

Fast modulator sizing

Table 3.3 summarizes the results for the sizing of the ΣΔ modulator. The specifications have been grouped into five categories; namely, modulator, front-end integrator, amplifier, comparators, and A/D/A converter. The most significant in-band modulator errors are reflected at the bottom of the table. The main considerations for this sizing are described next.

1. Recall that we denote this procedure as modulator sizing.

TABLE 3.3 Modulator sizing and main in-band error contributions.

MODULATOR	Topology	2-1-1
	Dual-quantization	1bit / 4bit
	Oversampling ratio	16
	Clock frequency	64MHz
	Differential reference voltage	2V
	Clock jitter	15ps (0.1%)
FRONT-END INTEGRATOR	Sampling capacitor	0.5pF
	Unit capacitor	0.5pF
	Capacitor standard deviation	0.12%
	Bottom parasitic capacitor	25%
	Capacitor non-linearity	25ppm/V
	Switch on-resistance	250Ω
AMPLIFIER	Open-loop DC gain	2500 (68dB)
	Gain-bandwidth product (1.6-pF load)	235MHz
	Slew rate (1.6-pF load)	380V/μs
	Differential output swing	±2V
	Input equivalent noise	$6\text{nV}/\sqrt{\text{Hz}}$
	DC-gain non-linearity	$20\%/\text{V}^2$
COMPARATORS	Hysteresis	30mV
	Offset	±10mV
	Resolution time	3.5ns
A/D/A CONVERTER	Resolution	4bit
	DAC *INL*	0.4%FS
Quantization Noise		–86.6dB
	Ideal quantization noise	–94.4dB
	Amplifier DC-gain leakage	–95.7dB
	Capacitor mismatch leakage	–88.9dB
	DAC non-linearity error	–95.9dB
Thermal Noise		–84.8dB
	kT/C noise	–86.9dB
	Amplifier noise	–88.9dB
Clock Jitter		–88.6dB
In-Band Error Power		–81.6dB
Dynamic Range		84.6dB
Effective Resolution		13.8bit

As shown in the first group of specifications, the 2-1-1 ΣΔM uses an $16 \times$ oversampling, which results in a clock frequency $f_s = 64\text{MHz}$, given the 2-MHz bandwidth intended for the A-to-D conversion. The last-stage quantizer resolution is nominally 4 bits, but it can be programmed to 2 and 3 bits. The reference voltage (V_{ref}) is 2V, which implies that references V_r^+ and V_r^- in Fig. 3.5 are $+1\text{V}$ and -1V, respectively. This leads to a full-scale range (Δ) of 4V using a 3.3-V supply, thanks to the differential implementation. Under linear analysis and ideal modulator operation, the in-band quantization error yields

$$P_Q \cong d_3^2 \cdot \frac{\Delta_3^2}{12} \cdot \frac{\pi^8}{9OSR^9} \tag{3.5}$$

where $\Delta_3 = \Delta/(2^B - 1)$ stands for the quantization step of the multi-bit third stage and the scaling coefficient d_3 equals 2 with the set of integrator weights used. Eq(3.5) leads to an ideal in-band quantization error of -94.4dB.

The second group of specifications in Table 3.3 includes the requirements for the capacitor and switches in the front-end SC integrator. The value of the sampling capacitor has been selected considering thermal noise and power dissipation. As stated in Section 2.4, in-band thermal noise power originated by the switches —kT/C noise— can be estimated as

$$P_{kT/C} \cong 2 \cdot \frac{2kT}{C_{S1}} \cdot \frac{1}{OSR} \tag{3.6}$$

where C_{S1} stands for the front-end sampling capacitor and the factor 2 accounts for the differential implementation. This contribution is approximately -86.9dB for a sampling capacitor of 0.5pF. The selected value, although small, still leaves some margin for other error mechanisms, while reducing the equivalent capacitive load of the front-end integrator, and thus the amplifier dynamic requirements and power dissipation. As we show later on, the reduction of capacitive load becomes an important task for high-speed operation in the intended digital technology, since capacitors are implemented using multi-metal sandwich structures that exhibit a large parasitic from the bottom plate to substrate (above 25% of the nominal capacitance).

Due to the sensitivity of cascades to noise leakage, a good capacitor matching is also required. As stated in Table 2.2, this leakage term is proportional to OSR^{-5} in the 2-1-1 ΣΔM. For a mismatch error contribution of approximately -89dB, the matching required is around 0.12%, which is feasible for multi-metal capacitors in the intended technology. On the other hand, the tolerated switch on-resistance (R_{on}) is around 250Ω. This value is large enough to implement switches as standard transmission gates with the nominal 3.3-V supply, with no need for clock-boosting techniques [Wu98] [Bult00].

The third group of specifications in Table 3.3 corresponds to the amplifier. Following a similar procedure to that in Section 2.1, noise leakage due to finite amplifier DC gain in the first stage—which is dominant— can be estimated as

$$P_{\mu} \cong \frac{\Delta^2}{12} \cdot \left(\frac{C_{S1}/C_{I1} + \sum_{i=1,2} C_{S2,i}/C_{I2}}{A_{DC}} \right)^2 \cdot \frac{\pi^2}{3OSR^3} \tag{3.7}$$

where A_{DC} stands for the open-loop amplifier DC gain, C_{S1} and C_{I1} stand for the sampling and integration capacitors in the first integrator, respectively, whereas $C_{S2,i}$ and C_{I2} stand for those in the second integrator, with two input branches. Note that eq(3.7) accounts for the value of integrator weights and capacitor sharing in the second integrator, for a more exact estimation of the contribution of the leaky integrators—expressions in Section 2.1 were approximated for comparison purposes among cascade architectures (see footnote 1 in Chapter 2). For a DC gain of 2500, the corresponding error contribution is around -95.7dB, which is comparable with the ideal quantization noise and considerably lower than that of capacitor mismatch.

The required gain-bandwidth product (GB) of the amplifier is estimated to be around 235MHz. This value leads to $GB \approx 3.7f_s$, which is enough for proper integrator settling and still provides some margin for the slowing-down effect caused by switch on-resistance on the integrator dynamic response. From eq(2.74), the effective amplifier GB can be recalculated as

$$GB_{eff}\big|_{\text{Hz}} \cong \frac{GB}{1 + GB/f_{on}} = \frac{GB}{1 + GB \cdot 2\pi \cdot 2R_{on}C_{S1}} \tag{3.8}$$

which leads to approximately 170MHz; i.e., $GB_{eff} \approx 2.7f_s$ given that $f_{on} \approx 10f_s$, with the selected values for R_{on} and C_{S1}. On the other hand, the amplifier slew rate, $SR = 380\text{V}/\mu\text{s}$, is large enough to avoid a dominant full-slew dynamic of the integrator.

Once the effective GB of the amplifier is determined, its contribution to the total thermal noise can be estimated using eq(2.109) as

$$P_{op} \cong \int_{-f_b}^{+f_b} \frac{2\pi \cdot GB_{eff}}{2f_s} \cdot S_{op}^t \cdot df = \frac{2\pi \cdot GB_{eff}}{2OSR} \cdot S_{op}^t \tag{3.9}$$

where GB_{eff} is in Hz and S_{op}^t stands for the input-referred thermal noise PSD of the amplifier (in V^2/Hz). From eq(3.9), the in-band contribution fir a reasonable amplifier input-referred noise of $6\text{nV}/\sqrt{\text{Hz}}$ is approximately -89dB, which is 2dB lower than the kT/C contribution.

The fourth and fifth group of specifications in Table 3.3 refer to the quantizers in the 2-1-1 cascade. Note that the main requirement for comparators is a low resolution time—around 3.5ns; i.e., below a quarter of the clock frequency—, whereas offset and hysteresis are not demanding. On the other hand, requirements on the multi-bit quantizer are basically imposed to the DAC non-linearity. Since multi-bit quantization is only used at the last stage, DAC errors will add to the last-stage input and will be therefore 3rd-order shaped, so that an INL as large as 0.4%FS can be tolerated without affecting the modulator linearity. Hence, power and area consuming correction or calibration techniques are avoided.

Finally, given the high-speed operation of the modulator, the jitter requirements on the clock signal must be also considered. Using eq(2.128), the in-band jitter noise can be estimated as

$$P_J \cong \frac{A_x^2}{2} \cdot \frac{(2\pi f_x \sigma_J)^2}{OSR} \cong \frac{(X_{OL} \cdot V_{ref})^2}{2} \cdot \frac{(2\pi f_x \sigma_J)^2}{OSR} \tag{3.10}$$

where A_x and f_x stand for the amplitude and frequency of the modulator input signal, respectively, and σ_J for the sigma of the clock jitter. Note from eq(3.10) that the maximum input amplitude considered corresponds to that at the modulator overload level. For an overload level around –5dBFS ($X_{OL} \approx 0.56 V_{ref}$), a jitter of 0.1% the clock period—i.e., 15ps—is enough for a maximum in-band contribution of approximately –88.6dB.

The former in-band errors are summarized at the bottom of Table 3.3. The main error source is thermal noise—especially kT/C noise—, which contributes –84.8dB. Quantization noise follows contributing –86.6dB, capacitor mismatch being dominant. According to these estimations, the total in-band error is –81.6dB, leading to a DR of 84.6dB (13.8bit).

Fine-tuning of blocks specs

At this point, the modulator sizing must be validated through more accurate time-domain simulations. This is mandatory given the limited complexity of the equations that can be realistically included in the design database—starting from quantization error, for which the additive white noise approximation is used. Such simulations cannot be carried out at the modulator level using electrical simulators, since transistor descriptions are not yet available and, in any case, CPU-time needed is prohibitive [Wolff97] [Mede99a]. Dedicated behavioral simulators are used instead. As shown in Fig. 3.1, the modulator design flow followed here incorporates the tool ASIDES [Mede99a], a behavioral simulator that supports non-linear difference equations and employs event-driven simulation for enhanced efficiency. The models developed in Chapter 2 for the effect of the different modulator non-idealities have been incorporated to this tool. This way, non-idealities with relatively complex models—such as that for the settling of integrators—can be more accurately evaluated, including also distortion generated by non-linear effects.

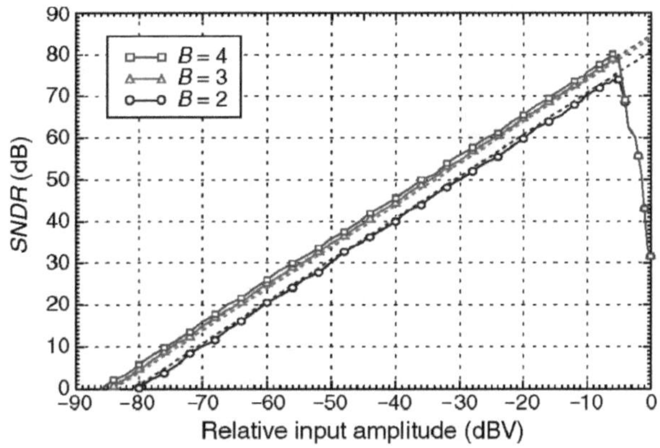

FIGURE 3.6 *SNDR* curves at 4MS/s obtained through behavioral simulation.

Fig. 3.6 shows the *SNDR* curves obtained through behavioral simulation of the 2-1-1 modulator with the sizing in Table 3.3. In addition to the nominal 4-bit resolution in the last stage, the curves for $B = 3$ and 2 are also depicted. For comparison purposes, Fig. 3.6 includes in dashed lines the estimated *SNDR* using the equations database. Note that results agree within 1dB outside the overload region (not considered in the equations database). According to behavioral simulation results, the modulator *DR* is approximately 85.5dB with 4-bit quantization in the last stage. For $B = 3$ and 2, *DR* is 84.5dB and 80.5dB, respectively. Only in the last case in-band quantization error dominates over thermal noise, so that a reduction in the modulator performance is clearly visible when the multi-bit resolution is decreased to 2 bits.

The latter issue can also be observed in Fig. 3.7*a*, that shows the modulator output spectrum obtained from a 64k-sample FFT for an input tone at 250kHz with amplitude of 0dBV (−6dBFS, near overload). Results correspond to 4- and 2-bit quantization in the last stage. In both cases, the cumulative error power is plotted for comparison purposes. For $B = 4$, the cumulative error error is white-noise limited (10-dB/dec slope) within the signal band (vertical line at 2MHz), whereas it is quantization-error limited outside the band (90-dB/dec slope, due to the 4th-order shaping). For $B = 2$, it increases at the upper part of the band, where quantization error is dominant.

Fig. 3.7*b* shows the large-input region of the *SNDR* curve for $B = 4$. Results correspond to a 30-run Monte Carlo simulation, assuming a mismatch of 0.12% in the unit capacitors. The estimated spread in *SNDR* due to capacitor mismatch is 2dB. On the other hand, note from Fig. 3.7*a* that, in spite of the large input amplitude, small 2nd- and 3rd-order harmonic components are generated, so that *SNDR* almost coincides with *SNR* for the full input range.

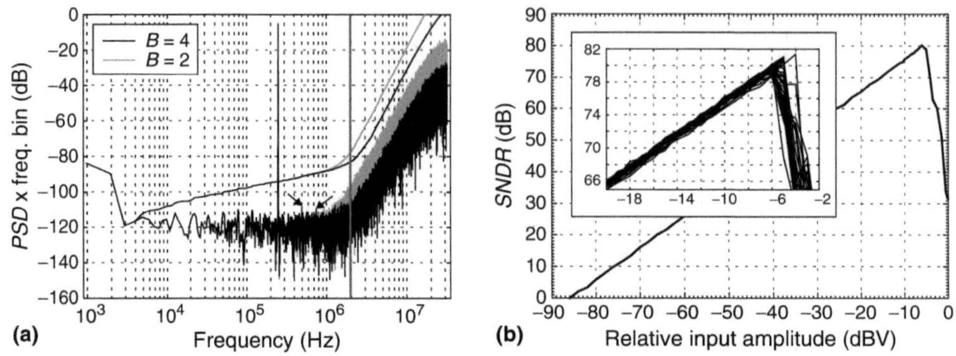

FIGURE 3.7 (a) Output spectrum for a 0dBV@250kHz input tone for $B = 4, 2$; (b) Detail of the *SNDR* curve under Monte Carlo simulation of capacitor mismatch ($\sigma_C = 0.12\%$, $B = 4$).

3.4.2 Integrator scaling

Thanks to oversampling, some specifications in Table 3.3 referring to the front-end integrator can be relaxed for the remaining integrators. The value of the sampling capacitor in those integrators can be progressively scaled down, since their contributions to the overall kT/C noise are attenuated within the signal band by the gain of the preceding integrators. Nevertheless, matching considerations preclude using very small capacitors—especially, in the second integrator, whose contribution to mismatch leakage is comparable with that of the first integrator. Taking this into account, the value of the unit capacitor is scaled from 0.5pF to 0.25pF in the second, third, and fourth integrators.

A more aggressive treatment can be applied to the required amplifier input noise, which allows a considerable reduction from the original specification of $6\text{nV}/\sqrt{\text{Hz}}$ at the modulator front-end. This value is scaled to $15\text{nV}/\sqrt{\text{Hz}}$ in the second integrator and $50\text{nV}/\sqrt{\text{Hz}}$ at the modulator back-end.

Similarly, the amplifier DC gain of the third and fourth integrators can be reduced from 2500 to only 500, since their contributions due to the in-band noise leakage are proportional to OSR^{-7} and OSR^{-9}, respectively. On the other hand, the DC gain of the first stage allows also some reduction. Since the total in-band shaped error is dominated by capacitor mismatch (see Table 3.3), a larger integrator leakage can be tolerated in the dominant first stage without degrading the modulator performance. For this reason, the amplifier DC gain in the second integrator is scaled to one half its original value. Although leakages of the first two integrators are of similar importance, that of the front-end integrator is kept to 2500 for a larger attenuation of the distortion generated by the DC-gain non-linearities (see Section 2.6.2).

The scaling applied to the specifications of the integrators are summarized in Table 3.4. The modulator performance after integrator scaling has been re-evaluated using behavioral simulation, exhibiting no degradation with respect to that shown at the end of the previous section.

TABLE 3.4 Scaling of the integrator specifications.

SPECs	1st Integ	2nd Integ	3rd Integ	4th Integ
Unit capacitor	0.5pF	0.25pF	0.25pF	
Open-loop DC gain	2500 (68dB)	1250 (62dB)	500 (54dB)	
Input equivalent noise	$6\text{nV}/\sqrt{\text{Hz}}$	$15\text{nV}/\sqrt{\text{Hz}}$	$50\text{nV}/\sqrt{\text{Hz}}$	

3.5 Design of the Building Blocks

This section describes the topology selection for the different building blocks in the ΣΔ modulator and their sizing at transistor level, in order to fulfill the formerly derived specifications.

3.5.1 Amplifiers

Major issues to be considered in the design phase of amplifiers are the requirements imposed on:

- ***Open-loop DC gain.*** As previously stated, the sensitivity of cascade architectures to integrator leakage demands an amplifier DC gain of 62dB in the first stage, although a reduction to 54dB can be allowed for OA3 and OA4. However, linearity issues force the DC gain of the front-end amplifier (OA1) to be increased to 68dB.
- ***Dynamics.*** The settling dynamics are critically dependent on the equivalent capacitive load that each integrator must drive. Moreover, as discussed in Section 2.3, capacitive loads change from sampling to integration. Given the SC implementation in Fig. 3.5, these loads can be estimated as

$$\begin{aligned} C_{eq,s} &= C_P + (C_L + C_{Sn,s})\left(1 + \frac{C_P}{C_I}\right) \\ C_{eq,i} &= C_S + C_p + (C_L + C_{Sn,i})\left(1 + \frac{C_S + C_P}{C_I}\right) \end{aligned} \qquad (3.11)$$

where C_S and C_I are the respective sampling and integrating capacitors of the integrator being considered, C_P and C_L refer to the permanent parasitics at the integrator input and output nodes, $C_{Sn,s}$ stands for capacitances loading the integrator only during sampling (e.g., sampling capacitors of the next integrator in the cascade), and $C_{Sn,i}$ refers to capacitors loading the integrator only during integration. This last variable is introduced to take into account the additional capacitive load seen by the last-stage amplifier during ϕ_2 due to the multi-bit ADC, which uses an SC input stage (explained in detail in Section 3.5.5). Note that C_L is determined by the amplifier output node parasitic and, eventually, the comparator input capacitance (for OA2 and OA3). On the other hand, bottom plate parasitics of the sampling and integration capacitors must also be considered. This is specially important in this implementation, because of the large bottom parasitic of the multi-metal capacitors used.

From the former issues, it turns out that an accurate estimation of integrator equivalent loads requires a previous knowledge of the capacitances involved in amplifiers and comparators, which can only be obtained after their complete design at transistor level. As an starting point, an initial estimation has been made with already known capacitances and assuming equal input and output capacitances in each amplifier (0.1pF and 0.2pF, respectively) and negligible comparator input capacitance. Table 3.5 summarizes the estimated equivalent loads for the four amplifiers. Note that the capacitive load is similar in the first three amplifiers (also from sampling to integration), but becomes considerably larger for OA4 during integration, because of the driving of the multi-bit ADC.

TABLE 3.5 Equivalent capacitive load for the amplifiers.

Amplifier	$C_{eq,s}$	$C_{eq,i}$
OA1	1.49pF	1.51pF
OA2	1.24pF	1.32pF
OA3	1.24pF	1.32pF
OA4	0.56pF	4.89pF

- ***Output swing.*** For a given reference voltage in a ΣΔ modulator, output swing demands on the amplifiers strongly depend on the value of integrator weights. For the 2-1-1 ΣΔM with the set of weights selected (see Fig. 3.4), the required output swing is reduced to only the reference voltage. Therefore, a differential output swing of only $\pm 2\text{V}$ is needed, which is not demanding when operating with a 3.3-V supply. The output swing in the last integrator must be however a bit higher for making use of the multi-bit quantizer full scale.

  The minimum output swing of each amplifier can be more exactly estimated using behavioral simulation. Fig. 3.8 depicts the histogram of the integrator output voltages for an input signal with amplitude of 0dBV (−6dBFS, near overload). Results show that the output swing required is

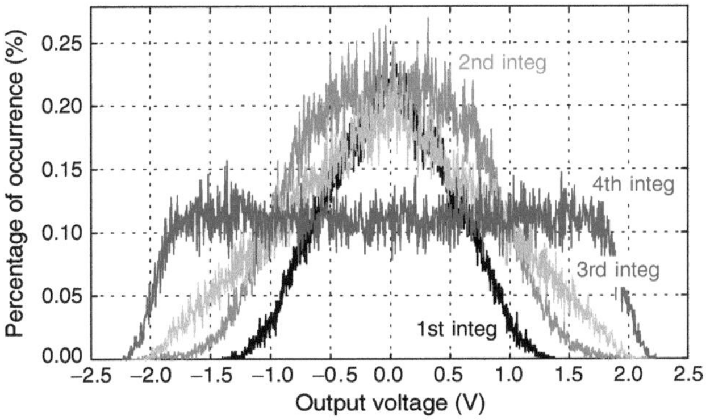

FIGURE 3.8 Histogram of the integrator outputs for a 0dBV input tone (64-k clock cycles).

around ±1.5V in the first amplifier (OA1), ±2.0V in the second (OA2) and third (OA3), and ±2.2V in the fourth (OA4).

Based on these estimations, behavioral simulations have been carried out to extract optimized dynamic requirements for the amplifiers that do not limit the modulator performance. Table 3.6 summarizes these dynamic specifications, together with the ones previously derived.

TABLE 3.6 Amplifier specifications.

	OA1	**OA2**	**OA3**	**OA4**
DC gain	2500 (68dB)	1250 (62dB)	500 (54dB)	500 (54dB)
GB	235MHz (1.6pF)	300MHz (1.4pF)	240MHz (1.4pF)	160MHz (4.5pF)
SR	380V/μs (1.6pF)	360V/μs (1.4pF)	285V/μs (1.4pF)	165V/μs (4.5pF)
Input noise	$6\text{nV}/\sqrt{\text{Hz}}$	$15\text{nV}/\sqrt{\text{Hz}}$	$50\text{nV}/\sqrt{\text{Hz}}$	$50\text{nV}/\sqrt{\text{Hz}}$
OS	±1.5V	±2.0V	±2.0V	±2.2V

In order to avoid over-sizing and minimize power consumption, the diversity of specifications shown in Table 3.6 recommends a dedicated design for each amplifier. This involves proper selection of the amplifier topology and optimum sizing of transistors. These tasks have been completed with the support of the optimization tool FRIDGE [Mede99a] (see Fig. 3.1). This tool combines statistical optimization with electrical evaluation for optimum sizing of cells at transistor level.

Front-end amplifier

The large DC gain required in the first integrator involves the design of its amplifier in the intended 3.3-V 0.35-μm technology. As a consequence of the output conductance degradation in short-channel transistors, high gain and high speed are difficult to achieve simultaneously with workable output

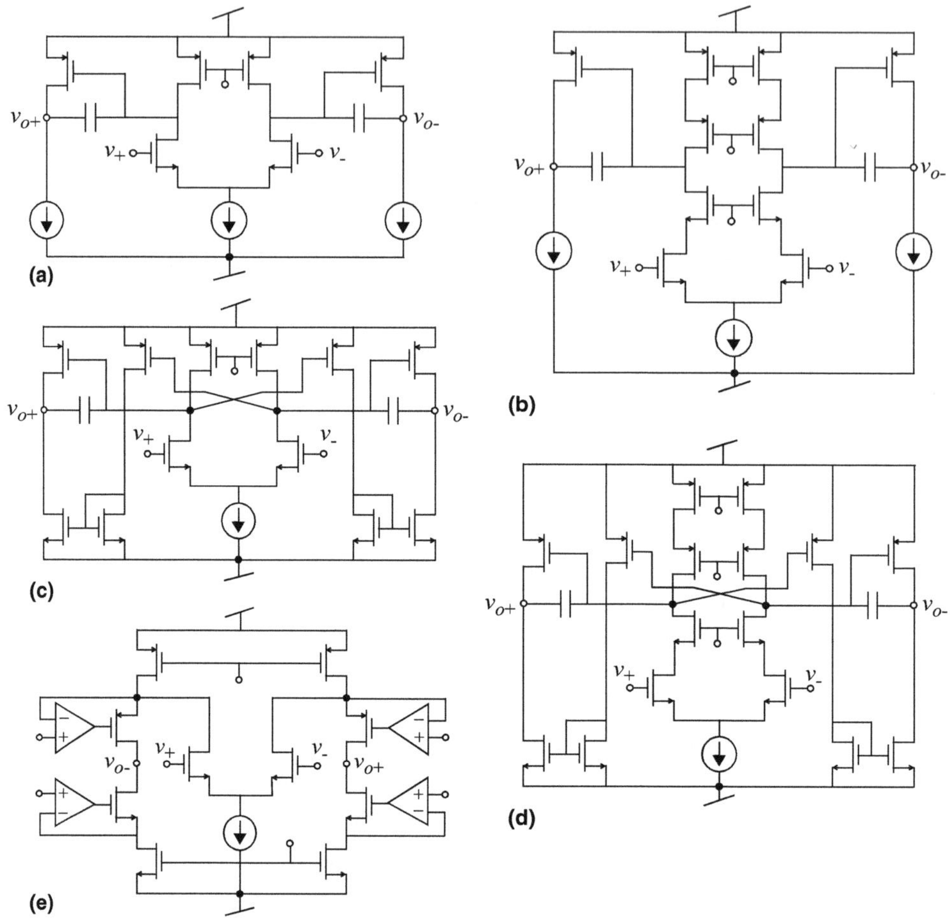

FIGURE 3.9 Two-stage amplifiers: (a) class A/A; (b) class A/A with telescopic first stage; (c) class A/AB; (d) class A/AB with telescopic first stage; (e) gain-boosted folded-cascode amplifier.

swing. A way to overcome this problem is resorting to multi-stage amplifier topologies. However, the required DC gain, though large, is small enough to be attained with just two stages. Thus, complex compensation schemes can be avoided. Fig. 3.9 shows possible two-stage candidates, namely:

- class A/A amplifier [Raza00] (Fig. 3.9*a*)
- class A/A amplifier, with a telescopic first stage [Raza00] (Fig. 3.9*b*)
- class A/AB amplifier [Rabii97] (Fig. 3.9*c*)
- class A/AB amplifier, with a telescopic first stage [Raza00] (Fig. 3.9*d*)
- gain-boosted folded-cascode amplifier [Bult90] (Fig. 3.9*e*)

The performance of these amplifiers has been confronted with the previously mentioned criteria of simplicity and reduced power consumption.

For a direct application of the main conclusions, our analysis, partially based on theoretical foundations, has been supported by design experiences in the 0.35-μm CMOS technology. In fact, several designs for each amplifier topology have been obtained using FRIDGE, trying to fulfil the specifications for OA1 with minimum power consumption. Minimizing occupation area played a secondary role, since the integrator dimensions are determined more by the size of the multi-metal capacitors than by the amplifier itself. Main conclusions are the following:

- The DC-gain requirement is too demanding for simple stages, such as the one in Fig. 3.9*a*. As the drain-to-source voltage increases, the degradation of the output conductance in minimum-length transistors obligates to resort to non-minimum L devices, thus compromising speed. Theoretically, a gain improvement by a factor 2 is obtained using a class AB second stage (Fig. 3.9*c*) and the same applies to the position of the lowest non-dominant pole. With this topology, a 68-dB DC gain, 250-MHz GB amplifier was obtained consuming 52mW. Power was reduced to 35mW while keeping performance by allowing certain gain in the 2nd-stage current mirror.
- The trade-off between gain and speed is better solved by including a cascode stage. Obviously, in order to accommodate a large output swing, this stage should be the first one (Fig. 3.9*b* and Fig. 3.9*d*). Thus, the achievable DC gain using minimum-length transistors increases considerably. For instance, a design of Fig. 3.9*b* with 80-dB DC gain and 250-MHz GB consumes 37mW, whereas a 93-dB DC gain, 250-MHz GB amplifier implemented with the topology of Fig. 3.9*d* consumes 30mW. Apart from consumption, the use of a class AB second stage improves the common-mode rejection figures. However, it also complicates the stabilization of the common-mode voltage, requiring two independent common-mode feedback (CMFB) nets for the first and second stages.
- The gain-boosted folded-cascode has been successfully employed with 5-V supplies [Bult90] [Marq98a], where the boosting amplifier can be a simple two-transistor common-source stage (see Fig. 3.10). In spite of its simplicity, this structure significantly involves the amplifier frequency response, because it introduces a pair of complex poles and a zero typically around GB. In order to maintain a good integrator transient response, the frequency of the non-dominant pole and zero must be well above GB and, in addition, the quality factor of the complex poles must be small. This implies a large power dissipation in the boosting amplifier. To illustrate this, a folded-

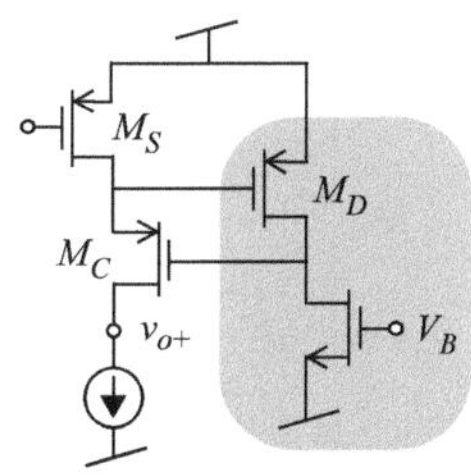

FIGURE 3.10 Gain-boosting with a simple common-source stage.

cascode has been sized with the boosting strategy in Fig. 3.10. It fulfils the DC-gain and dynamic requirements of OA1 with 26mW, from which 75% is dissipated on the boosting amplifiers. However, the output swing is limited because the voltage across M_S is forced to the gate-to-source voltage of M_D. In a 3.3-V technology, such arrangement leads to an unacceptably low output swing. To avoid this limitation one has to resort to differential OTAs to boost the cascode stage [Geer99], which further complicates the frequency response and augments the power consumption, making it comparable with that achieved with less complex architectures.

The discussion above drives our attention to amplifiers in Fig. 3.9*b* and Fig. 3.9*d*. Although that in Fig. 3.9*c* is capable of achieving the required performance with comparable power consumption, the DC gain obtained is just that in Table 3.6. Increasing the DC gain in order to have a safety margin implies larger power dissipation, because of the formerly mentioned limitations. As shown, this problem can be solved by using a telescopic first stage.

The class A/A amplifier with Miller compensation has been adopted for its simplicity. The complete amplifier schematic is shown in Fig. 3.11. A nulling-resistor technique to compensate for the first non-dominant pole (output pole) would reduce power dissipation, but the compensation must be assured for process and temperature variations and, more important, for varying amplifier loads (from sampling to integration). The value of the resistor, usually implemented with a MOSFET, may also vary with the output voltage coupled through the compensation capacitor, which degrades the integrator transient response. Although it is possible to compensate for these effects [Johns97] [Raza00], the required extra power and design complexity may mask the initial advantages of the topology.

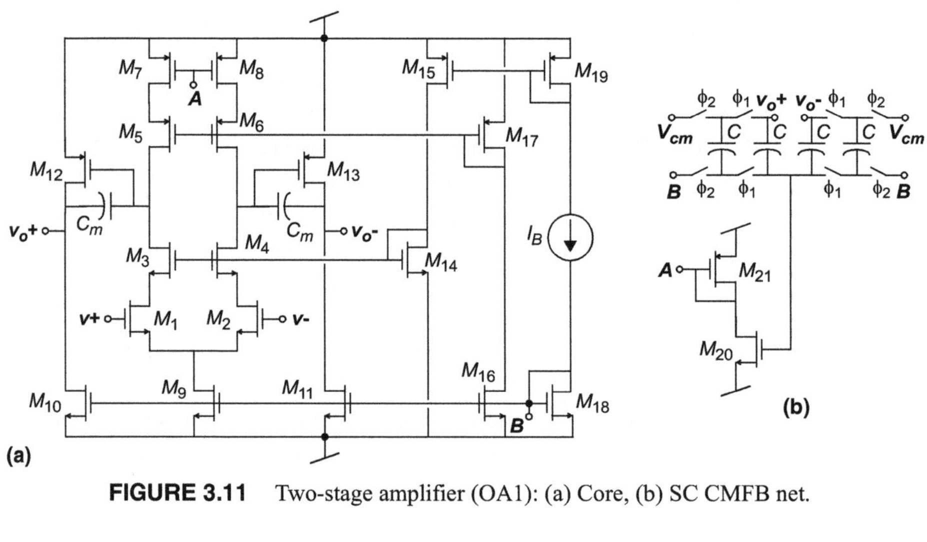

FIGURE 3.11 Two-stage amplifier (OA1): (a) Core, (b) SC CMFB net.

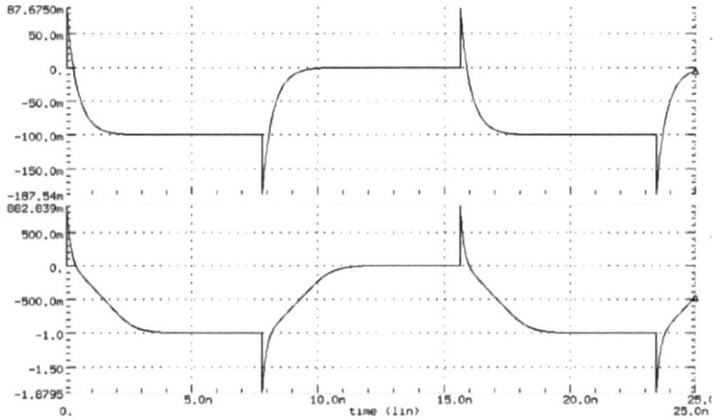

FIGURE 3.12 Simulated small- and large-signal transient response of the front-end integrator.

The amplifier biasing stage is implemented by transistors $M_{14} - M_{19}$. The CMFB net is of dynamic type, which has no voltage range issues and static power dissipation. Note that an inverting stage (M_{20}, M_{21}) is added to the basic SC net in order to implement the negative common-mode feedback.

OA1 has been optimized with FRIDGE to fulfil the requirements in Table 3.6. As a result, the bias current I_B is set to $52\mu A$; compensation and CMFB capacitors are 0.61pF and 86fF, respectively. Electrical simulation results are given in Table 3.7. Note that the obtained DC gain is considerably larger than required, as a result of the telescopic first stage. However, further fine-tuning of the amplifier sizing does not imply significant reduction of power consumption, which is limited by speed requirements. The lowest non-dominant pole in the open-loop frequency response of the amplifier is typically at 830MHz, which corresponds to the output pole $(\sim g_{m12}/C_m)$. Thus, reducing as much as possible the compensation capacitor significantly improves the power consumption. Fig. 3.12 shows the small- and large-signal transient response of the front-end integrator. Note that it behaves like a first-order system. In spite of the onset of slew rate in the large-signal step, the integrator settles properly in the available time slot.

Remaining amplifiers

Because of the medium/low DC gain required for these amplifiers, a cascode single-stage can be used. Among these structures, the telescopic offers the best speed/power figure, since it does not suffer from mirror poles, the signal path can be formed only with nMOS transistors, and just a bias current is required. Also, the typical differential output swing of this topology with 3.3-V supply is around the required value (4V). The only problem is that, to accommodate such an output swing, the input and output common-mode levels must be different. Although this can be easily done in an SC integrator, it requires an additional well-controlled voltage reference. To avoid this

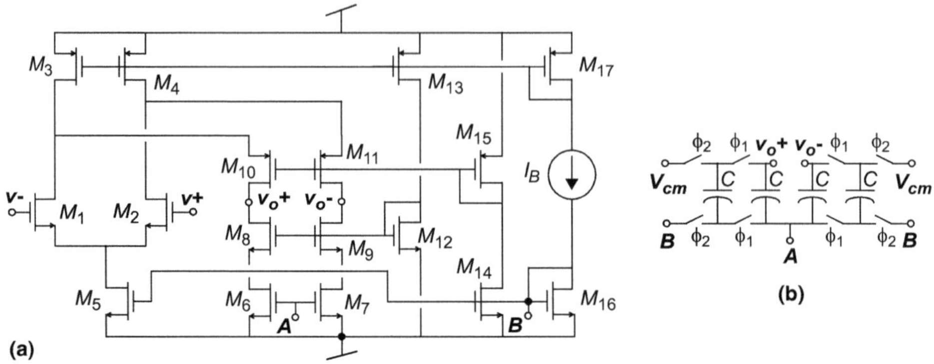

FIGURE 3.13 Folded-cascode amplifier (OA2, OA3, OA4): (a) Core, (b) SC CMFB net.

problem, a folded-cascode amplifier can be used. The price to pay is a larger power dissipation than that obtained with a telescopic amplifier.

Two folded-cascode amplifiers have been sized using FRIDGE. Both amplifiers share the topology shown in Fig. 3.13. One of the designed versions is intended to fulfil the specifications of both OA2 and OA4, what can be accomplished with the same transistor sizes just by changing the bias current (120μA and 175μA, respectively). OA3 uses a different sizing and a bias current of 100μA. The corresponding CMFB nets use 86-fF capacitors. Electrical simulation results for the amplifiers are summarized in Table 3.7. Note that the power consumption of OA4 is larger than that of OA2 and OA3 as a consequence of its larger loading capacitance.

TABLE 3.7 Electrical simulation results for the amplifiers.

	OA1	OA2	OA3	OA4
DC gain	80.0dB	62.8dB	55.7dB	62dB
GB	250MHz (1.6pF)	311MHz (1.4pF)	261MHz (1.4pF)	167MHz (4.5pF)
PM	67° (1.6pF)	65° (1.4pF)	71° (1.4pF)	79° (4.5pF)
SR	390V/μs (1.6pF)	410V/μs (1.4pF)	320V/μs (1.4pF)	185V/μs (4.5pF)
Input noise	$5\text{nV}/\sqrt{\text{Hz}}$	$3.2\text{nV}/\sqrt{\text{Hz}}$	$4.1\text{nV}/\sqrt{\text{Hz}}$	$3.8\text{nV}/\sqrt{\text{Hz}}$
OS	±2.5V	±3.1V	±2.9V	±3.1V
Power	38.5mW	4.5mW	4.0mW	6.6mW

Fig. 3.14 shows the small- and large-signal transient response of the fourth integrator. Note that the large-signal step is slew-rate limited, with a final linear response. According to the settling model, such a characteristic can be tolerated, thus reducing power dissipation.

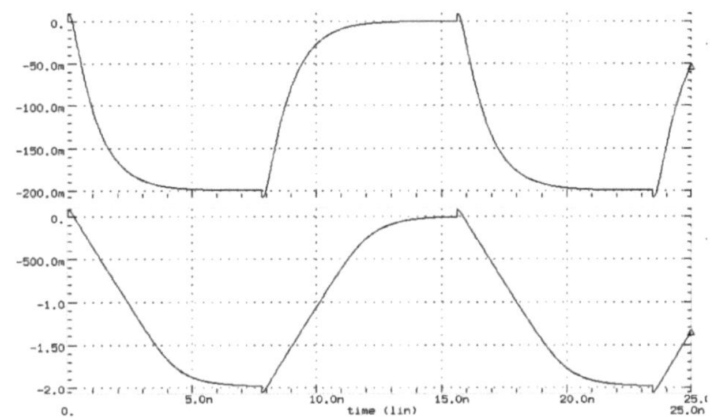

FIGURE 3.14 Simulated small- and large-signal transient response of the fourth integrator.

3.5.2 Comparators

The single-bit quantizers at the end of the first and second stages of the modulator demand a low resolution time, whereas hysteresis as large as 30mV can be tolerated. The same applies for comparators in the last-stage A/D converter, which is implemented using a flash structure. These requirements recommend the use of a dynamic comparator based on a regenerative latch [Rodr99]. Fig. 3.15 shows the architecture employed for the implementation of comparators [Yuka85], which has been widely used in ΣΔ modulator design. In practice, this topology is capable of achieving resolutions about that required with no pre-amplifying stage.

An alternative latched comparator [Yin94b] [Marq98a] employs a pMOS-input differential pair to perform the voltage-to-current conversion at nodes X1 and X2 (instead of M_1 and M_2), and a switch short-circuiting both nodes to keep M_3 and M_4 ON during reset. This improves the common-mode rejection ratio of the comparator and may speed up the response. However, it consumes

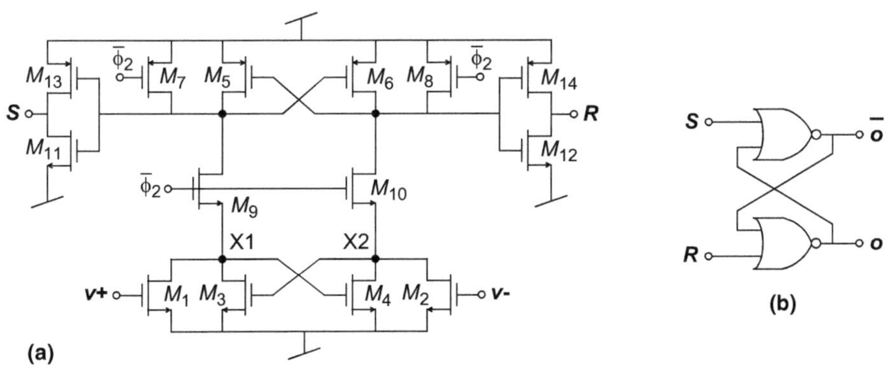

FIGURE 3.15 Comparator: (a) Regenerative latch, (b) SR latch.

during the reset phase and the power consumption cannot be too small in order to have sufficient transconductance at the differential pair.

Because of this, the simplified regenerative latch illustrated in Fig. 3.15*a* has been adopted. It is activated by $\bar{\phi}_2$ at the end of the integration phase, solving the difference of the integrators outputs. At the beginning of the next integration phase, the outputs of the latch are forced to the low state and an SR flip-flop, depicted in Fig. 3.15*b*, latches the comparator output until the next activation. The main features of the comparator, obtained by electrical simulation, are summarized in Table 3.8.

TABLE 3.8 Electrical simulation results for the comparator.

Hysteresis	< 10mV
Resolution time, LH	2.5ns
Resolution time, HL	2.3ns
Power consumption	0.65mW

3.5.3 Switches

Due to the high operation speed of the ΣΔ modulator, the value of the finite switch on-resistance (R_{on}) is mainly constricted by dynamic considerations. As shown in Section 2.3.5, the influence of R_{on} in combination with the finite amplifier dynamics must be carefully evaluated, since it leads to a further degradation of the integrator response.

Fig. 3.16 illustrates this effect, showing the transient evolution of the front-end integrator output voltage during a clock cycle, for several values of R_{on}. Note that, as the switch on-resistance increases, charge-transfer at the beginning of integration and sampling is not instantaneous anymore and the integrator dynamic is slowed down. Using behavioral and electrical simulations, it has

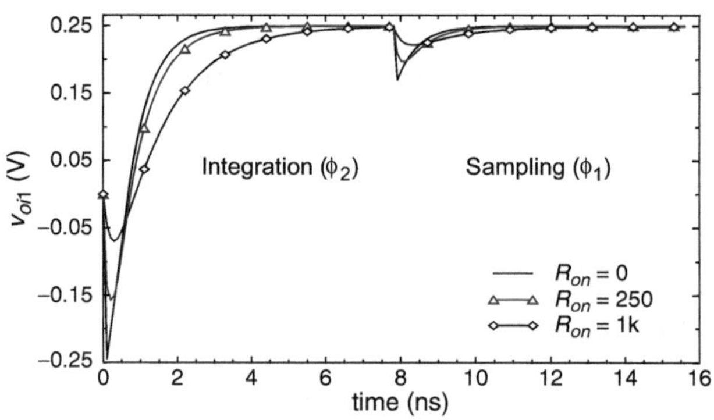

FIGURE 3.16 Transient evolution of 1st-integrator output voltage in a clock period.

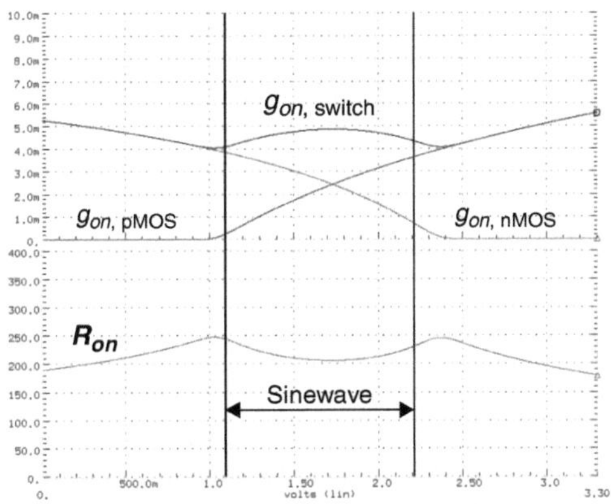

FIGURE 3.17 DC curve of the implemented analog CMOS switch.

been evaluated that switch on-resistances around 250Ω can be tolerated, with minor degradation of the modulator performance.

The required R_{on} can be achieved using standard transmission gates operated at the nominal 3.3-V supply. The use of clock-boosting strategies, low-V_t transistors, or similar techniques can be thus avoided [Wu98] [Bult00]. The analog CMOS switch is implemented with an nMOS transistor of aspect ratio $8/0.35$ and a pMOS transistor of $29.5/0.35$. Fig. 3.17 shows the characteristic of the switch as a function of the signal level. Note that the applied switch sizing tries to compensate the difference in the transconductance parameter of the nMOS and pMOS transistors—i.e., $K_N W_N \approx K_P W_P$, with W_N and W_P being the width of the nMOS and pMOS transistors, respectively [†2]. As stated in Section 2.6.3, the non-linearity exhibited by the switch characteristic can be therefore reduced, in order to limit distortion.

3.5.4 Capacitors

Integrator weights are physically implemented through arrangements of unit capacitors. For matching and thermal noise considerations, the value of the unit capacitor is selected to 0.5pF for the first integrator and 0.25pF for the rest. The total number of unit capacitors required for the differential implementation of the 2-1-1 ΣΔ modulator is only 2×16, smaller than for other sets of coefficients—e.g., 2×29 capacitors for the weights in [Marq98a] and [Geer99], and 2×35 for those in [Yin94b]. Table 3.9 summarizes how the integrators weights are obtained through unit capacitor combinations.

2. The switch charge injection plays a secondary role here, because the delayed-phase switching (see Section 3.5.6) makes it signal independent.

TABLE 3.9 Implementation of integrator weights.

1st Integrator	2nd Integrator	3rd Integrator	4th Integrator
$g_1 = \dfrac{C_{1,1}}{C_{1,2}+C_{1,3}+C_{1,4}+C_{1,5}}$	$g_2 = \dfrac{C_{2,1}+C_{2,2}}{C_{2,3}+C_{2,4}}$	$g_3 = \dfrac{C_{3,1}+C_{3,2}}{C_{3,3}+C_{3,4}}$	$g_4 = \dfrac{C_{4,1}+C_{4,2}}{C_{4,3}}$
$g_1' = \dfrac{C_{1,1}}{C_{1,2}+C_{1,3}+C_{1,4}+C_{1,5}}$	$g_2' = \dfrac{C_{2,2}}{C_{2,3}+C_{2,4}}$	$g_3' = \dfrac{C_{3,1}}{C_{3,3}+C_{3,4}}$	$g_4' = \dfrac{C_{4,1}}{C_{4,3}}$
–	–	$g_3'' = \dfrac{C_{3,2}}{C_{3,3}+C_{3,4}}$	$g_4'' = \dfrac{C_{4,2}}{C_{4,3}}$

Unit capacitors are implemented using a multi-metal sandwich structure, employing the five metal layers available in the 0.35-μm digital CMOS technology. A side view of the sandwich capacitor is illustrated in Fig. 3.18*a*, whereas Fig. 3.18*b* shows its top view. Metal-1, metal-3, and metal-5 are connected together, constituting the bottom plate. Metal-2 and metal-4 build up the top plate. Using this structure, the 0.5-pF unit capacitor is approximately $52\mu m \times 52\mu m$ size and the 0.25-pF unit is $36\mu m \times 36\mu m$. The bottom-plate parasitic capacitor, specially important for evaluating the integrator settling requirements, is above 25% of the nominal capacitance.

In practice, the actual values of integrators weights may differ from the nominal ones due to capacitor mismatch. These deviations result into incomplete cancellation of the quantization errors of the first and second modulator stages, thus degrading the modulator performance. The in-band quantization error, taking into account noise leakage due to capacitor mismatch, can be estimated for the 2-1-1 ΣΔM to be—see eq(2.31)

$$P_Q(\varepsilon_g) \cong \frac{\Delta^2}{12} \cdot \varepsilon_1^2 \cdot \frac{\pi^4}{5OSR^5} + d_3^2 \cdot \frac{\Delta_3^2}{12} \cdot \frac{(1+\varepsilon_{g_4''})^2 \pi^8}{9OSR^9} \tag{3.12}$$

where $\Delta_3 = \Delta/(2^B - 1)$ stands for the quantization step of the multi-bit third stage, the scaling coefficient d_3 equals 2, $\varepsilon_1 = |\varepsilon_{g_3''} - \varepsilon_{g_3}| + \varepsilon_{g_2} + \varepsilon_{g_1'}$ and ε_{g_i} stands for the error in the corresponding integrator weight.

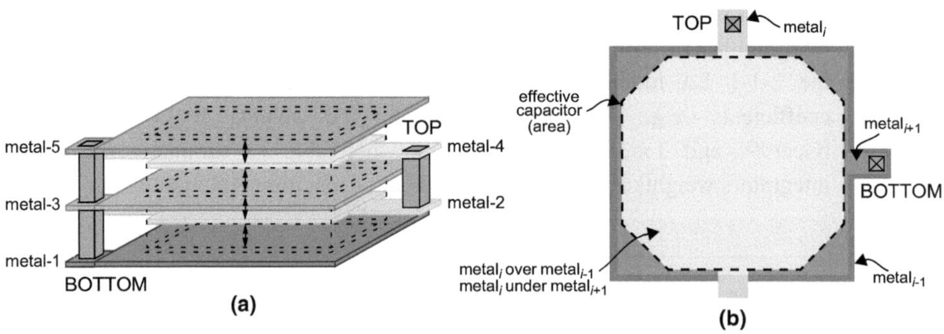

FIGURE 3.18 Multi-metal sandwich capacitor: (a) Side view, (b) Top view.

Using eq(2.22) and footnote 4 in Section 2.2.2, the former error terms can be calculated for the actual implementation of weights in Table 3.9, leading to

$$\begin{aligned} \varepsilon_{g_1'} &= \frac{3}{\sqrt{2}}\sqrt{\frac{1}{1}+\frac{1}{4}}\sigma_{C_1} & \varepsilon_{g_2} &= \frac{3}{\sqrt{2}}\sqrt{\frac{1}{2}+\frac{1}{2}}\sigma_{C_2} \\ \varepsilon_{g_3''}-\varepsilon_{g_3} &= \frac{3}{\sqrt{2}}\sqrt{\frac{1}{1}-\frac{1}{2}}\sigma_{C_3} & \varepsilon_{g_4''} &= \frac{3}{\sqrt{2}}\sqrt{\frac{1}{1}+\frac{1}{1}}\sigma_{C_4} \end{aligned} \tag{3.13}$$

with σ_{C_i} being the standard deviation of the unit capacitor in each integrator.

Equations (3.12) and (3.13) have been used to evaluate the impact of mismatch in each integrator on the modulator performance. Fig. 3.19 shows the modulator *ENOB* against capacitor mismatch, considering the influence of each integrator separately. The same equation database as in Section 3.4 is used, with modulator parameters in Table 3.3. For each integrator, the standard deviation on its unit capacitor is considered to vary from 0.1% to 1%, whereas that of the remaining integrators is fixed to 0.12%[†3]. Note from Fig. 3.19 that larger sensitivity to mismatch is placed in the first stage of the cascade. The influence of the first and second integrators is quite similar and both are dominant in regard to noise leakage. Mismatches in the third integrator are of less importance, whereas those in the fourth integrator are almost negligible.

Fig. 3.20 shows the symbolic layout of capacitors implementing the integrators weights in the ΣΔ modulator. Note that only capacitors in the first

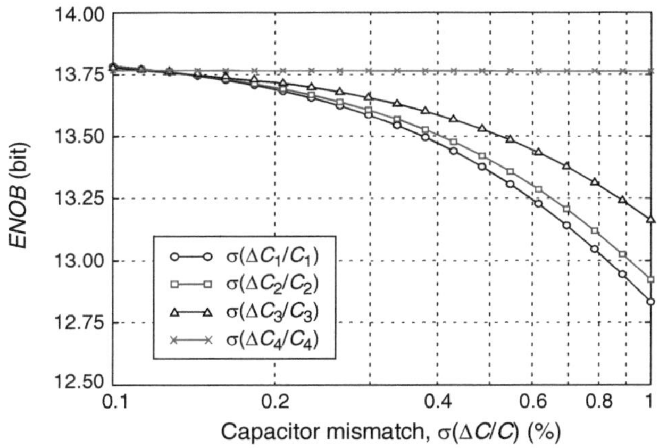

FIGURE 3.19 *ENOB* vs. capacitor mismatch ($\sigma_{C_i} = \sigma(\Delta C_i/C_i) = 0.12\%$ when not varying).

3. This value has been extracted from the characterization of capacitor matching in the intended technology, where multi-metal capacitors with different structures and sizes exhibit a sigma around 0.05% ~ 0.25%. For 0.5-pF and 0.25-pF capacitors with the structure in Fig.3.18, the matching is in the range of 0.05% ~ 0.15%.

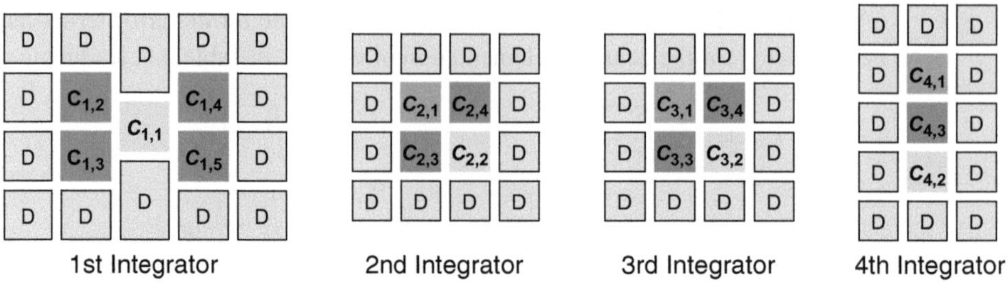

FIGURE 3.20 Representation of the layout of unit capacitors (D stands for dummy).

integrator actually use a common-centroid topology, whereas the matching of the remaining weights is just based on closely-placed unit capacitors.

3.5.5 Programmable A/D/A converter

The last-stage multi-bit quantizer is implemented by means of a selectable-resolution A/D/A converter, capable of 2-, 3-, or 4-bit quantization. The structure of this programmable A/D/A converter basically corresponds to that of a 4-bit architecture, whose digital output code can be adapted for 2 or 3 bits of resolution. The SC implementation of the 4-bit A/D/A converter is illustrated in Fig. 3.21.

A/D converter The A/D converter uses a fully-differential flash architecture [Lewis87] [Bran91b] and compares the differential input voltage v_i^+, v_i^- (4th-integrator output) with reference voltages v_r^+, v_r^- generated in a resistive-ladder DAC. During ϕ_1 references v_r^+, v_r^- are stored in the input capacitors, which are then used to compute the difference $(v_i^+ - v_i^-) - (v_r^+ - v_r^-)$ during ϕ_2. At the end of ϕ_2, comparators are activated to solve the sign of that difference. The thermometer output code of the 15 comparators is then translated to a 1-of-16 code using AND gates.

Comparators in the ADC are identical to those used in the 1st- and 2nd-stage single-bit quantizers (see Fig. 3.15), with no need for a pre-amplifying stage. Multi-metal sandwich capacitors of value $C = 0.25\text{pF}$ are used and the analog switches are identical to those in the SC integrators ($8/0.35$ for the nMOS and $29.5/0.35$ for the pMOS transistor).

The timing scheme of the switches has been adapted to reduce the capacitive load to the fourth integrator. Nevertheless, it suffers from input-dependent charge injection from switches controlled by ϕ_2. This problem has been overcome making these switches considerably smaller ($1/0.35$ for both nMOS and pMOS), with no degradation of the converter performance.

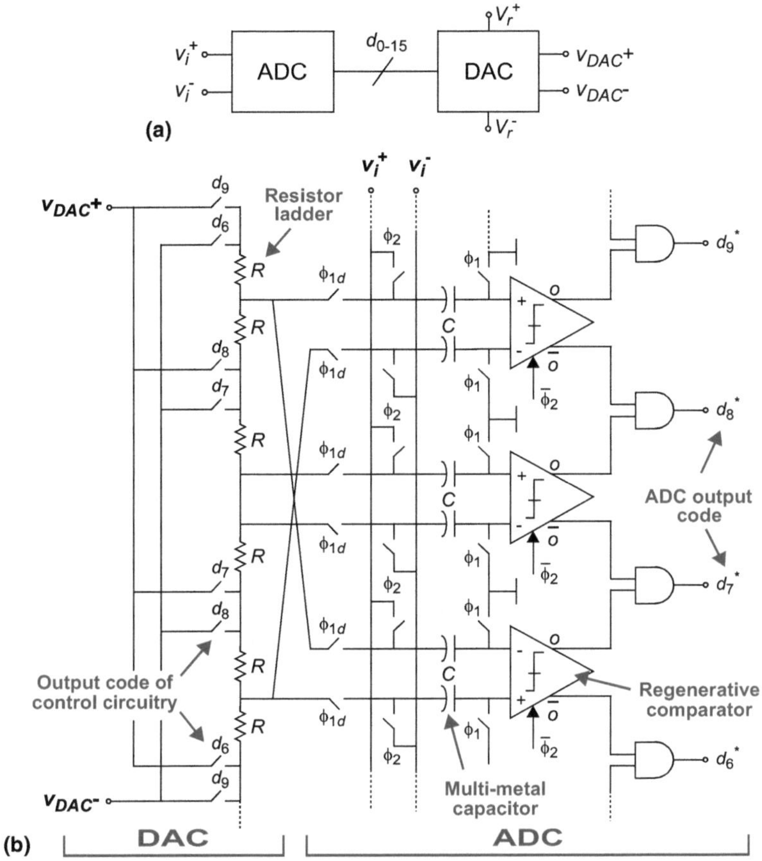

FIGURE 3.21 A/D/A converter: (a) Block diagram, (b) Partial view of the SC implementation.

D/A converter

It consists of a simple resistor ladder that:

- provides the reference voltages for the operation of the ADC, and
- generates the analog output of the overall A/D/A converter, through the selection of the voltages generated in the ladder by the 1-of-16 output code of the ADC ($d_{0\text{-}15}$).

The ladder consists of 30 unit resistors of value $R = 50\Omega$ connected between reference voltages $V_r^+ = +1\text{V}$ and $V_r^- = -1\text{V}$, providing a differential full scale of $\pm 2\text{V}$ with a 1.33-mA current consumption. The value selected for R ensures a small settling error in the voltage references transferred to the ADC (during ϕ_1) and in the sampling capacitors of the fourth integrator (during ϕ_2). Resistors are implemented using unsalicided $p+$ poly layers due to their high resistivity, good matching, and low dependence on voltage and temperature variations.

Control circuitry The 4-bit A/D/A converter can be adapted to provide 2- or 3-bit resolution using a digital control circuitry—see Fig. 3.22. The selection of the desired resolution is done with two signals, S_{3b} and S_{2b}, as shown in Table 3.10.

TABLE 3.10 Selection of resolution in the programmable A/D/A converter.

S_{3b} S_{2b}	Output code	Resolution
0 0	1-of-16	4 bits
0 1	1-of-4	2 bits
1 0	1-of-8	3 bits
1 1	not allowed	

The functionality of the control circuitry is shown in Fig. 3.23, for both the 4bit-to-2bit and the 4bit-to-3bit conversion. Conceptually, it consists in the OR operation of the appropriate digital outputs d_{0-15}* of the 4-bit ADC in order to accommodate them to a 1-of-4 or 1-of-8 code. For instance:

- digital output d_2 of the programmable ADC must be given by $d_8* + d_9* + d_{10}* + d_{11}* + d_{12}*$ if the selected resolution is 2 bits. It must activate analog switches controlled by d_{10} in the 4-bit DAC, while forcing switches controlled by d_8, d_9, d_{11}, and d_{12} to be off.

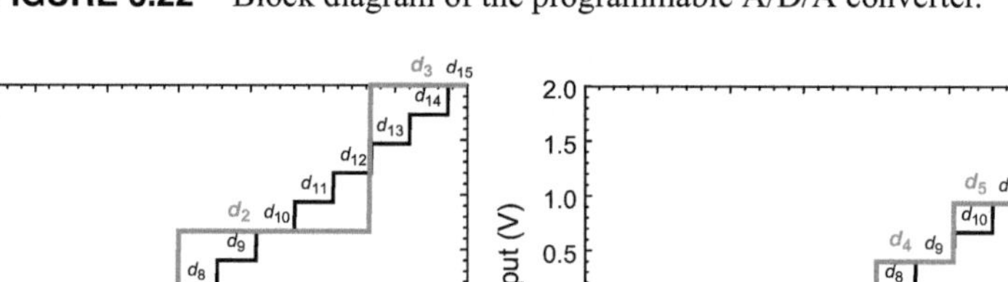

FIGURE 3.22 Block diagram of the programmable A/D/A converter.

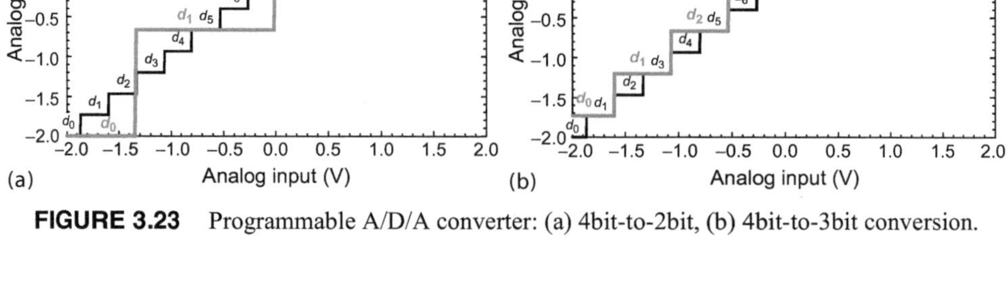

FIGURE 3.23 Programmable A/D/A converter: (a) 4bit-to-2bit, (b) 4bit-to-3bit conversion.

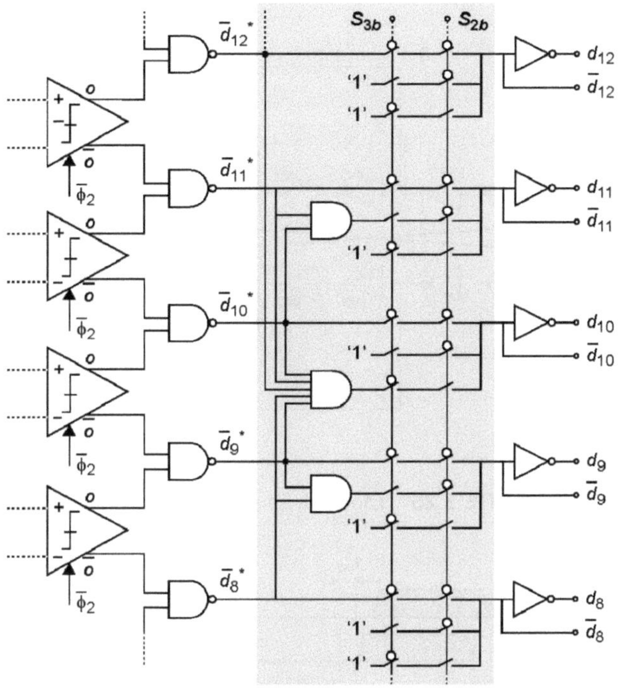

FIGURE 3.24 Partial view of the control circuitry.

- digital output d_5 must be given by $d_{10}^* + d_{11}^*$ if the selected resolution is 3 bits, and must activate switches controlled by d_{11} in the 4-bit DAC, while switches controlled by d_{10} are forced to be off.

Note from Fig. 3.23*a* that the 4bit-to-2bit conversion leads by construction to a 2-bit A/D/A converter with full scale of ±2V, given that $(2^4 - 1)/(2^2 - 1) = 5$ is an integer. In the case of the 4bit-to-3bit conversion (Fig. 3.23*b*), $(2^4 - 1)/(2^3 - 1) = 2.14$ is not an integer. The 3-bit A/D/A converter has by construction a full scale of ±1.87V and an offset error of 133mV (0.25LSB). Nevertheless, cascade multi-bit ΣΔMs present very low sensitivity to offset errors in the DAC [Mede99a], so that the overall modulator performance is not degraded, as confirmed through behavioral simulation.

The implementation of the control circuitry in the programmable A/D/A converter is illustrated in Fig. 3.24.

3.5.6 Clock phase generator

Fig. 3.25 shows the schematic of the clock phase generator. Two non-overlapped clock phases—ϕ_1, ϕ_2— are obtained from an external clock signal. Delayed versions of these phases—ϕ_{1d}, ϕ_{2d}—are generated in order to attenuate

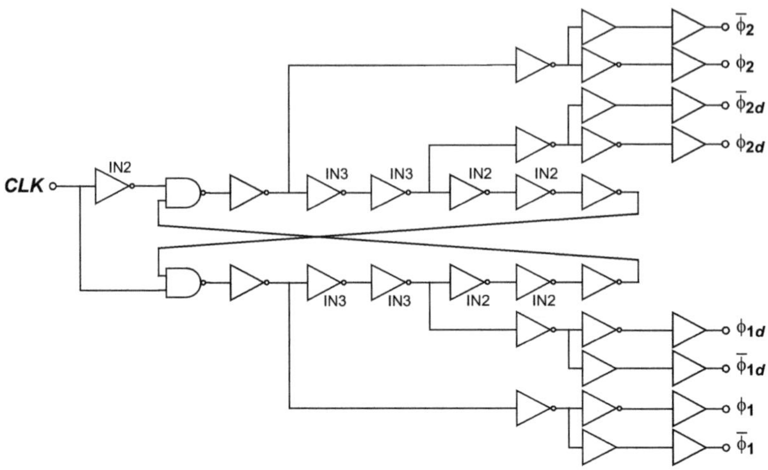

FIGURE 3.25 Clock phase generator and drivers.

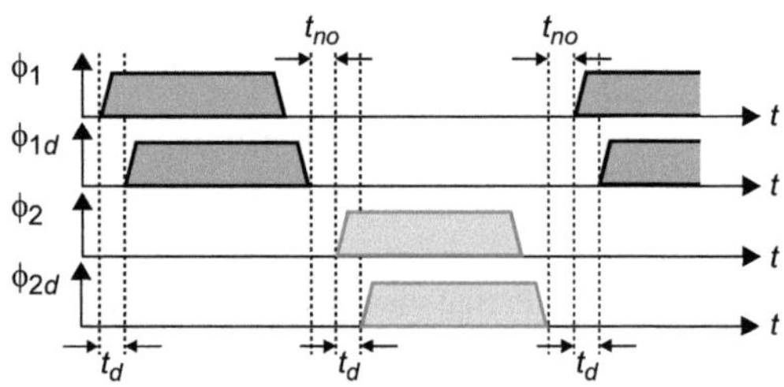

FIGURE 3.26 Timing of the clock phases in the ΣΔ modulator.

signal-dependent charge injection [Lee85]. The complementary versions of all these phases are also generated—$\overline{\phi}_1$, $\overline{\phi}_2$, $\overline{\phi}_{1d}$, $\overline{\phi}_{2d}$—for the control of the CMOS switches. All phases are properly buffered at the output of the clock driver, to account for the different capacitive load of the signals.

Fig. 3.26 illustrates the timing of the generated clock phases. The non-overlapping time t_{no} and the phase delay t_d are approximately 0.35ns and 0.31ns, respectively.

3.6 Layout and Prototyping

The ΣΔ modulator has been implemented in a standard digital 0.35-μm CMOS technology that uses an epitaxial process with heavily-doped bulk. High-performance mixed-signal circuits are specially challenging in this kind of processes, because of the great impact of the on-chip switching activity of the digital circuitry on the analog section.

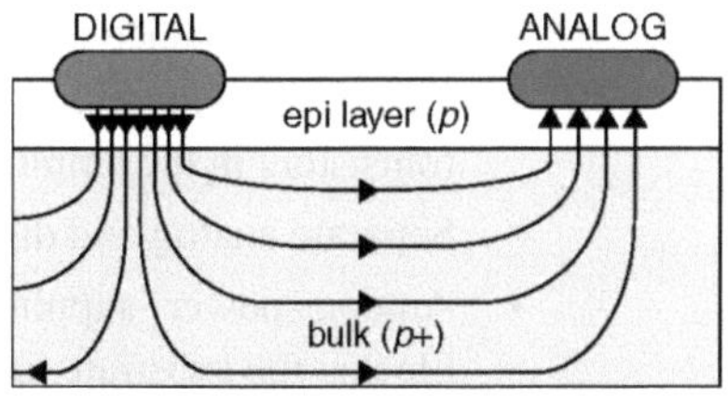

FIGURE 3.27 Current lines in an epitaxial process with low-resistive bulk.

Due to the conductive nature of the deep substrate, injected disturbances can propagate through the chip through this low-impedance path (see Fig. 3.27), instead of the epitaxial layer, that presents higher resistivity [Felde99] [Libe00]. This means that traditional layout techniques are of little effectiveness in low-resistive bulk epitaxial processes. Guard-rings used for shielding analog parts provide a conductive path only for surface currents (in the epi layer). On the other hand, the separation of digital and analog blocks at considerable distance usually does not provide sufficient attenuation of the switching noise injected to analog devices, since the substrate can be considered as a single node, at least to a first approximation.

In spite of the limited effectiveness of these common strategies, they have been incorporated in the prototype. The layout of the 2-1-1 dual-quantization ΣΔM is shown in Fig. 3.28 and considers the following issues:

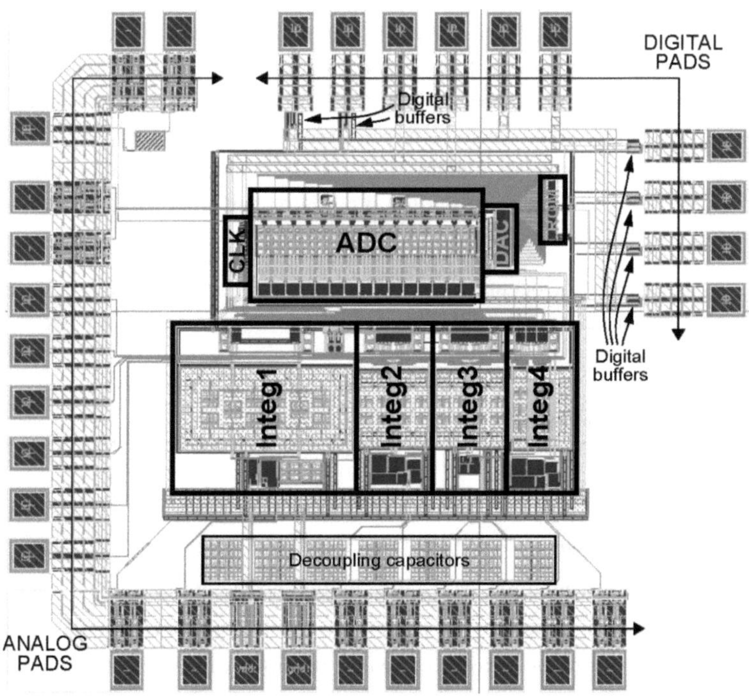

FIGURE 3.28 Layout of the prototype.

- Increased distance among sensitive analog blocks and noisy digital ones.
- Centroid layout techniques with unit transistors are employed for matched transistors in the amplifiers and in the regenerative latches.
- Separate analog and digital power supplies are used.
- Analog power supplies—V_{DDA}, V_{SSA}—are employed in the analog blocks, the substrate, and the guard-rings. Separated routing lines are used for that purpose.
- Digital power supplies—V_{DDD}, V_{SSD}—are used in the modulator digital blocks: clock phase generator, SR latches, 4-bit ROM, etc.
- Powerful digital buffers have been designed for the high-speed outputs of the cascade stages (Y_1, Y_2, $Y_{3,\,0\text{-}3}$). Dedicated power supplies—V_{DDD2}, V_{SSD2}—are used for them.
- Pads with ESD protection are used for signals driving transistor gates (S_{3b}, S_{2b}, CLK, and amplifier bias currents), whereas no protections are used for the modulator differential input voltage and references V_r^+, V_r^-.
- The modulator digital outputs (Y_1, Y_2, $Y_{3,\,0\text{-}3}$) and the digital power supplies (V_{DDD},V_{SSD} and V_{DDD2},V_{SSD2}) also use pads with no ESD protection. Since no biasing is required for these pads, they have been be placed outside the pad ring.

Given the low resistivity of the substrate, some degradation of the prototype performance is foreseeable, due to the switching noise introduced by the digital circuitry—especially, by the digital buffers that drive the output of each modulator stage off the chip. In order to palliate its impact, the prototype accounts for a possible reduction of the digital supplies below their nominal value (3.3V) without affecting the modulator performance, as a direct strategy to reduce the power of the switching activity.

The prototype occupies an area of 1.32mm^2 without pads (4.30mm^2 pads included) and has been packaged in a 44-pin ceramic quad flat pack.

3.7 Experimental Results

A two-layer printed circuit board (PCB) was designed for testing the ΣΔ modulator. Its schematic is illustrated in Fig. 3.29. The PCB includes:

- Separate ground planes for the analog and the digital signals.
- Decoupling capacitors in the supply and the reference voltages lines.
- Termination resistors for impedance coupling in the digital output lines.
- ESD protection for sensitive input pins.
- A 1st-order RC anti-aliasing filter at the modulator differential input.
- Independent biasing control of the amplifiers.

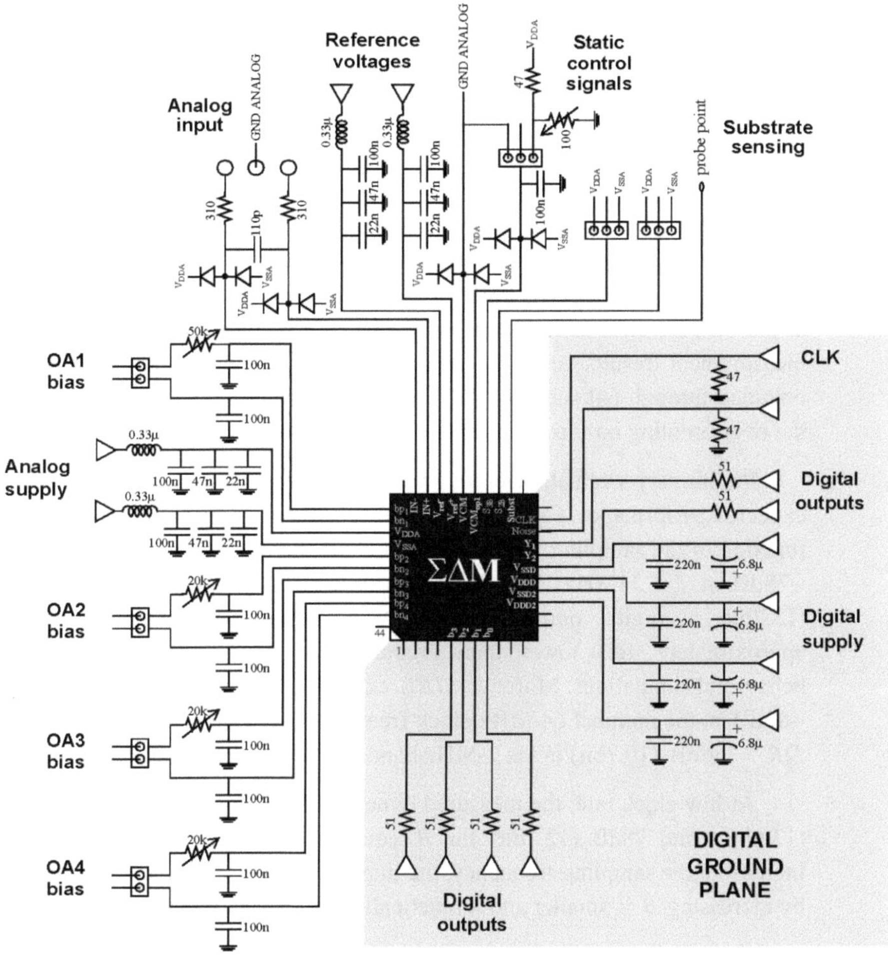

FIGURE 3.29 Schematic of the PCB for the test of the prototype.

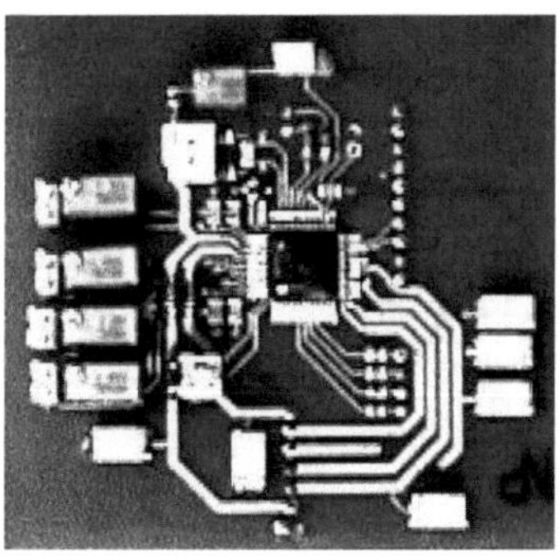

FIGURE 3.30 PCB for the test of the prototype.

Fig. 3.30 shows a photograph of the PCB. In order to avoid socket parasitics, samples were directly soldered to the PCB. Nine modulator samples have been evaluated, all being operative and providing similar performance.

The test set-up for the ΣΔ modulator comprises a high-linearity sinusoidal source with differential output (Tek SG5010) and a digital test-unit (HP82000). The latter generates the off-chip clock signal, acquires the output bit-streams of the modulator stages, and provides the supply and reference voltages. The test set-up is controlled with C-routines from a work-station. Bit-streams are transferred from the digital tester to the work-station, where MATLAB is used to perform the error cancellation logic. Henceforth, unless otherwise specified, measurement results are computed from the modulator output spectrum, obtained through 64k-sample FFT of its overall output bit-stream. By default, the oversampling ratio is 16 and the multi-bit resolution is 4 bits.

Preliminary evaluation of the prototype showed a degradation from the expected performance. Fig. 3.31 depicts the measured in-band error (IBE) as a function of the sampling frequency (f_s). Note that the measured IBE is about -78dB at $f_s \leq 24\text{MHz}$. This leads to a modulator dynamic range of 81dB †4 (13.2bit) at digital output rate $DOR \leq 1.5\text{MS/s}$. This dynamic range is approximately 4dB lower than predicted by the equation database and behavioral simulations. Moreover, IBE exhibits a rapid increase with f_s, being -63dB at the nominal 64-MHz clock frequency. This leads to a dynamic range $DR = 66\text{dB}$ (10.7bit) in the 2-MHz band, which is 3-bit lower than expected.

At low clock rate, the measured dynamic ranges are 81dB (13.2bit), 80dB (13.0bit), and 78dB (12.7bit) for B equal 4, 3, and 2, respectively. When increasing the sampling frequency, the improvement in dynamic range obtained by increasing B is smaller and it practically vanishes at nominal clock rate.

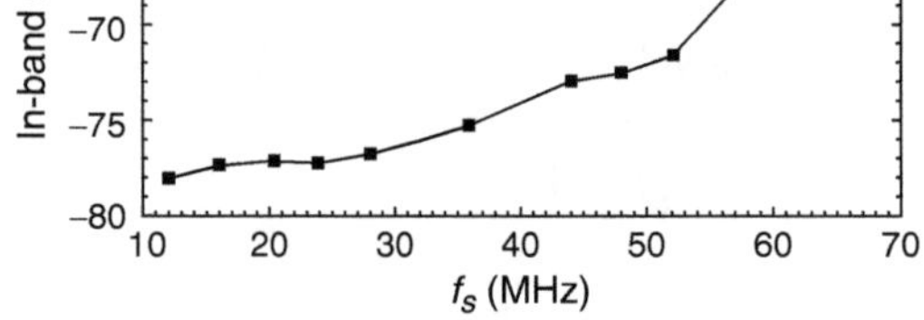

FIGURE 3.31 Measured in-band error power vs. clock frequency ($OSR = 16$, $B = 4$).

4. The full-scale amplitude of the differential input signal is 2V (3-dB power).

Next, tests applied trying to identify the source of performance degradation will be discussed.

3.7.1 Performance of the A/D/A converter

First, the functionality of the A/D/A converter was measured from specific samples including the ADC, the DAC, and the clock driver. The same test set-up formerly described was used.

The ADC performance was measured using the code-histogram method for sinewaves [Doer84] [IEEE01]. An input tone with 2.16-V amplitude and 103-kHz frequency was applied to the ADC, operating with 4-bit resolution at 64-MHz clock frequency. The digital output codes for 100 input periods were acquired with the digital tester. From the histogram of the 16 digital output codes, the analog voltages corresponding to the code transitions were estimated.

The DAC performance was measured forcing each of the 16 digital input codes with the digital tester and measuring the corresponding analog output voltage with a high-accuracy multimeter (HP34401A).

Measured performances are summarized in Table 3.11 and show that the A/D/A converter meets the specifications required by the ΣΔ modulator. In all measured samples, DAC *INL* is well below 0.4%FS. The programmability to 2 and 3 bits of resolution proved also to be fully operative.

TABLE 3.11 Measured performance of the A/D/A converter.

	ADC		DAC	
	%FS	LSB	%FS	LSB
Offset error	–2.017	–0.303	0.313	0.047
Gain error	4.595	0.689	–0.623	–0.094
DNL	2.873	0.431	0.125	0.019
INL	1.531	0.230	0.104	0.016

3.7.2 Influence of jitter noise

In eq(2.128) in-band jitter error was estimated to be proportional to the square of both the amplitude A_x and frequency f_x of a sinusoidal input. These dependencies were used to evaluate the influence of clock jitter, considering it a possible source of performance degradation at high-speed operation. Measurements were performed at $f_s = 64\text{MHz}$ for variations in A_x with fixed f_x, and vice versa. This is illustrated in Fig. 3.32, which shows the modulator baseband spectrum for a -15dBV input tone at frequencies varying from

10kHz to 160kHz (maximum frequency of the Tek SG5010 signal generator). No change in the noise floor is observed, although eq(2.128) predicts a 24-dB increase of IBE from 10kHz to 160kHz, if jitter noise were the dominant error.

Since no variations could be observed for the measured IBE in the cases considered, jitter noise was finally discarded as a limiting source of error.

3.7.3 Influence of settling errors

Defective settling of the integrators was initially foreseen as a possible cause of performance degradation at high clock frequencies, especially considering:

- on the one hand, the difficulty to accurately estimate parasitic capacitances in the final implementation during the early design phases of the modulator, and,
- on the other, the use of a novel technology whose characterization had not been yet confirmed by silicon results.

For the reasons above, we adopted a current biasing scheme that can be independently controlled for each amplifier at the PCB level.

Fig. 3.33 shows the measured IBE at f_s = 64MHz when varying the bias current of each amplifier, while keeping the rest at their nominal values (52μA, 120μA, 100μA, and 175μA for OA1 to OA4, respectively). Note that IBE significantly decreases when increasing the bias currents—especially, for the second and third amplifiers. At first glance, this contradicts the results in Section 3.5.1 for the optimization of the amplifier dynamics using behavioral simulations, where settling errors were not limiting the modulator performance.

For a better understanding of this discrepancy, the modulator schematic and its corresponding extracted layout were electrically simulated using

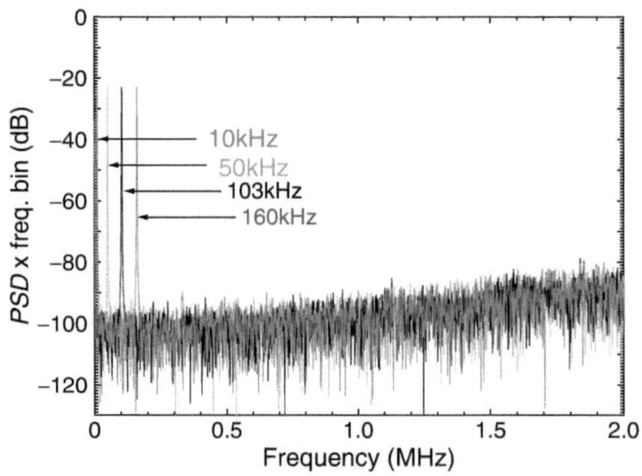

FIGURE 3.32 Baseband output spectrum for a –15dBV input tone at different frequencies.

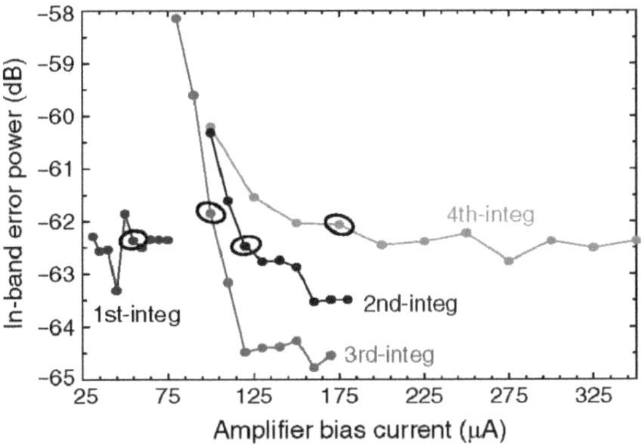

FIGURE 3.33 Measured in-band error power versus amplifier bias current ($f_s = 64\mathrm{MHz}$).

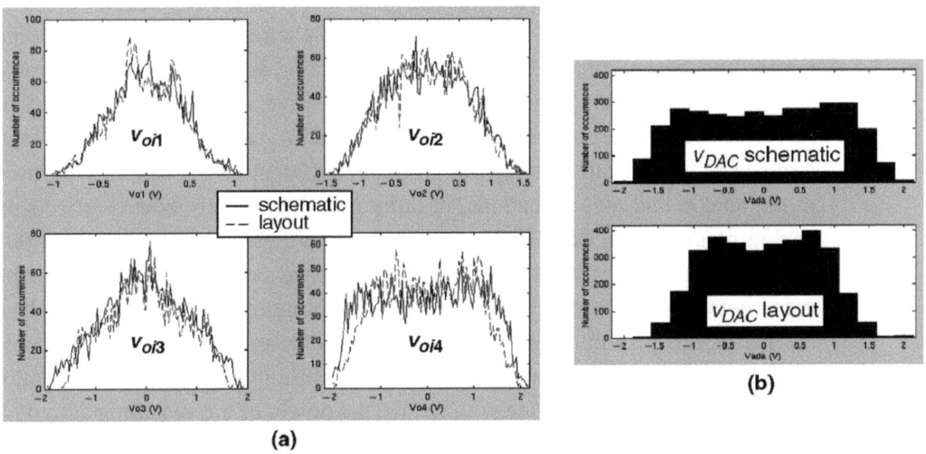

FIGURE 3.34 Electrical simulation of the modulator at $f_s = 64\mathrm{MHz}$: (a) Histogram of the integrator outputs, (b) Histogram of the A/D/A converter output.

HSPICE. Fig. 3.34 compares both simulation results for a –6dBV@250kHz input tone at $f_s = 64\mathrm{MHz}$. The histograms of the four integrator outputs are plotted in Fig. 3.34*a*, whereas Fig. 3.34*b* shows the output of the multi-bit A/D/A converter. Note that there are differences between the two simulations, especially in the swings of the third and fourth integrator outputs, and therefore also in the multi-bit converter output. These discrepancies in the histograms are caused by parasitic capacitances, which are included in the extracted layout, but not in the schematic. Parasitic capacitances in the final layout were thus underestimated during the early design phases.

Originally, the bottom parasitic of multi-metal capacitors was estimated to be around 25% of the nominal capacitance. This figure was revised in the extracted layout, showing that it can reach 40% as a consequence of the large fringing capacitance in the intended technology. Given that parasitics are larger than expected, the equivalent capacitive loads for the amplifiers are also larger. Consequently, the amplifier dynamics must be faster for a non-limiting settling at high clock rates. This explains the measured IBE improvement when increasing the amplifier bias currents. The inclusion of the extracted parasitics in behavioral simulations predicts the need of increased bias currents. Their final values are fixed to $50\mu A$, $175\mu A$, $175\mu A$, and $350\mu A$ for OA1 to OA4, respectively. With these bias currents, the measured IBE is reduced by 3dB at the nominal 64-MHz clock rate ($IBE = -66dB$).

3.7.4 Influence of switching noise

The switching activity of the digital section was considered a potential cause of performance degradation. This possibility was already taken into account during the design phase, so that the chip floorplan includes considerable distance among noisy digital blocks and sensitive analog ones, as well as shielding to reduce crosstalk effects and guard rings to provide a low-impedance return path to noisy digital lines. Different power supplies are used in the analog and digital sections, in order to reduce supply bounce effects on analog blocks. Also, dedicated power supplies are used for the digital buffers driving the high-speed bit-streams of the three stages off the chip. The pad ring is neither closed, in order to avoid a direct coupling of the switching signals at the pad level. The final design of the ΣΔ modulator accounts for a possible reduction of the digital supplies below 3.3V without affecting the modulator performance, as a direct strategy to reduce the power of the switching activity.

The influence of switching noise was first evaluated through the reduction of both digital supplies (digital core and digital output buffers). Results show that the modulator performance is not appreciably affected by the reduction of the voltage supplied to the internal digital blocks. However, it is significantly improved when the supply of the high-speed digital buffers is reduced. This is illustrated in Fig. 3.35, which shows the measured IBE versus the clock rate for two different operating conditions: nominal ones and reduced supply for the digital output buffers plus increased amplifier bias currents. Note that, in the latter case, the performance degradation at high clock rates is considerably decreased. At the nominal rate, $IBE = -70.5dB$; i.e., a 7.5-dB reduction is obtained in comparison with the nominal operation conditions. As formerly stated, the increased bias currents lead to a 3-dB reduction, whereas an additional 4.5-dB reduction is obtained when decreasing the supply of the output buffers to 1.2V, which is the lowest value for proper operation.

The performance of the ΣΔ modulator was re-evaluated under the latter operating conditions. Fig. 3.36 shows the measured in-band output spectra for a

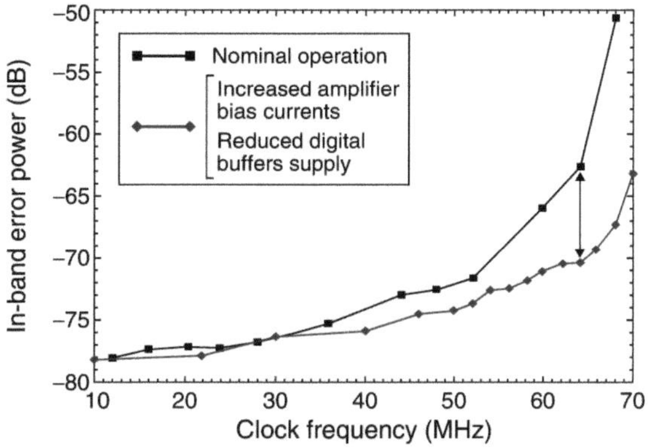

FIGURE 3.35 Measured in-band error power vs. clock frequency ($OSR = 16$, $B = 4$).

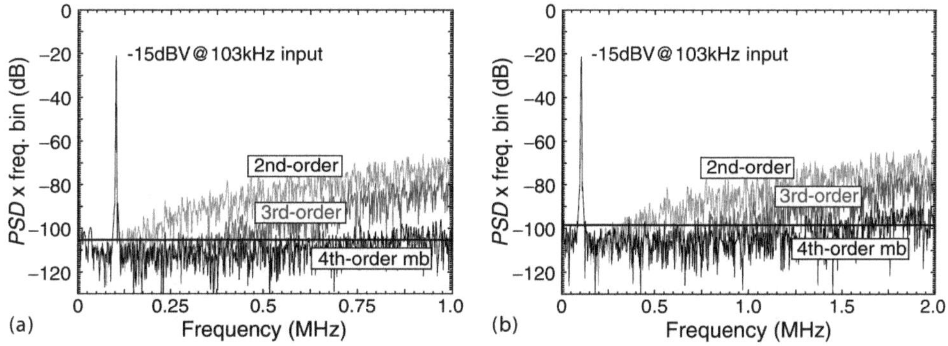

FIGURE 3.36 Measured output spectrum at (a) 32-MHz and (b) 64-MHz clock rate.

–15dBV@103kHz input tone. Sampling frequencies are 32 MHz (Fig. 3.36*a*) and 64MHz (Fig. 3.36*b*). The spectra labelled '4th-order mb' correspond to the overall modulator output, whereas the ones labelled '2nd-order' and '3rd-order' correspond to the first-stage output (2nd-order ΣΔM) and to the combination of the first and second stages (2-1 ΣΔM), respectively. Note from the above figures that the performance of the first and second modulator stages is similar at both clock rates. Nevertheless, in Fig. 3.36*b* the noise floor is higher and a larger contribution of the last-stage shaped error is also noticeable at the upper part of the signal band. This means that, as the sampling frequency increases, the performance of the multi-bit last stage is degraded and non-shaped errors increase.

In order to further investigate this degradation, experimental data were collected at 32-MHz and 64-MHz clock rate and processed in MATLAB considering several oversampling ratios. Fig. 3.37 shows the computed *SNDR* against *OSR* for a –15dBV input test signal. For comparison purposes, the corresponding curves obtained through behavioral simulation are also included.

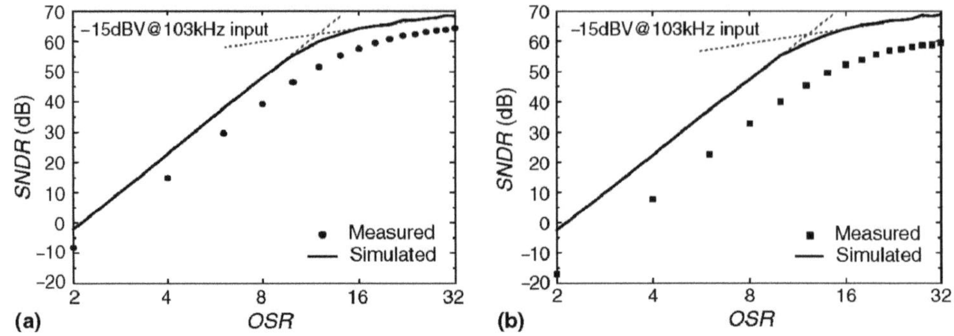

FIGURE 3.37 *SNDR* versus *OCR* when operating at: (a) f_s = 32MHz, (b) f_s = 64MHz.

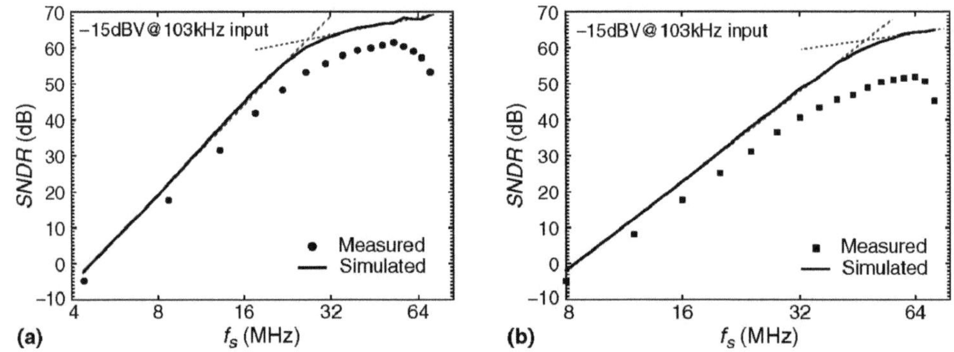

FIGURE 3.38 *SNDR* versus clock frequency in the: (a) 1-MHz band, (b) 2-MHz band.

Both simulated curves are almost equal, since settling errors do not limit performance. A priori, the *SNDR* should be limited by 4th-order shaped errors (90dB/dec) for low *OSR* (wide signal band), whereas it should be limited by white noise (10dB/dec) for high *OSR* (narrow signal band). The slopes of the measured curves in Fig. 3.37*a* and Fig. 3.37*b* are in agreement with this, what indicates that the shaping order of quantization errors is not degraded in the prototype—thus, noise leakages are not limiting performance—and that non-shaped errors limit the low-frequency region of the spectra. However, there is a shift of the two measured curves from the simulated ones, which is more significant the larger the clock rate. Therefore, both non-shaped errors and 4th-order shaped quantization errors in the last stage increase as the clock rate does.

A similar conclusion can be derived from Fig. 3.38, which shows the measured *SNDR* versus clock frequency for the same test input, computed in fixed signal bands of 1MHz (Fig. 3.38*a*) and 2MHz (Fig. 3.38*b*). The simulated curves are also included. Note that now, there is no shifting of the experimental curves from simulated ones—as occurred in Fig. 3.37—, but a variation of their slopes, since the degradation effects increase with the sampling frequency.

Given that the measured in-band error power decreases if the output buffers supply is lowered, it is more than reasonable to conclude that they are the source of most of the on-chip switching activity that degrades the modulator performance. In fact, the analog blocks closest to the digital buffers are the fourth integrator in the cascade and the multi-bit quantizer (see Fig. 3.28), which explains the degradation of their performance as the clock rate increases.

3.8 Performance Summary

Fig. 3.39 shows the modulator *SNDR* curves measured at 32-MHz and 64-MHz sampling frequency, for noise computed in the 1-MHz and 2-MHz signal band, respectively. The respective dynamic ranges are 79.5dB (12.9bit) and 73.5dB (11.9bit).

Table 3.12 summarizes the measured performance of the prototype in both signal bands. The corresponding values of the figures-of-merit FOM_1 and FOM_2—defined in equations (1.70) and (1.71), respectively— are also included. The total power dissipation at 64-MHz clock rate is 78.3mW, from which 60.2mW (77%) are consumed in the analog blocks, 4.5mW (6%) in the digital section, and 13.6mW (17%) in the digital output buffers.

Note that, in spite of the expected performance degradation caused by the on-chip switching activity, the prototype provides an A-to-D conversion of approximately 13bit@2MS/s and 12bit@4MS/s. Moreover, this is achieved in a standard digital CMOS technology that suffers from poor device matching (especially, in multi-metal capacitors) and large substrate conductivity—disadvantages that are not found in analog or mixed-signal oriented processes.

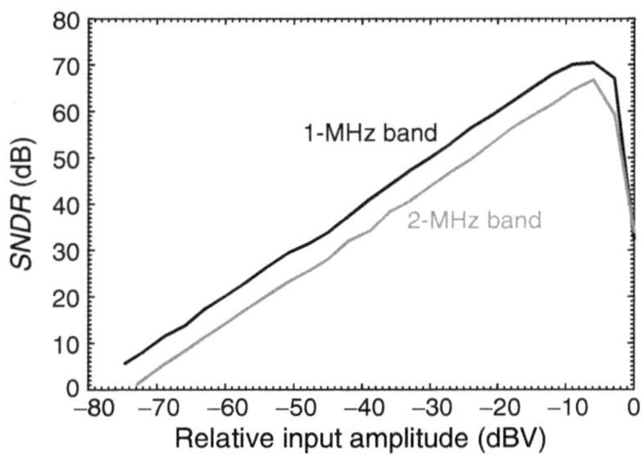

FIGURE 3.39 Measured *SNDR* curves in the 1-MHz and 2-MHz signal bands. Respective clock rates are 32MHz and 64MHz. ($OSR = 16$, $B = 4$, and 103-kHz input tone).

TABLE 3.12 Summary of measured prototype performance.

	1-MHz band	2-MHz band
Topology	2-1-1(4b)	
Oversampling ratio	16	
Reference voltage	2V	
Clock frequency	32MHz	64MHz
Digital output rate	2MS/s	4MS/s
Dynamic range	79.5dB	73.5dB
ENOB	12.9bit	11.9bit
SNDR peak	71dB	67dB
THD	–80dB	
Power consumption	73.7mW	78.3mW
FOM_1	4.82	5.12
FOM_2	3.96	1.86
Active area	1.32mm^2 (pads excluded)	
Supply voltage	3.3V (1.2V in digital O 's)	
Technology	0.35-μm STD CMOS (1P5M)	

3.9 Performance Comparison with the State of the Art

Table 3.13 shows the state of the art in high-speed low-pass ΣΔMs implemented in CMOS processes. Considered prototypes achieve resolutions larger than 11 bits at rates above 1MS/s. Their features are summarized including information about the technology and the modulator architecture. The list of abbreviations used can be found in Section 1.7 on page 56. For comparison purposes, the presented prototype is included at the bottom of the table.

It must be noted that the prototype here exhibits the largest full-scale/supply ratio, which is $2 \cdot 2/3.3 \cong 1.2$. The closest reported values are 1.1 in [Jiang02] (1.8-V supply) and [Feld98] (3.3-V supply), and 1.0 in [Grilo02] (2.7-V supply) and [Vleu01] (2.5-V supply). From the viewpoint of the practical implementation of a ΣΔM, the larger the value of this ratio, the more demanding the requirements of output swing on amplifiers and of linearity on amplifiers and switches, what obviously impacts the power consumption needed to fulfill them. Although this ratio thus affects the global features of the designs, it is not taken into account in the following comparison with the state of the art—albeit it would benefit the prototype—, in order to use the two figures-of-merit (FOM_1 and FOM_2) commonly established in literature.

Fig. 3.40*a* to Fig. 3.40*d* depict the main performance parameters of each high-speed ΣΔM against its digital output rate. The performance of the ICs are

TABLE 3.13 State-of-the-art high-speed low-pass ΣΔ prototypes in CMOS technologies.

REFs	*ENOB* (bit)	*DOR* (S/s)	*OSR*	Architecture	Technology	V_{ref} (V)	V_{supply} (V)	Power (W)	FOM_1	FOM_2
[Bran91b]	12	2.1M	24	2-1(3b)	1μm STD	1.5	5	41m	4.77	2.14
[Yin94b]	15.82	1.5M	64	2-1-1	2μm BiCMOS	2.5	5	180m	2.07	69.61
[Broo97]	14.5	2.5M	8	2(5b)-0(12b) [5-b flash with DDS, 12-b pipeline]	0.6μm MS [2P]	1	5 & 3	500m	8.63	6.70
[Feld98]	12.5	1.4M	16	2-2-2(1.5b)	0.7μm MS [2P]	1.8	3.3	81m	9.99	1.45
[Marq98a]	14.8	2M	24	2-1-1	1μm MS [2P]	2	5	230m	4.03	17.66
[Geer99]	15	2.2M	24	2-1-1	0.5μm MS [2P]	0.9	3.3	200m	2.77	29.48
[Mede99b]	13	2.2M	16	2-1-1(3b)	0.7μm STD	2	5	55m	3.05	6.70
[Fuji00]	15	2.5M	8	2(4b)-1(4b)-1(4b) [Bi-DWA in all stages]	0.5μm MS [2P, low-Vt]	2	5	105m	1.28	63.81
[Geer00]	15.8	2.5M	24	3rd-or(4b) [DWA]	0.65μm MS [2P]	1	5	295m	2.07	68.85
	11.5	12.5M	8					380m	10.50	0.69
[Mori00]	13	2.2M	24	2-2-2	0.35μm MS [2P]	1.2	3.3	150m	8.32	2.46
	12			2-2(5b)		1		99m	10.99	0.93
[Lamp01]	13.5	1.56M	32	2-2(3b) [LR]	0.35μm –	–	2.5	50m	2.76	10.47
[Vleu01]	15.5	4M	16	2(5b)-2(3b)-1(3b) [2S, P-DWA]	0.5μm MS [2P]	1.25	2.5	150m	0.81	142.94
[Grilo02]	13	1M	32	2nd-or(4b) [1st-or DEM]	0.35μm BiCMOS	1.4	2.7	11.88m	1.45	14.10
[Gupta02]	14.6	2.2M	29	2-1-1(2b) [2S]	0.35μm STD	1.5	3.3	180m	3.29	18.81
[Jiang02]	13.8	4M	8	5th-or(4b) [hybrid FIR-IIR, DWA]	0.18μm STD	1	1.8	149m	2.61	13.63
[Kuo02]	13.7	1.25M	12	4th-or(4b) [FB-FF, I-DWA]	0.25μm MS [MiM]	0.675	2.5	100m	6.01	5.53
	13.0	2M						105m	6.41	3.19
[Reut02]	14	2.5M	32	5th-or(1.5b) [FFS, LR]	0.25μm STD	–	2.5	24m	0.59	69.79
[Veld02]	11.3	4M	40	4th-or(1.5b) [RC-active/GmC, FFS]	0.18μm STD	–	1.8	6.6m	0.65	9.62
[Lee03]	14.16	1M	64	2-2	0.35μm MS [2P]	0.9	1.8	150m	8.20	5.57
	12	2M	32						18.31	0.56
[Mill03]	12.8	1.25M	18	2nd-or(6b) [mDWA]	0.18μm MS [dual-gate, MiM]	1.2	2.7	30m	3.30	5.51
	11.7	3.84M	12					50m	4.00	2.04
This Work	12.9	2M	16	2-1-1(4b)	0.35μm STD	2	3.3	73.7m	4.82	3.96
	11.9	4M						78.3m	5.12	1.86

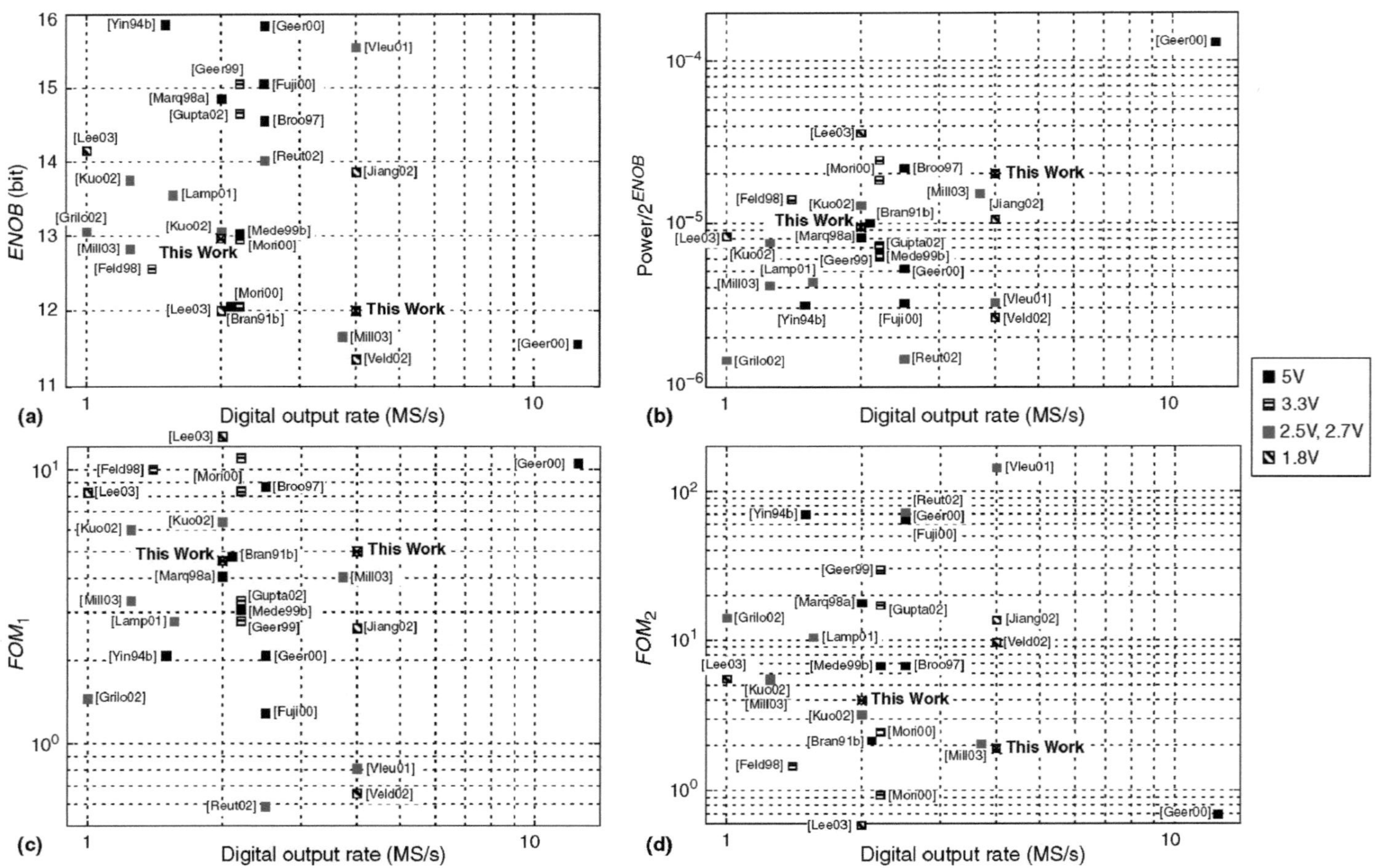

FIGURE 3.40 State of the art in high-speed ΣΔMs: (a) $ENOB$, (b) Normalized power consumption, (c) FOM_1, and (d) FOM_2 vs. digital output

summarized in terms of the achieved $ENOB$, the normalized power consumption, and the values of the figures-of-merit.

Note that the overall features of the prototype compare well with the state of the art, in spite of the degradation of its dynamic range due to the switching noise in the conductive substrate. In deed, the prototype outperforms the 3.3-V ICs reported [Feld98] and [Mori00] regarding resolution, power consumption, and figures-of-merit, although both are implemented in mixed-signal technologies, which benefit from poly-poly capacitors (better matching) and lightly-doped bulks (reduced switching noise). It also outperforms the global features of the ΣΔM in [Bran91b] and improves the normalized power and FOM_1 of [Broo97], although both ICs operate from a 5-V supply.

3.10 Summary

This chapter presents the implementation of a high-speed ΣΔ modulator targeted to achieve an A-to-D conversion of 14-bit resolution over a signal bandwidth of 2MHz. The topology selected is a 4th-order 3-stage cascade (2-1-1 ΣΔM) employing dual-quantization. Given that multi-bit quantization is only used in the modulator last stage, quantization error power can be reduced with no compromise of the modulator linearity, avoiding therefore the need of correction/calibration mechanisms.

The requirements of the building blocks in an SC implementation are derived with the support of dedicated CAD tools, which make extensive use of the models and equations developed in Chapter 2 for the effect of non-idealities on the performance of ΣΔ modulators. The topology of the different blocks is described and their designs are presented at transistor level.

The prototype has been implemented in a 0.35-μm standard digital CMOS technology operated at 3.3-V supply, with no use of extra process steps, higher on-chip voltages, or low-Vt transistors. Multi-metal structures with thick-oxide are used for capacitors and standard transmission gates for the switches.

The resolution measured for the prototype exhibits a degradation due to switching noise problems associated to the conductive substrate. In spite of this, the prototype achieves 13bit@2MS/s and 12bit@4MS/s, with reduced power budgets of 74mW and 78mW, respectively.

The measured performance of the designed ΣΔ modulator is comparable with state-of-the-art high-speed ΣΔMs reported in open literature and proves the feasibility of their implementation in deep-submicron digital CMOS technologies.

CHAPTER 4

A ΣΔ Modulator in 2.5-V 0.25-μm CMOS for ADSL/ADSL+

WIRELINE SOLUTIONS FOR BROADBAND ACCESS to the Internet and home networking are continuously evolving to provide increasing data rates and more functionality. Asymmetric digital subscriber line (ADSL) is an example of such applications and extensions of this technology, like ADSL+ (with doubled number of channels) or VDSL (providing video-rate reception), are just round the corner. This motivates the increasing demand for highly-linear fast analog front-ends capable to achieve more than 12-bit accuracy for signal bandwidths ranging from 1MHz to 20MHz [Casi01], with the ultimate goal of incorporating them to system-on-chip (SoC) commercial solutions for that applications implemented in modern VLSI CMOS processes. The ΣΔ ADC presented in this chapter is an example of this industrial interest.

The design of an ADC oriented to these applications becomes a difficult task, beyond the high-accuracy and high-speed requirements themselves. The use of a deep-submicron CMOS process will force the ADC to operate at low voltage supply, using transistors whose threshold voltages are comparatively high, avoiding extra process steps to improve the features of the devices, but maintaining a reduced power budget. At the same time, the ADC must coexist with large noisy digital circuits in the same die (the DSP of the SoC), with no performance degradation.

In this hostile environment, while ADC topologies traditionally oriented to high-speed applications—such as pipeline [Guil01]—have to resort to area and power consuming calibration methods to improve their linearity, the high intrinsic linearity exhibited by ΣΔ ADCs, employing relatively simple analog circuitry, has stimulated the interest for embedding them into SoCs.

Nevertheless, only low-oversampling ΣΔMs are feasible in broadband applications, because the sampling frequency must be limited so that the speed requirements of the analog circuitry are achievable in a CMOS process. Thus, high-order shaping and/or multi-bit quantization must be used in order to enhance resolution. Among the alternatives to that purpose, the combined use of multi-stage ΣΔ architectures and dual-quantization techniques has demonstrated to be a feasible solution to circumvent both the stability and linearity problems involved, while leading to efficient and robust silicon implementations [Bran91b] [Mede99b] [Mori00] [Lamp01] [Rio01a] [Gupta02].

Yet, the viability of SC ΣΔ cascades in deep-submicron CMOS depends on two key process features: the supply voltage and the performance of capacitors.

- The supply voltage affects the selection of the reference voltage, which defines the available dynamic range and also imposes the output swing requirements in the integrators. This impacts the selection of the amplifier topology, and its capability to trade DC gain, speed, and output swing must be carefully considered—especially in the front-end stage of the ΣΔM. Single-stage amplifiers traditionally used in high-speed applications are not adequate in low-voltage implementations, because cascode devices will be required to achieve enough DC gain, and the attainable output swing will be small. Among the alternatives, two-stage amplifiers offer the possibility to yield a large DC gain, still providing large enough output swings. This allows to increase the value of the reference voltage, thus augmenting the modulator full scale to useful levels.

  Besides the amplifiers, the performance of the switches with low supply voltages needs also careful control, especially for dynamic distortion considerations [Yu99]. For broadband ΣΔ modulators, solutions can be found in clock-boosting strategies, low-Vt transistors, or high-voltage devices available in double-oxide processes [Wu98] [Bult00]. However, these techniques often lead to an increase in circuit complexity, power dissipation, or process cost—apart from potential reliability problems—, and should be avoided whenever possible.

- The second relevant technology feature refers to the quality of the capacitor structures. Typical capacitor matching requirements in cascade ΣΔMs range from 0.05% to 0.2% standard deviation. Also, capacitor parasitics have an important impact on the settling of integrators—and thus on the power consumption—, so that small parasitics are desirable for the efficient implementation of high-frequency ΣΔMs. Finally, capacitor linearity must be also considered, although its requirements are usually relaxed provided that symmetrical fully-differential circuitry is used.

  A first option is to use multi-metal sandwich capacitors, in spite of their large bottom-plate parasitics. However, their capacitance density is not high and it does not scale with the technology shrinking [Apar02], what results in considerable area occupation. Moreover, they sometimes suffer from gradients in the process of polishing of inter-metal dielectric [Lei98], which reflects in poor matching properties. Metal-insulator-metal (MiM) capacitors are now becoming available in many CMOS processes; they exhibit excellent matching and linearity, and small bottom parasitics.

In this scenario, this chapter presents a high-speed cascade ΣΔM targeted to be incorporated in a customer premises equipment (CPE) modem for digital subscriber line applications (both ADSL and ADSL+). The prototype is implemented in a 0.25-μm CMOS process with MiM capacitors and has a differential full scale of 3V operating from a single 2.5-V supply, with no need for higher on-chip voltages or low-Vt transistors. The potential problems associated to this low-voltage implementation have been circumvented while keeping the original philosophy of simple, robust, non-calibrated analog circuitry of ΣΔMs. Two-stage amplifiers are used at the front-end in order to achieve both large DC gain and output swing. Standard CMOS transmission gates are used for the switches, thus avoiding the use of clock-boosting.

Section 4.1 discusses the selection of the cascade topology, with special emphasis on the deep-submicron CMOS process to be used. The specifications for the main modulator blocks are obtained in Sections 4.2 and 4.3, whereas the complete design is presented in Sections 4.4 and 4.5. Finally, Section 4.6 shows the measured performance of the prototype.

4.1 Topology Selection

The topology of the ΣΔM has been selected among cascade candidates using single- or multi-bit quantizers. Given the broadband application considered, the oversampling ratio must be restricted to a small value, and these architectures offer a high-order shaping without involving the modulator stability. Furthermore, an enhanced, linear operation can be achieved incorporating a multi-bit quantizer only in the last stage of the cascade, while the remaining stages are single-bit; this dual-quantization scheme avoids correction/calibration methods in the multi-bit DAC.

The potentialities of these architectures have led us to propose a family of cascade ΣΔMs that can be easily expanded to any order, while preserving a low systematic loss of resolution (only 1bit) and a high overload level. It comprises a 2nd-order loop followed by 1st-order stages, which are identical thanks to the selected integrator coefficients (see Appendix A). Cascades belonging to this family can be described by three design parameters: the modulator order (L), the oversampling ratio (OSR), and the number of bits in the last-stage quantizer (B). Thus, a triad $\{L, OSR, B\}$ is used to codify them.

An analytical procedure has been developed to estimate their power consumption. Appendix B details the underlying expressions, which contemplate both architecture and technological features, together with simplifying assumptions inspired in practical design solutions. The aim here is not only to draw conclusions about architectural choices, but also to track their evolution under technology scaling. To that purpose, the comparison of the performance they can achieve is made from a twofold perspective:

- For a given technology, the performance of the cascade modulators is compared for varying converter specifications, and
- For given converter specifications, the architectures are evaluated in the technology road map.

The comparison is made according to the following figure-of-merit [Good96]

$$FOM_1 = \frac{\text{Power(W)}}{2^{ENOB(\text{bit})} \cdot DOR(\text{S/s})} 10^{12} \quad (4.1)$$

where DOR stands for the digital output rate, i.e., the Nyquist rate.

In a first comparison step, the triads $\{L, OSR, B\}$ describing specific cascades have been evaluated along the curve in the resolution-speed plane shown in Fig.4.1 (dashed line). Although this particular resolution-speed relationship is arbitrary, it fits the usual requirements for wireline telecom ADCs: ISDN (Integrated Services Digital Network), ADSL, VDSL, etc., which have been placed in the figure for illustration. For each section of the resolution-speed curve, the cascade architecture with the minimum FOM_1 has been noted down. Note that, as the output rate increases, the oversampling ratio decreases and, simultaneously, the increased number of bits in the multi-bit quantizer shows up to compensate for the oversampling reduction. Note that the 4.4-MS/s DOR employed in ADSL+ falls into the region led by the architecture $\{4, 16, 3\}$; i.e., a 4th-order 2-1-1 cascade with 3-bit quantization in the last stage and using 16 × oversampling, which will be the choice for our design.

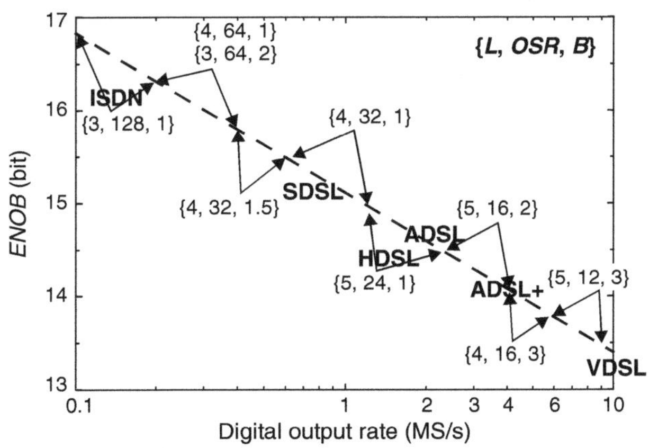

FIGURE 4.1 Most efficient cascade ΣΔM for each region of the resolution-speed plane, considering a 2.5-V 0.25-μm CMOS technology. ($\sigma_C = 0.1\%$, $INL = 0.25\%\text{FS}$, $A_{DC} = 3000$, and two-stage amplifiers have been assumed).

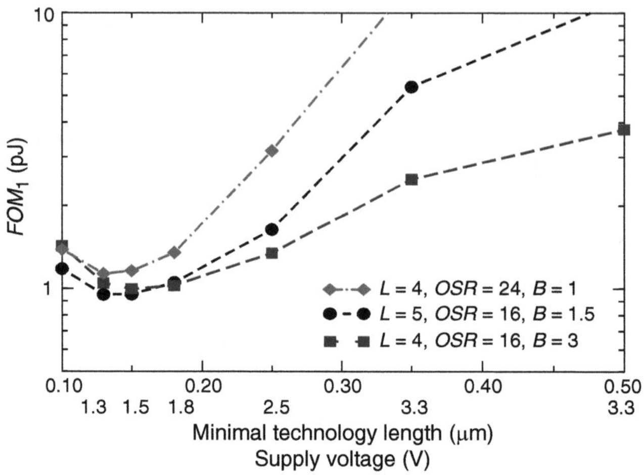

FIGURE 4.2 Estimated evolution of FOM_1 under technology scaling for three cascade architectures obtaining 14bit@4.4MS/s.

In a second step, we estimate how the performance of cascade ΣΔMs is going to evolve under technology changes, taking advantage of the fact that some technology features enter the formulation of their power consumption. Fig. 4.2 shows the estimated evolution of FOM_1 for three cascade topologies; namely, $\{4, 24, 1\}$, $\{5, 16, 1.5\}$, and $\{4, 16, 3\}$, aimed at obtaining 14 bits at 4.4MS/s. Two facts are noticeable:

- Despite the reduction of the supply voltage, the overall power dissipation does not decrease below $0.18\mu\text{m}$. This is basically due to the reduction in the supply voltage, which imposes a reduction in the reference voltage and, hence, a compensating increase in the sampling capacitors. Since the incomplete settling error power must be also kept constant, this mechanism leads to an increased current absorption, which makes the power consumption increase below $0.18\mu\text{m}$. The location of the inflection point depends on the converter specifications. For instance, if for the same speed, the resolution is to be increased, the inflection point moves to the right in Fig. 4.2.
- Fig. 4.2 also illustrates the dynamic nature of the architecture selection in Fig. 4.1. Note that the $\{4, 16, 3\}$ cascade ΣΔM outperforms for $0.25\mu\text{m}$ and above, but it does not below $0.18\mu\text{m}$. The reason behind is that the multi-bit modulator has a fixed amount of power contributed by the last-stage quantizer that is not present in the single-bit implementation ($\{4, 24, 1\}$ ΣΔM). In addition, the latter takes advantage of the faster technologies to compensate for the increased oversampling ratio with respect to the multi-bit modulator.

4.2 Switched-Capacitor Implementation

Fig. 4.3 shows the fully-differential SC schematic of the selected $\{4, 16, 3\}$ ΣΔM. The first stage of the cascade includes two integrators—with one and two input branches, respectively—and switches controlled by the comparator outputs to feed the quantized signal back. The second stage uses an integrator with only two input branches to implement weights g_3, g_3', and g_3'', since the values used (see Fig. A.4) allow to distribute g_3 between the two SC branches. The same applies for g_4 in the fourth integrator. This integrator drives the 3-bit ADC and the loop is closed by a 3-bit DAC. The 1-of-8 output code of the ADC is converted into binary by a ROM that generates the corresponding bit-streams.

The modulator operation is controlled by two non-overlapped clock phases. The integrator input signals are sampled during phase ϕ_1. During phase ϕ_2 the algebraic operations are performed and results are accumulated in the feedback capacitors. In order to attenuate signal-dependent charge injection, delayed versions of the two phases (ϕ_{1d}, ϕ_{2d}) are also provided. This delay is incorporated only to the falling edges of the signals (switches turn-off), while the rising edges are synchronized in order to increase the effective time slot for the modulator operations [Marq98a]. The comparators and the last-stage ADC are activated at the end of ϕ_2—using $\bar{\phi}_{2d}$ as strobe—to avoid any possible interference due to the transient response of the integrators at the beginning of sampling. This timing guarantees a single delay per clock cycle.

4.3 Specifications for the Building Blocks

The design of the ΣΔ modulator has been faced following the top-down methodology described in Section 3.1. Table 4.1 summarizes the modulator sizing achieving 13bit@4.4MS/s. Five groups of specifications are enclosed: modulator, front-end integrator, amplifier, comparator, and A/D/A converter. In this procedure, the worst-case performance has been evaluated in the presence of variations in the process (for instance, changes in device parameters), temperature, and supply. Table 4.2 shows a summary of the most significant contributions to the in-band error power. Main considerations made for this sizing are described next.

The first step of the modulator sizing is the selection of the reference voltage. In this selection, both the overloading characteristics of the modulator and the type of signal being converted must be considered. In our case, the overload level is nearly –5dBFS (see Fig. A.2), while the largest input is the –15dB discrete multi-tone (DMT) signal shown in Fig. 4.4*a*. Note that,

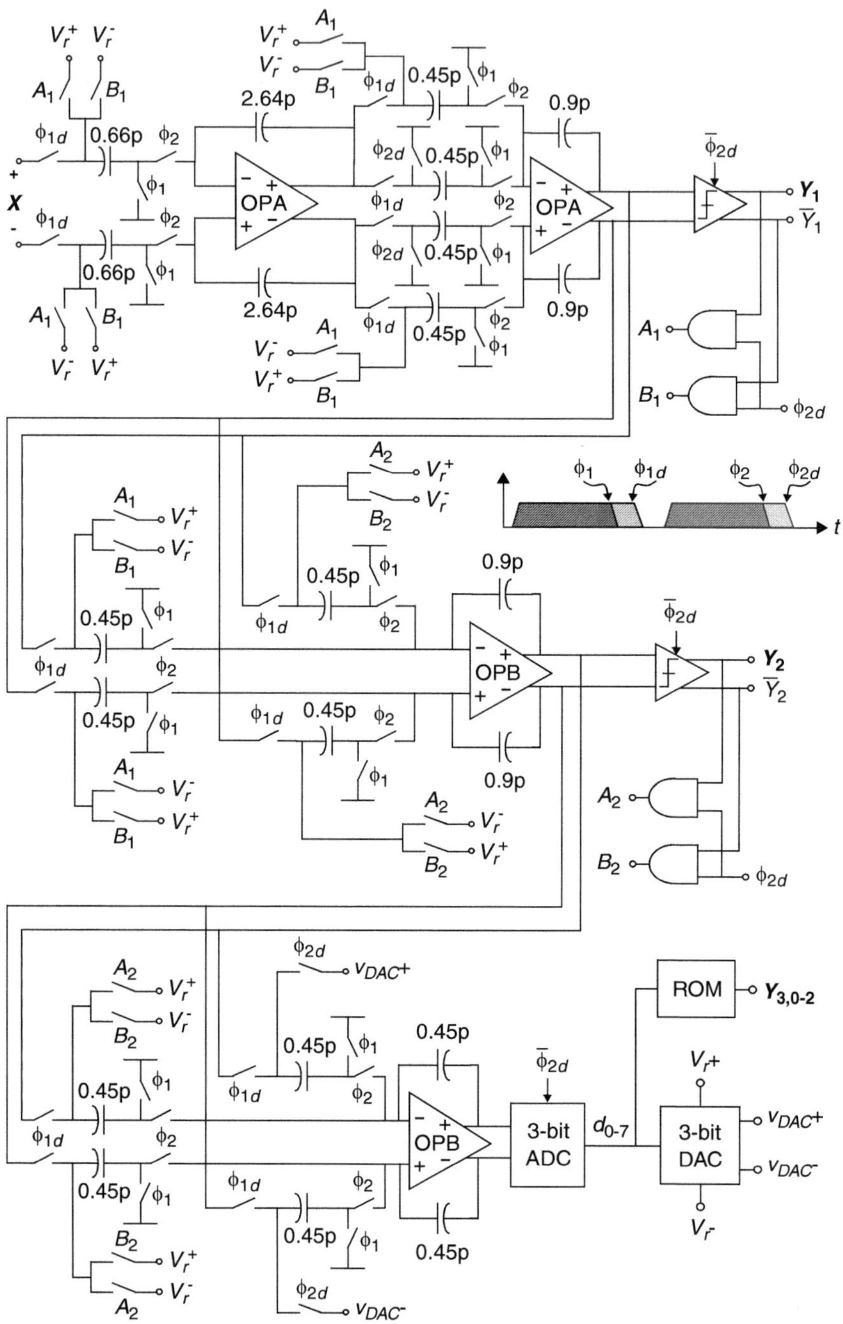

FIGURE 4.3 SC implementation of the 2-1-1 multi-bit ΣΔ modulator.

TABLE 4.1 Modulator sizing.

MODULATOR	Topology	2-1-1
	Dual-quantization	1bit / 3bit
	Oversampling ratio	16
	Clock frequency	70.4MHz
	Differential reference voltage	1.5V
	Clock jitter	15ps (0.1%)
FRONT-END INTEGRATOR	Sampling capacitor	0.66pF
	Unit capacitor	0.66pF
	Capacitor standard deviation (1-pF MiM cap)	0.05%
	Capacitor tolerance	±20%
	Bottom parasitic capacitor	1%
	Switch on-resistance	150Ω
AMPLIFIER	Open-loop DC gain	3000 (70dB)
	Gain-bandwidth product (1.5-pF load)	265MHz
	Slew rate (1.5-pF load)	800V/μs
	Differential output swing	±1.8V
	Input equivalent noise	$6\text{nV}/\sqrt{\text{Hz}}$
COMPARATORS	Hysteresis	20mV
	Offset	±10mV
	Resolution time	3ns
	Input capacitance	0.2pF
A/D/A CONVERTER	Resolution	3bit
	DAC *INL*	0.5%FS

although its power is not too high, large peaks appear from time to time, thus yielding the high crest factor [Gust00] peculiar to DMT signals (5.4 in our case). Fortunately, the duration of these peaks is short enough not to overload the modulator. In order to illustrate this, Fig. 4.4*b* shows behavioral simulation results of the modulator *SNDR* for such an input signal as a function of the reference voltage. In spite of the presence of signal peaks of approximately 1V, the modulator *SNDR* is correct up to $V_{ref} = 1.3\text{V}$ [†1]. In order to provide a safety margin, $V_{ref} = 1.5\text{V}$ was taken. In a two-stage fully-differential amplifier supplied with 2.5V, this reference voltage provides a margin of 500mV for the saturation voltage of the output transistors. As shown in Fig. 4.3, V_{ref} is implemented using differential references, so that $V_{ref} = V_r^+ - V_r^-$.

In Table 4.2 the in-band power of quantization error has been split up in its four contributions associated to: the ideal quantization error [first term in eq(A.4)], finite DC gain [first term in eq(A.5)], capacitor mismatch [second

TABLE 4.2 Main in-band error contributions.

		Nominal	Worst-Case
Quantization Noise		–88.1dB	–86.2dB
	Ideal quantization noise	–90.3dB	
	Amplifier DC-gain leakage	–99.8dB	
	Capacitor mismatching leakage (σ_C = 0.05% | σ_C = 0.1% for 1pF)	–95.4dB	–89.4dB
	DAC non-linearity error	–96.4dB	
Thermal Noise		–84.8dB	–82.2dB
	kT/C noise	–88.1dB	–86.0dB
	Amplifier noise	–87.5dB	–84.5dB
Clock Jitter		–90.1dB	
IN-BAND ERROR POWER		–82.3dB	–80.3dB
DYNAMIC RANGE		82.8dB (13.5bit)	80.8dB (13.1bit)

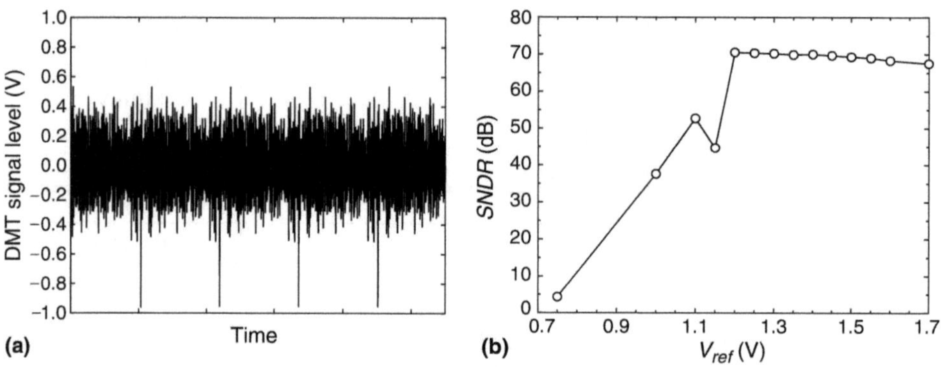

FIGURE 4.4 (a) Time-domain representation of a –15dB DMT signal; (b) *SNDR* of the converted DMT signal as a function of the reference voltage.

term in eq(A.5)], and last-stage DAC non-linearity [second term in eq(A.4)]. Note that the quantization error leakage is dominated by capacitor mismatch. Although MiM capacitors exhibit good matching—$\sigma_C = 0.05\%$ for 1-pF caps—, the use of small unit capacitors (0.66pF) for dynamic considerations increases the sensitivity of the cascade, so that we have assumed twice that value for σ_C. The contribution of the 3-bit DAC non-linearity is 6dB below the ideal quantization noise for $INL = 0.5\%\text{FS}$, which is easily achievable

1. Note that this would never be the case for a 1-V amplitude input sinewave, since it would be inside the overload region of the modulator with 1.3-V reference (see Fig. A.2).

without calibration. The noise leakage due to the amplifier DC gain is almost negligible for $A_{DC} = 3000$. However, as later discussed, this value is not be further relaxed to avoid excessive distortion due to the DC-gain non-linearity.

A small sampling capacitor ($C_S = 0.66\text{pF}$) is used in order to reduce the capacitive load of the integrators and, hence, their power dissipation. So, white circuit noise becomes dominant. An approximate expression for its in-band error power is

$$P_{CN} = P_{kT/C} + P_{op} = \frac{4kT}{C_S} \cdot \frac{1}{OSR} + \frac{2\pi \cdot GB_{eff}}{2OSR} S^t_{op} \tag{4.2}$$

where S^t_{op} stands for the input-referred thermal noise PSD of the amplifier (in V^2/Hz) and GB_{eff} is the effective amplifier gain-bandwidth product (in Hz), which during integration that can be approximated to

$$GB_{eff} \cong \frac{GB}{1 + GB/f_{on}} = \frac{GB}{1 + GB \cdot 2\pi \cdot 2R_{on}C_S} \tag{4.3}$$

with GB being the amplifier gain-bandwidth product (in Hz), R_{on} the switch on-resistance, and f_{on} the pole associated to the RC constant of the SC branch during integration.

The first contribution in eq(4.2) yields a worst-case value of -86.0dB— for maximum temperature ($+110°\text{C}$) and -20% tolerance in the capacitor value. On the other hand, for GB and R_{on} fixed according to settling considerations to 265MHz and 150Ω, respectively, the effective bandwidth is 250MHz. An equivalent thermal noise at the amplifier input of $6\text{nV}/\sqrt{\text{Hz}}$ is therefore enough to obtain a noise contribution similar to that of the kT/C noise (-87.5dB). Besides, the worst-case amplifier white noise contribution corresponds to the largest GB_{eff}, which varies along the design corners. Assuming that it can be as large as twice its nominal value (i.e., 500MHz), the worst-case amplifier noise contribution yields -84.5dB.

As stated in Section 2.3.3, the limited amplifier GB introduces basically a gain error in the integrator transfer function. This error is specially important in the integrators of the first stage of the cascade, because the quantization error of this stage will leak to the modulator output. For the architecture considered operating with $OSR = 16$, the amplifier must fulfill $GB(\text{Hz}) > 2.5 f_s$ to avoid degradation of the modulator performance due to incomplete settling, with f_s being the sampling frequency.

If the finite on-resistance R_{on} of the switch is also considered, the effective amplifier response is slowed-down, as stated in eq(4.3). This effect is illustrated in Fig. 4.5*a*, that shows behavioral simulation results for the in-band error power as a function of the normalized amplifier GB, for different values of R_{on}. The

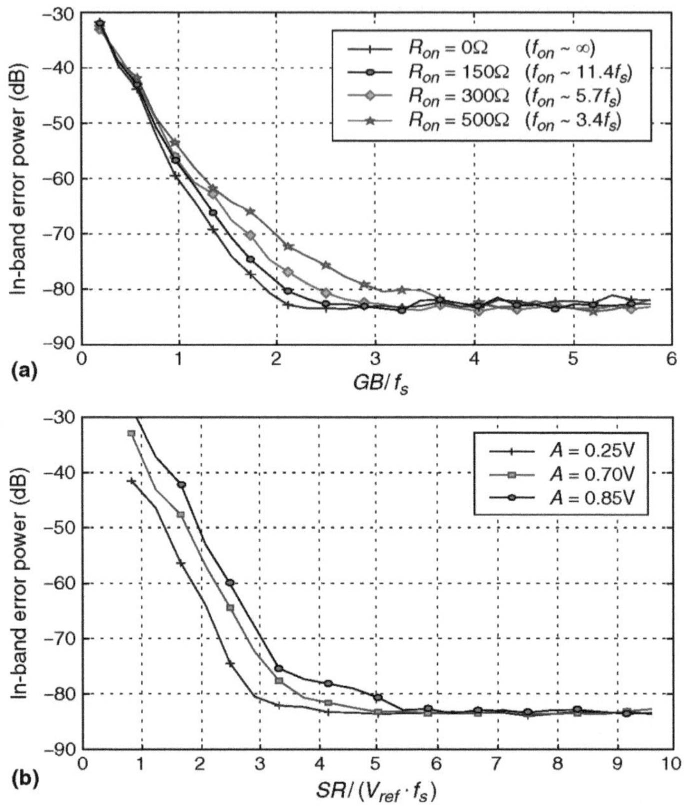

FIGURE 4.5 (a) In-band error vs. normalized amplifier GB for different values of the switch on-resistance; (b) In-band error vs. normalized amplifier SR for different input amplitudes.

corresponding values of the normalized RC-pole are also depicted. Note that, as the RC-pole decreases, the amplifier GB must be increased in order to compensate for the slow-down effect. A switch resistance of 150Ω is fixed for this design. On the one hand, as will be shown further on, this resistance can be obtained using standard CMOS transmission gates, without clock-boosting. On the other, the amplifier GB must be increased just to $GB(\text{Hz}) > 3.2f_s$ in order to maintain the modulator performance. Assuming that approximately 85% of the clock cycle is left for the integrator operation (after ensuring non-overlapping and delay in the clock phase signals), the required GB is approximately 265MHz.

The required amplifier slew rate (SR) is established guarantying that the slew-rate limited evolution at the beginning of integration and sampling is fast enough for the subsequent linear dynamic to settle to the desired accuracy. For this modulator a normalized slew rate $SR/(V_{ref} \cdot f_s) = 5$ is sufficient to ensure correct performance. However, since the operation of the front-end integrator is partially slew-rate limited, the dynamic will be also partially non-linear and appreciable distortion may arise. This effect is illustrated in Fig. 4.5*b*, where behavioral simulation results are shown for the modulator in-band error power as a function of the normalized amplifier SR, for different amplitudes of a sinewave input. Note that, for the correct conversion of an input sinewave of maximum amplitude (0.85V), the normalized SR must be increased up to 6.5. Assuming that 85% of the clock cycle is left for the integrator operation, the required SR is approximately $800\text{V}/\mu\text{s}$.

Thanks to oversampling, some specifications in Table 4.1 referring to the front-end integrator can be relaxed for the rest of integrators. Specifically, the value of the sampling capacitor in those integrators can be progressively scaled down, since their contributions to the overall kT/C noise are attenuated in the signal band. Nevertheless, matching considerations and reliability preclude using very small capacitors. In this design the scaling of the nominal C_S (0.66pF) is limited to 32%, which means that 0.45-pF unit capacitors are used in the rest of integrators. On the contrary, the input-referred white noise of the amplifiers at the modulator back-end can be considerably increased without jeopardizing performance.

A more aggressive reduction can be applied to the other circuit requirements. For instance, the amplifier DC gain of the third and fourth integrators can be reduced to 600, because the in-band powers of the respective quantization error leakages are proportional to OSR^{-5} and OSR^{-7}, and the effect of their non-linearity is negligible in comparison with that of the front-end integrator. Moreover, the slew rate can be relaxed to $350\text{V}/\mu\text{s}$, as their settling behaviors are not so important.

Table 4.3 summarizes the specifications for the four integrators in the cascade after scaling.

TABLE 4.3 Scaling of the integrator specifications (1.5-pF load).

SPECs	1st Integ	2nd Integ	3rd Integ	4th Integ
Unit capacitor	0.66pF	0.45pF	0.45pF	
Open-loop DC gain	3000 (70dB)		600 (56dB)	
Gain-bandwidth product	265MHz		210MHz	
Slew rate	$800\text{V}/\mu\text{s}$		$350\text{V}/\mu\text{s}$	
Input equivalent noise	$6\text{nV}/\sqrt{\text{Hz}}$		$50\text{nV}/\sqrt{\text{Hz}}$	
Differential output swing	±1.80V		±1.60V	

4.4 Design of the Building Blocks

4.4.1 Amplifiers

The trade-off among DC gain, dynamics, and output swing, always present in an amplifier [Raza00] [Malo01], becomes tighter in low-voltage implementations. It has been already shown that the selection of the reference voltage and the topology of the front-end amplifier are interrelated in deep-submicron cascade ΣΔMs, the reason being that large enough V_{ref} requires two-stage amplifiers in order to achieve the DC-gain and dynamic requirements. Fortunately, this is not the case for amplifiers at the modulator back-end, whose DC gain can be largely relaxed, so that a single-stage topology may be enough. Thus, in order to avoid over-sizing and optimize the power consumption, two different amplifiers have been designed: a high DC gain, high-speed amplifier for the first stage (OPA), and a modest DC gain, high-speed amplifier for the third and fourth integrators (OPB).

The circuit-level sizing tool FRIDGE [Mede99a] was used to explore the potential of different topologies, ranging from single-stage telescopic and folded-cascode amplifiers to two-stage multi-path compensated amplifiers, some including gain-boosting stages. The search criteria were the achievement of the specifications for OPA and OPB with minimum power dissipation and reduced circuit complexity.

Front-end amplifiers

OPA is implemented using a two-stage two-path compensated architecture, shown in Fig. 4.6. It uses a telescopic first-stage and both Miller and Ahuja compensation [Ahuja83] through capacitors C_c and C_{ac}, respectively. The common-mode feedback (CMFB) nets employed in the first and the second stage are dynamic, because they have no static consumption and help to circumvent voltage range problems.

A p-type input scheme has been preferred, the main reason being the possibility of cancelling the body effect in the pMOS devices—one of the mechanisms for substrate noise coupling [Arag99] [Char01]. Another reason for this choice is that, in the target technology, $1/f$ noise of nMOS devices is considerably larger than that of pMOS ones. Although $1/f$ noise usually plays a secondary role in telecom converters, since it normally does not aliases and the low-frequency region of the spectrum is commonly out of the signal band, the $1/f$ noise *PSD* of very small devices can be huge [Kung88] and sometimes poorly modelled [Zhou01], thus deserving special attention in deep-submicron implementations. This trend precludes using

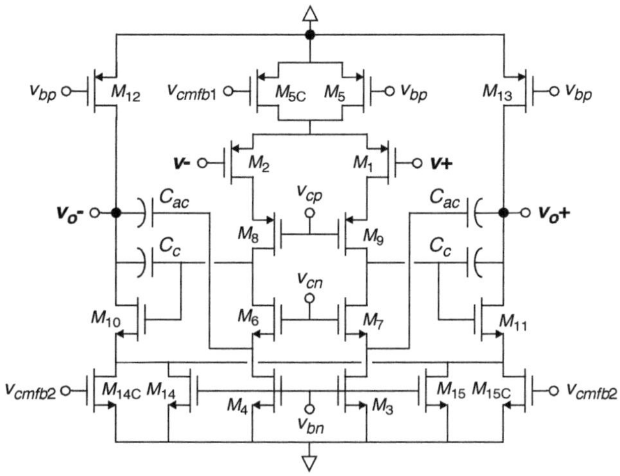

FIGURE 4.6 Two-stage two-path compensated amplifier (OPA).

minimum-length transistors, even more noticeably than if only matching considerations are taken into account. In our case, the devices contributing most to the input-referred amplifier noise are M_1, M_2 and M_3, M_4. In order make the $1/f$ noise contribution negligible, the length of those devices was increased up to $0.5\mu m$ for the pMOS and $2\mu m$ for the nMOS. In the worst case the in-band error power due to the $1/f$ noise of the front-end amplifier is $-103.6dB$, low enough not to degrade the performance.

Table 4.4 summarizes the target values imposed to the parameters of OPA during the optimization process in FRIDGE. Those featured after full amplifier sizing are included in the table. The second column corresponds to electrical simulation results in typical operation conditions. Results in the third column correspond to the worst-case value of each parameter in a corner analysis, considering fast and slow device models, $\pm 5\%$ variation in the 2.5-V supply, and temperatures in the range $[-40°C, +110°C]$.

TABLE 4.4 Electrical simulation results for OPA (1.5-pF load).

	Target	Typical	Worst-case
Open-loop DC gain	>70dB	78.6dB	73.5dB
Gain-bandwidth product	>265MHz	446.8MHz	331.5MHz
Phase margin	>55°	64.0°	57.9°
Slew rate	>800V/μs	1059V/μs	883V/μs
Differential output swing	>±1.80V	±2.09V	±1.86V
Input capacitance	<200fF	126fF	129fF
Input equivalent noise	$<6nV/\sqrt{Hz}$	$5.1nV/\sqrt{Hz}$	$5.5nV/\sqrt{Hz}$
Power consumption	<20mW	17.2mW	19.4mW

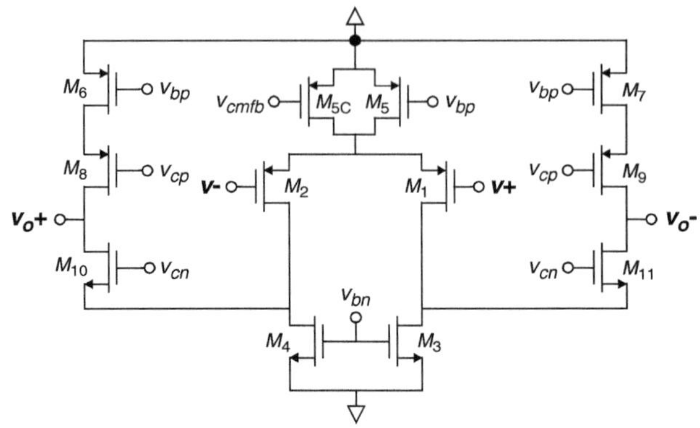

FIGURE 4.7 Folded-cascode amplifier (OPB).

Back-end amplifiers

A single-stage folded-cascode architecture was selected for amplifier OPB, which is enough to accomplish its moderate DC-gain requirement with reduced power dissipation. The amplifier schematic is depicted in Fig. 4.7.

As in OPA, a p-type differential pair is used to cancel the body effect in the input devices. A switched-capacitor CMFB net is also employed.

Table 4.5 summarizes the specifications pursued for OPB during its optimization in FRIDGE. The amplifier features, obtained by electrical simulation after full sizing, are included in the second column (typical conditions) and in the third column (worst-case value in the corner analysis).

TABLE 4.5 Electrical simulation results for OPB (1.5-pF load).

	Target	Typical	Worst-case
Open-loop DC gain	>56dB	58.0dB	56.8dB
Gain-bandwidth product	>210MHz	393.5MHz	331.7MHz
Phase margin	>60°	70.3°	67.7°
Slew rate	>350V/μs	377V/μs	373V/μs
Differential output swing	>±1.60V	±1.97V	±1.72V
Input capacitance	<350fF	300fF	343fF
Input equivalent noise	$<50\text{nV}/\sqrt{\text{Hz}}$	$4.1\text{nV}/\sqrt{\text{Hz}}$	$5.1\text{nV}/\sqrt{\text{Hz}}$
Power consumption	<7.5mW	6.6mW	6.9mW

Non-linearities

Besides the former aspects concerning the amplifier design itself, two non-idealities deserve special attention in this low-voltage implementation, since they can critically affect the performance of the ΣΔ modulator:

- ***DC gain non-linearity.*** When the amplifier output voltage swings, the drain-to-source voltage of the output transistors changes, and so does

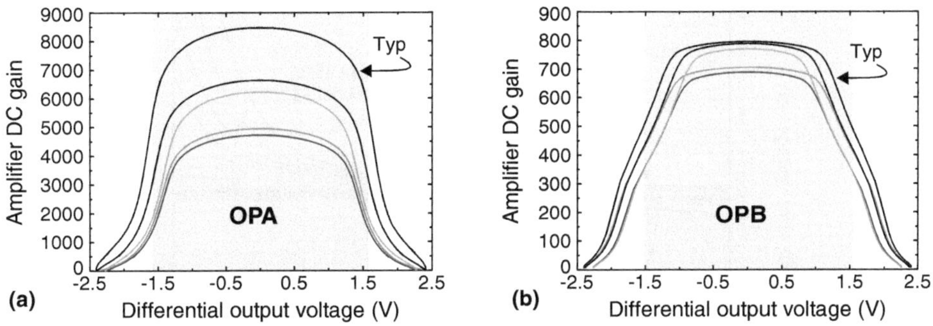

FIGURE 4.8 DC gain non-linearity for: (a) OPA, (b) OPB at several process corners.

the output impedance. This effect (illustrated in Fig. 4.8 for OPA and OPB for different design corners) translates into a dependence of the open-loop DC gain on the output voltage, so that the DC gain reaches its maximum at the central point and decreases as the output approaches the rails.

Such a non-linearity can be modeled by a second-order polynomial dependence of the gain on the output voltage (see Section 2.6.2), but this is only valid for weak non-linearities; i.e., for small voltage excursions around the central point. On the contrary, in this 2.5-V implementation it is expected that small-gain regions of the DC curves are often visited during the normal operation of the modulator (shadowed areas in Fig. 4.8).

In order to accurately account for this non-linearity in behavioral simulations, we can resort to a table look-up procedure from amplifier DC curves obtained by electrical simulation, whose data are included in ASIDES [Mede99a] through a fast-convergence iterative procedure, illustrated in Fig. 4.9.

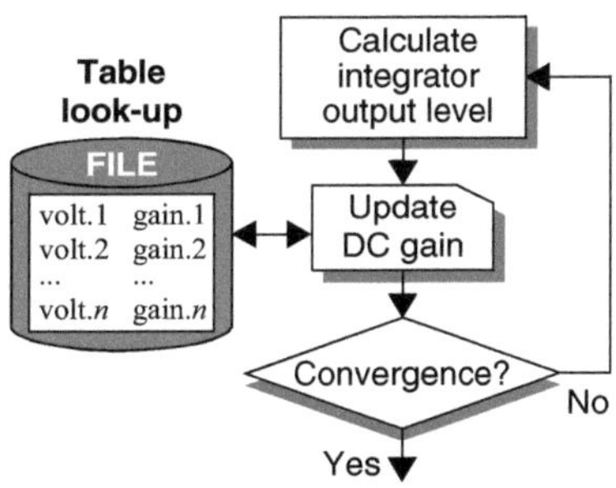

FIGURE 4.9 Flow graph to account for the DC-gain non-linearity.

- ***Non-linear settling.*** From the point of view of settling errors, linear integrator dynamics are not mandatory, but slew-rate limited dynamics can be allowed as long as they are followed by a linear dynamic that settles to the desired accuracy. This trade-off between bandwidth and slew rate has been exploited in the design of amplifiers to fulfill the final settling requirements with reduced power consumption. Besides settling itself, distortion arising from partially slew-rate limited dynamics needs also to be precisely estimated in behavioral simulations.

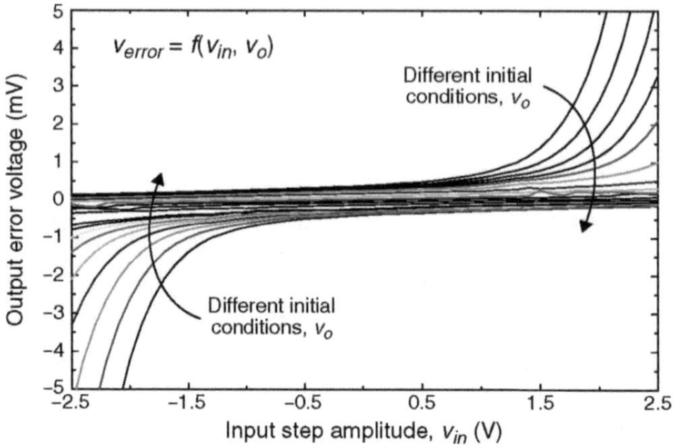

FIGURE 4.10 Two-dimensional table look-up for the integrator transient response.

To this purpose, ASIDES includes the non-linear model for the transient response of integrators with single-stage amplifiers (such as OPB) presented in Section 2.3.1. However, integrators in the modulator first stage employ a two-stage amplifier with non-constant slew rate (OPA). In order to accurately account for them in behavioral simulations, a table loop-up procedure has also been employed from integrator transient responses obtained by electrical simulation. As illustrated in Fig. 4.10, the files used for the table look-up contain information of the error voltages in the integrator outputs at the end of one integration-sampling process (one clock cycle), for a given initial condition and input step amplitude.

4.4.2 Comparators

The resolution specifications for the comparators in the first and second stage are not very demanding: offset and hysteresis smaller than 10mV and 20mV, respectively. However, the maximum comparison time is only 3ns—a quarter of the worst-case clock period. For this reason, the latched comparator in Fig. 4.11 has been adopted. It includes a differential-pair input transconductor [Yin92], which attenuates the impact of common-mode interferences, a CMOS regenerative stage, and a SR latch. In this circuit, the small voltage imbalance created across the nMOS switch controlled by ϕ_{2d} during the reset phase is rail-to-rail regenerated during the positive-feedback comparison phase. The latter starts when $\bar{\phi}_{2d}$ goes high, thus making the latch react before the integrator output changes at the beginning of ϕ_1. This strategy avoids using an extra SC stage at the comparator front-end. Differenced supply paths are used for the pre-amplifier and the regenerative latch in order to reduce the sensitivity to digital switching noise and supply bounce.

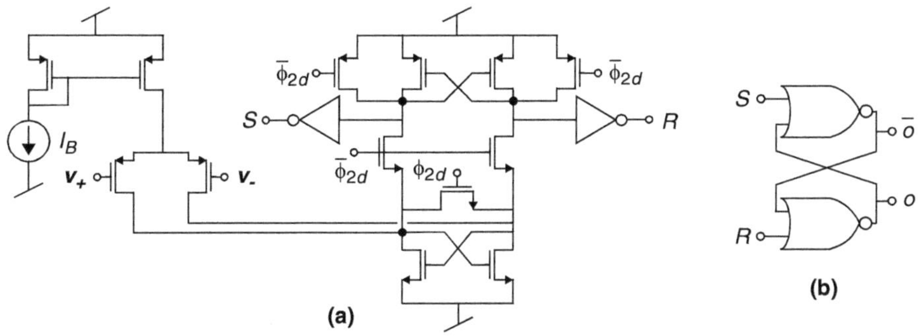

FIGURE 4.11 Comparator: (a) Pre-amplifier and regenerative latch, (b) SR latch.

Monte Carlo and corner analysis have been used to characterize the comparator after full sizing. Table 4.6 summarizes its worst-case performance.

TABLE 4.6 Worst-case electrical simulation results for the comparator.

Hysteresis	127.5μV	Offset	6.3mV
Resolution time, LH	3.9ns	Resolution time, HL	2.8ns
Input capacitance	0.1pF	Power consumption	0.3mW

4.4.3 Switches

The design of the CMOS switches has been tackled with two main considerations in mind. First, the on-resistance heavily affects the integrator dynamic, slowing down its transient response. Second, the switch on-resistance can be highly dependent on voltage in low-voltage implementations. The sampling process with such a non-linear resistance causes dynamic distortion [Yu99] at the ΣΔM front-end, the more evident the larger the signal frequency. Among the solutions to these problems, resorting to larger aspect ratios increases parasitics and power dissipation, whereas including clock-boosting [Wu98] [Bult00] increases complexity and leads to a less robust design.

According to settling considerations, resistances in the range of 150Ω can be tolerated in combination with the amplifier dynamics. In our process, such a value can be obtained using standard-threshold CMOS transmission gates, with no need for clock boosters. The sizes of the pMOS and nMOS devices—$36.5/0.25$ and $8.5/0.25$, respectively—have been selected to equalize their transconductances, keeping the resistance of the transmission gate as linear as possible. Fig. 4.12*a* shows its nominal DC curve.

In order to evaluate the distortion, the non-linear sampling has been extensively simulated using the differential circuitry in Fig. 4.12*b*. Note that the distortion will be mainly determined by switches S_{1p} and S_{1n} (connected to the

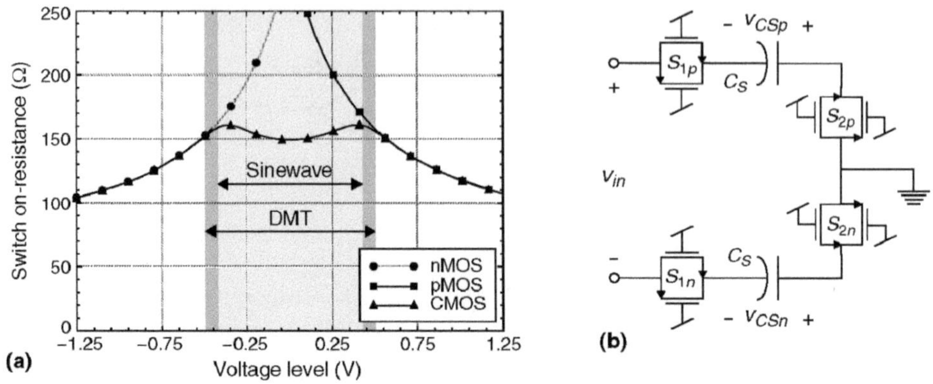

FIGURE 4.12 Switch on-resistance: (a) DC curve, (b) Circuit for evaluating distortion.

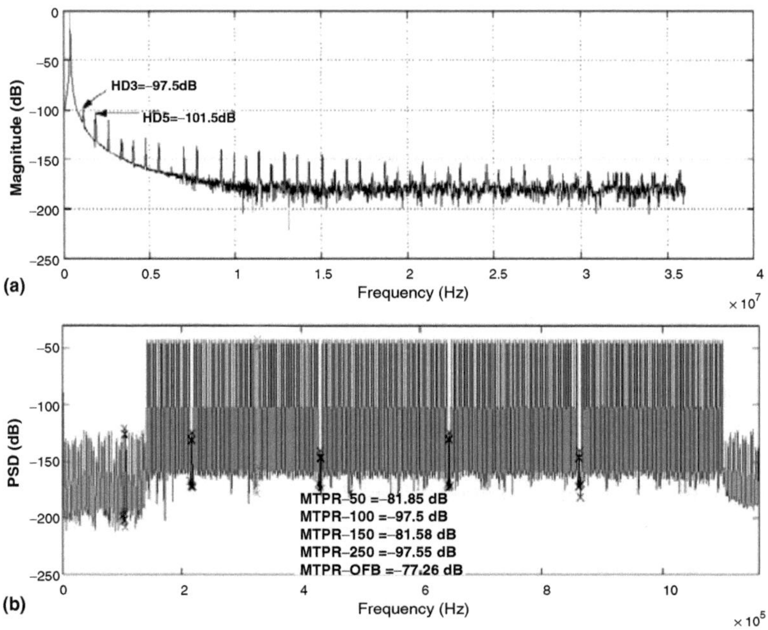

FIGURE 4.13 Worst-case dynamic distortion introduced by the switches for a: (a) $0.85V_{pd}$ @ 366kHz sinewave input signal, (b) –15dB DMT input signal.

input), whereas S_{2p} and S_{2n} are connected to the central voltage that is constant. Electrical simulations have been performed at $f_s = 70.4\text{MHz}$ to compute the first five in-band harmonics for a $0.85V_{pd}$ @366kHz input sinewave. Also the DMT signal in Fig. 4.4*a* has been considered. Fig. 4.13 shows the worst-case results obtained for both type of inputs during the corner analysis. The worst-case total harmonic distortion (THD) is –96dB for the input sinewave and the maximum multi-tone power ratio ($MTPR$) [Gust00] of

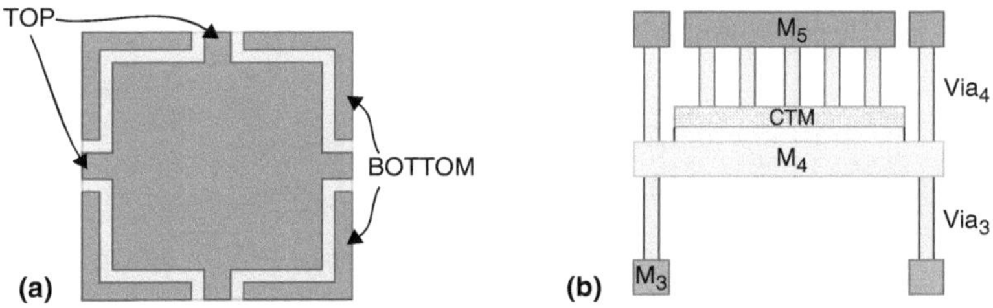

FIGURE 4.14 Metal-insulator-metal capacitor: (a) Top view, (b) Side view.

the converted DMT signal is $-81\,\text{dB}$. Both figures are small enough for our application, so that clock-boosting is not required.

4.4.4 Capacitors

Capacitor ratios implementing the integrator weights of the SC ΣΔM use unit elements. Taking into account thermal noise and mismatch, the value of the unit capacitor is fixed to 0.66pF for the front-end integrator and to 0.45pF for the rest. The total number of unit elements required for the differential implementation of the modulator is only 2×16. They have been arranged and laid out in a way similar to that shown in Table 3.9 and Fig. 3.20.

Capacitors use metal-insulator-metal (MiM) structures available in the intended technology, which allows thin inter-metal oxide between metal-4 and metal-5. Side and top views of a MiM capacitor are illustrated in Fig. 4.14. With these structures, the 0.66-pF unit capacitor occupies approximately $27\mu\text{m} \times 27\mu\text{m}$ and the 0.45-pF unit is $22\mu\text{m} \times 22\mu\text{m}$. Their bottom-plate parasitics are only around 1% of the nominal capacitance, what significantly helps to limit the equivalent capacitive load of the integrators. The mismatch exhibited by MiM capacitors is approximately 0.1% for 1-pF capacitance.

4.4.5 A/D/A converter

The multi-bit quantizer in the modulator last stage converts the 4th-integrator output into digital with 3-bit word-length ($Y_{3,\,0\text{-}2}$ in Fig. 4.3) and then back into an analog representation ($v_{DAC}+$, $v_{DAC}-$) to close the 3rd-stage loop. The 3-bit A/D/A converter is implemented with a flash ADC and a resistive-ladder DAC. A partial view of it is shown in Fig. 4.15.

A/D converter The ADC has a fully-differential flash architecture and compares the differential output of the fourth integrator v_i^+, v_i^- with voltage references v_r^+, v_r^- generated in the DAC. A static input scheme has been adopted, instead of

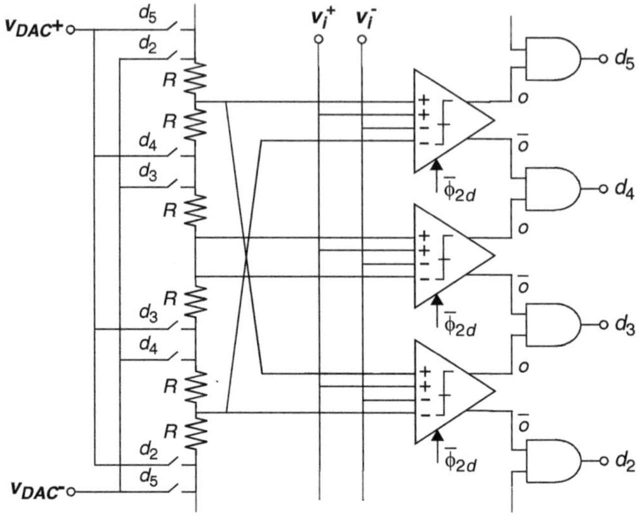

FIGURE 4.15 Partial view of the A/D/A converter.

the dynamic one used in the 0.35-μm prototype in Chapter 3 (see Fig. 3.21). This way the use of storage capacitors at the ADC input is avoided, resulting in a considerable area saving.

Comparators in the flash ADC are similar to those used in the modulator first and second stage (see Fig. 4.11). Given that, in this case, the differential input signal is compared with a differential voltage reference, an extra differential pair is added to the pre-amplifying stage. The size of the input transistors is also decreased in order to reduce the capacitive load to the fourth integrator (connected to the ADC during both integration and sampling).

During the reset phase ϕ_{2d} the difference $(v_i^+ - v_i^-) - (v_r^+ - v_r^-)$ is amplified, to be regenerated by the latches during $\bar{\phi}_{2d}$. The thermometer code provided by the seven comparators in the ADC is then translated into a 1-of-8 output code using AND gates.

Electrical simulations under Monte Carlo and corner analysis have been carried out for characterizing the worst-case comparator performance. Results are summarized in Table 4.7.

TABLE 4.7 Worst-case electrical simulation results for comparators in the ADC.

Hysteresis	34.7μV	Offset	9.9mV
Resolution time, LH	2.7ns	Resolution time, HL	1.7ns
Input capacitance	50fF	Power consumption	0.3mW

D/A converter

The resistive-ladder DAC is used to:

- generate the voltage references for the ADC operation, and
- generate the analog output of the A/D/A converter, by selecting one of the voltage references generated in the ladder through the 1-of-8 output code of the ADC ($d_{0\text{-}7}$).

The ladder consists of 14 segments of 50-Ω unit resistors connected between reference voltages $V_r^+ = 2.0\text{V}$ and $V_r^- = 0.5\text{V}$, thus providing a differential full scale of $\pm 1.5\text{V}$ with 2.1-mA current consumption. The reference voltages to the DAC are obtained directly from the on-chip analog supply by simple voltage division, whereas those provided to the SC integrators are obtained from a dedicated on-chip voltage generator. The underlying reasons for this will be exposed later on in Section 4.4.7.

The value of R is selected so that the settling error in the generated voltage references is low enough for a correct comparison in the ADC. Additionally, the settling error of the voltage sampled by capacitors in the fourth integrator during ϕ_2 is also small enough.

Resistors have been implemented using unsalicided $n+$ poly, due to its low dependence on voltage and temperature variations. According to the technological data of the intended process, standard deviations ranging from 0.5% to 1% can be expected for resistances around $1\text{k}\Omega$ using this layer. Thus, in order to guarantee that $INL \leq 0.5\%\text{FS}$, each of the 50-Ω resistors is obtained by connecting larger devices in parallel (eleven 550-Ω resistors).

4.4.6 Clock phase generator

Fig. 4.16 shows the clock driver that generates the non-overlapped clock phases—ϕ_1, ϕ_2—from a master clock signal CLK. Delayed versions of the phases—ϕ_{1d}, ϕ_{2d}—are also generated in order to attenuate signal-

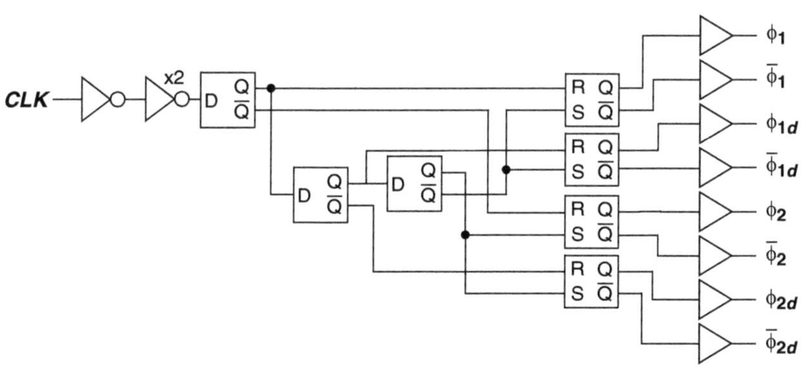

FIGURE 4.16 Clock phase generator and drivers.

dependent charge injection [Lee85]. As shown in Fig. 4.3, the delay is incorporated only to the turn-off of the switches (falling edges of the signals), while their rising edges are synchronized in order to increase the time slot available for sampling and integration [Marq98a]. Complementary versions of the phases are also generated to control the CMOS switches. All signals are properly driven at the output using a buffer tree that equalizes the differences in capacitive load among the phases.

After ensuring reliable non-overlapping time and phase delay, the worst-case effective phase eye is 6ns, which means that approximately 85% of the clock period is left for the modulator operation.

4.4.7 Auxiliary blocks

Besides the ΣΔ modulator itself, the implementation of the prototype includes several on-chip auxiliary blocks, which are described next.

Reference voltage generator

The reference voltages required for the modulator operation—namely, $V_{ref} = V_r^+ - V_r^- = 1.5\text{V}$ and the central voltage $V_{cm} = 1.25\text{V}$—can be obtained from the 2.5-V supply as

$$V_r^+ - V_r^- = \frac{3}{5}(V_{DD} - V_{SS}) \qquad V_{cm} = \frac{1}{2}(V_{DD} - V_{SS}) \tag{4.4}$$

The circuit in Fig. 4.17 has been used to implement the required ratios ($3/5$ and $1/2$) in robust manner and generate these references on-chip.

The analog ground V_{cm} is obtained from a simple resistive ladder, whereas an asymmetric OTA (amplifier V_{cm}-OP; see Fig. 4.18*a*) is used to buffer the voltage to the ΣΔM. References V_r^+, V_r^- are obtained from a fully-differential amplifier in inverting configuration with gain $3/2$. When

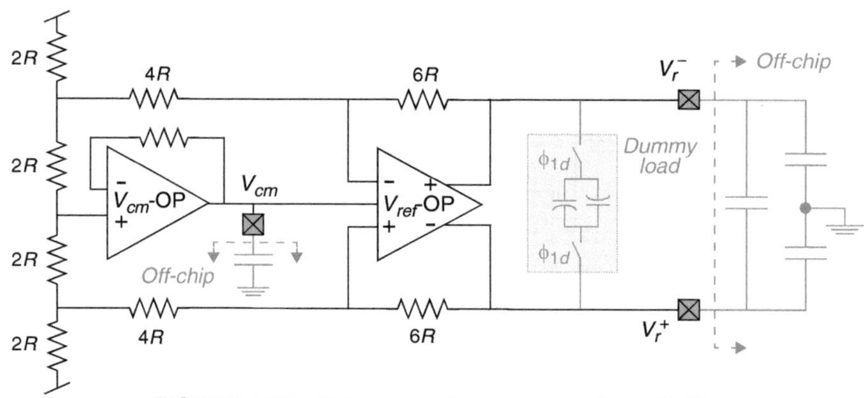

FIGURE 4.17 Reference voltage generator ($R = 3\text{k}\Omega$).

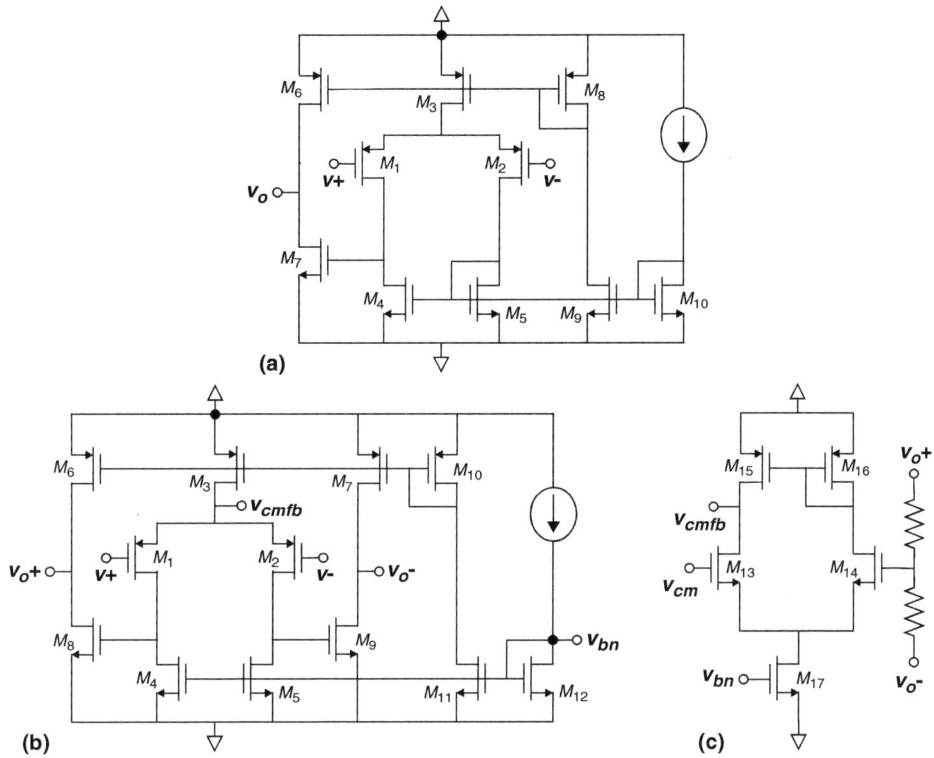

FIGURE 4.18 (a) Single-ended OTA (amplifier V_{cm}-OP); (b) Fully-differential OTA and (c) Static CMFB net (amplifier V_{ref}-OP).

connected to the resistive ladder, the 3/5-ratio is implemented. The fully-differential amplifier V_{ref}-OP consists of a symmetric OTA with static CMFB net (see Fig. 4.18*b* and Fig. 4.18*c*).

The main design considerations for the generation of these references are a fast settling [Pies02] and a low output impedance in the V_r^+, V_r^- lines in order to avoid dynamic distortion at the integrators [Ribn91b]. The latter prevents using the same reference voltages in the resistive-ladder DAC (see Section 4.4.5). A 7-Ω maximum output impedance is obtained along the signal band through the combined use of an on-chip resistive amplifier and two big external capacitors. An extra external capacitor is connected between the reference voltages, valued according to the (pad + wire + lead + pin) parasitics, so that the spurious components around half the sampling frequency are removed from the differential reference voltage.

References V_r^+, V_r^- are connected to the integrator sampling capacitors during integration (ϕ_{2d}), but disconnected from them during sampling (ϕ_{1d}), what leads to a ringing in that signals due to charge-redistribution effects. As

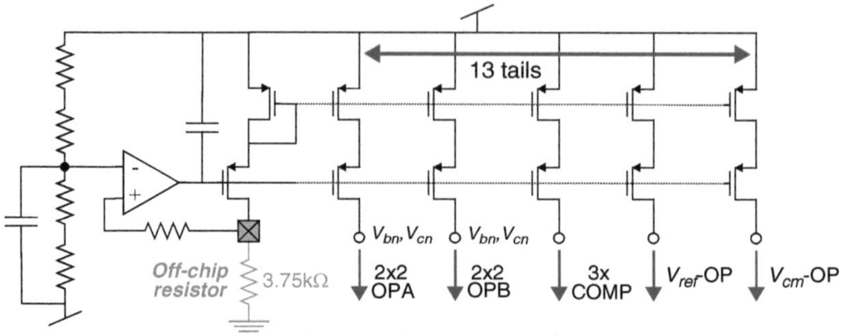

FIGURE 4.19 Master current generator.

shown in Fig. 4.17, a dummy capacitive load, which connects to the generator during ϕ_{1d}, has been included to compensate for this effect.

Master current generator The bias currents needed in the ΣΔ modulator are all internally generated, as shown in Fig. 4.19, from a single master current of 330μA, provided by an external 3.75-kΩ resistor. This current is mirrored and properly scaled to bias the amplifiers in the integrators, the pre-amplifying stages of the comparators, and the two amplifiers in the reference voltage generator. A single tail current is used for biasing the comparators in the ADC.

Anti-aliasing filter The anti-aliasing filter at the modulator input has been also included on-chip. It consists of a simple 2nd-order *RC* filter, whose bandwidth can be programmed to accomplish either ADSL (up to 1.1MHz) or ADSL+ (up to 2.2MHz) band requirements.

4.5 Layout and Prototyping

The prototype floorplan, illustrated in Fig. 4.20, was carefully studied and the following measures, valid for non-epi resistive substrates, were adopted [Arag99] [Char01]:

- Increased distance among analog and digital blocks, in order to attenuate the impact of the switching activity.
- Use of separate analog, mixed, and digital supplies, which are distributed on-chip through distinguished low-impedance paths.
- Analog supplies—V_{DDA}, V_{SSA}—are employed for the current biasing of analog blocks (mainly, amplifiers and pre-amplifying stages of comparators) and in the substrate and wells of the analog section.

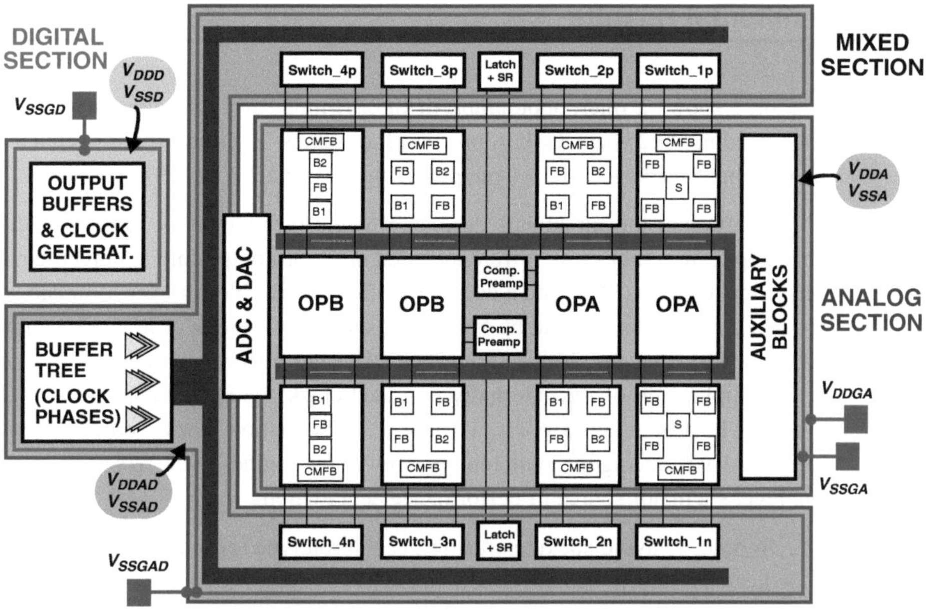

FIGURE 4.20 Illustration of the prototype floorplan.

- Mixed supplies—V_{DDAD}, V_{SSAD}— are used in the integrator switches, the dynamic CMFB nets of the amplifiers, the regenerative latches of comparators, and the buffer tree driving the clock phases.
- Digital supplies—V_{DDD}, V_{SSD}— are used in the clock phase generator and the buffers driving the output of the modulator stages.
- Placement of guard-rings (with dedicated pads and pins) surrounding the different chip sections, in order to avoid the spreading of switching noise and provide a quiet substrate for the sensitive analog blocks.
- Preserved layout symmetry and extensive use of common-centroid techniques, aimed at gaining insensitivity to common-mode interferences.
- An U-shaped bus distributes the clock signals along the die, including shielding to reduce cross-talk and provide a low-impedance return path.
- An U-shaped bus structure is used to distribute the supplies, references, and central voltage along the analog section.
- Extensive use of on-chip decoupling, including a mixed on/off-chip decoupling scheme for the analog supply [Inge97].
- The pad ring is divided into four parts (analog, mixed, digital, and digital IOs) by blocking diode cells for power separation and reduced cross-talk.
- Use of multiple bonding techniques, in order to reduce wire inductance and reduce supply bounce effects.

4.6 Experimental Results

Fig. 4.21 shows a microphotograph of the ΣΔ modulator and auxiliary blocks fabricated in a 0.25-μm CMOS process. It occupies 2.78mm^2 (pads excluded) and has been packaged in a 44-pin plastic quad flat pack.

Apart from the modulator described here, other blocks pertaining to the final application (not shown) were included in the prototype chip, among them a phase-locked loop (PLL) and a decimation filter. These blocks were arranged so that the ΣΔ modulator could be tested as a stand-alone block or in combination with the PLL and the digital filter. The PLL can be programmed to 2 × or 4 × multiplication of the externally provided clock signal, whereas the digital filter can be programmed to ↓8, ↓16, or ↓32 decimation. These configurations are aimed at reducing the switching activity of the digital IO buffers at the pad-pin level—a major source of performance degradation.

In order to avoid socket parasitics, each prototype sample is mounted onto a dedicated 4-layer printed circuit board (PCB), shown in Fig. 4.22, which includes typical measures for signal integrity, such as separate analog, mixed, and digital ground planes, intensive decoupling and filtering, proper impedance termination, etc. [Berr99]. Ten modulator samples have been tested, all being operative and exhibiting similar performance.

The input signal is generated by a high-resolution (–100dB *THD*) sinusoidal source with floating differential output (Tek SG5010), its common-mode voltage referenced to the on-chip generated central voltage V_{cm}. The output samples, either from the modulator bit-streams or from the decimation filter, are acquired by a digital tester (HP82000) that also provides the master

FIGURE 4.21 Microphotograph of the prototype in 0.25-μm CMOS.

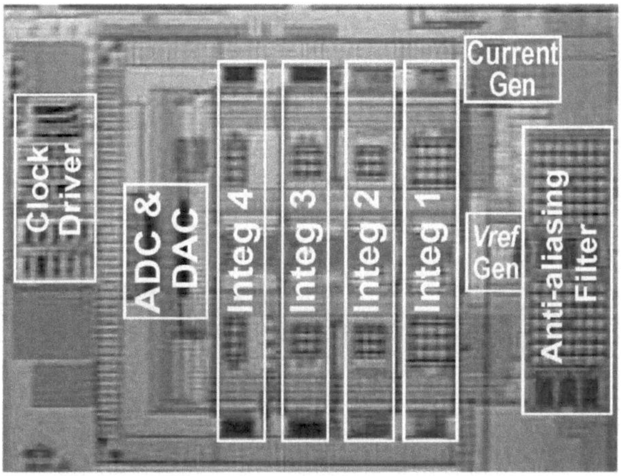

FIGURE 4.22 4-layer PCB used for testing.

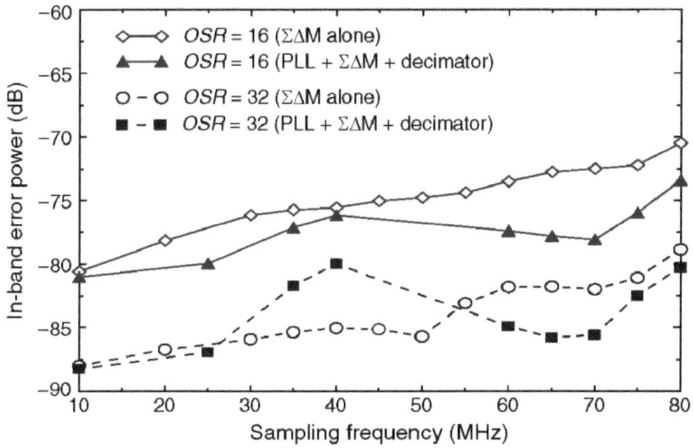

FIGURE 4.23 In-band error power vs. modulator sampling frequency.

clock stimulus and the supply voltage. The test set-up is controlled with C-routines from a work-station, where MATLAB is used to process the transferred output samples. When the ΣΔ modulator is tested as a stand-alone block, the error cancellation logic and the decimation are also implemented by software.

Fig. 4.23 shows the total in-band error power as a function of the modulator sampling frequency, for the nominal oversampling ratio ($OSR = 16$) and twice this value. Two curves are plotted for each value of OSR, corresponding to clock-rate acquired (ΣΔM alone) output samples and decimated (PLL + ΣΔM + decimator) output samples. Note that in the former case, the in-band error power increases as the sampling frequency does. This effect, explained by the increasing switching noise injected by the IO buffers †2, causes a degradation of around 9dB in performance at the nominal sampling frequency (70.4MHz). Nevertheless, when both the PLL and the decimator are used—so that the

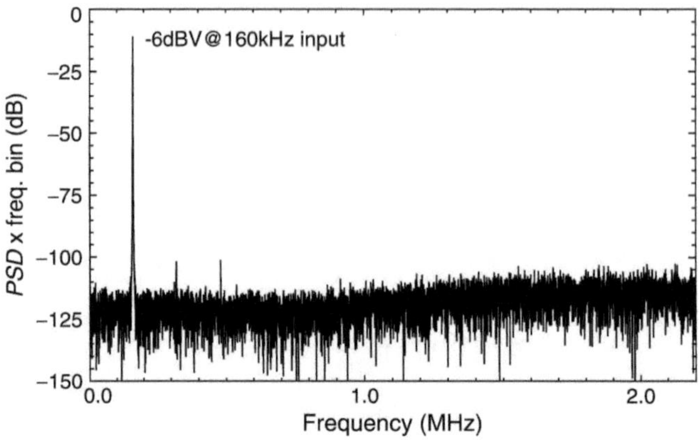

FIGURE 4.24 Baseband output spectrum for a -6dBV@160kHz input sinewave measured at nominal sampling frequency.

switching frequency of input and output buffers are divided by 4 and 16, respectively—, the loss of performance is reduced up to 3dB. Note also that, although the digital filter activity generates an increase of the in-band error power at intermediate sampling frequencies, its impact is largely suppressed at the nominal rate, thus demonstrating the validity of the decoupling schemes used, especially at the reference voltages. A similar behavior is obtained for 32 oversampling ratio.

Fig. 4.24 shows a 16k-sample FFT of the decimated converter output for a -6dBV@160kHz input sinewave. Despite the large signal level, no significant harmonic distortion is observed. In fact, in Fig. 4.24 the spurious-free dynamic range ($SFDR$) is 90dB, whereas the THD computed up to the fifth harmonic is -87dB, so that the $SNDR$ almost coincides with the SNR.

Fig. 4.25 shows the measured $SNDR$ curves for both $OSR = 16$ and $OSR = 32$, the error power being computed in the ADSL+ band (from 30kHz to 2.2MHz) and in the ADSL band (from 30kHz to 1.1MHz), respectively. The measured dynamic range is 78dB (12.7bit) for $OSR = 16$ and 85dB (13.8bit) for $OSR = 32$, with $SNDR$ peaks of 72.5dB and 80dB, respectively.

The good linearity of the converter also manifests as low integral and differential non-linearity (INL and DNL, respectively). Both curves are shown

2. The switching activity can be considered the responsible for the degradation at high sampling frequencies. Indeed, the measured performance loss is smaller the higher the temperature —considered range is $[-40°\text{C}, +110°\text{C}]$—, what can be explained by the slow-down effect of temperature on the digital circuitry, thus reducing the power of high-frequency components associated to the switching activity. Moreover, the case of maximum temperature is also the worst case for the integrator settling, so that settling errors can be discarded as a relevant source of degradation. Also, an increase of 30% in the master bias current does not modify the measured in-band error versus sampling frequency.

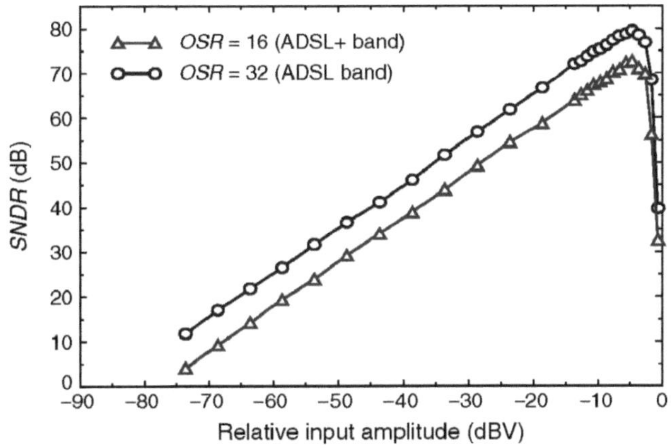

FIGURE 4.25 Measured *SNDR* curves in the 1.1-MHz (ADSL) and 2.2-MHz (ADSL+) bands.

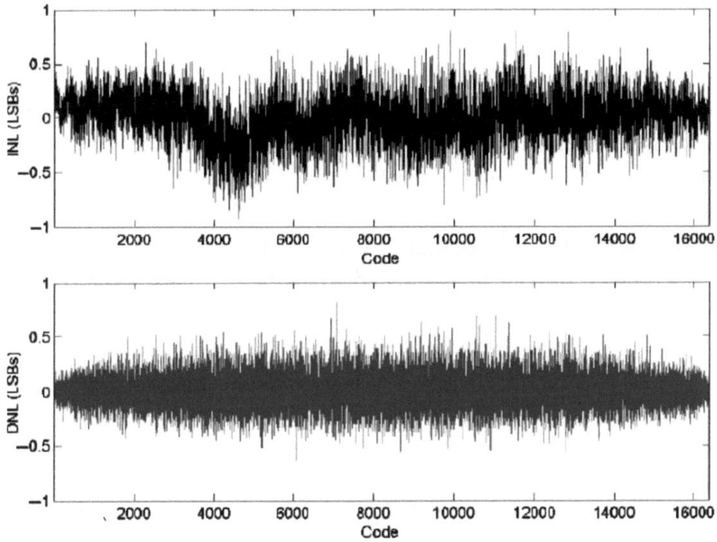

FIGURE 4.26 Measured *INL* and *DNL*. Vertical-axis units are LSBs of 14 bits.

in Fig. 4.26. They have been obtained applying the code-histogram method [Doer84] [IEEE01] to 89 output data records, each one containing 8192 consecutive output samples for a 0.8V@59.62kHz input sinewave. The input signal frequency is selected so that an integer number of signal periods fits into the record length (8192). Due to the uncertainty in the actual signal frequency, it may happen that the output vector does not contain an integer number of signal periods, thus corrupting the code histogram and, hence, the *INL*/*DNL* measurements. In order to avoid this problem, the first of the output records is examined and truncated to include an integer number of signal periods. The rest

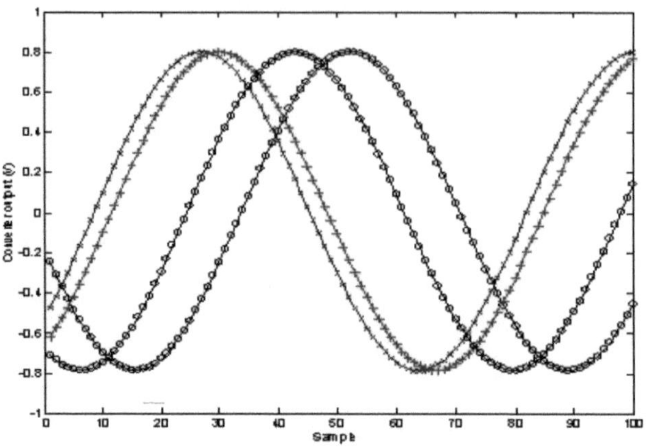

FIGURE 4.27 First 100 samples of several converter output records for a 0.8V@59.62kHz input sinewave.

of records are also truncated by the same amount, so that all the records have the same number of samples. Thus, the only difference among the data records is the phase of the sinewave, which can not be controlled (see Fig. 4.27). However, this situation helps to hit the converter output codes in a more uniform way, thus requiring fewer samples than in the case in which all the samples were taken consecutively. Then the output code histogram is obtained and the actual width of each code bin can be estimated from it. The units for *INL* and *DNL* in Fig. 4.26 are LSBs of 14 bits in a full scale of $\pm 0.8\text{V}$; i.e., $1\text{LSB} = (2 \cdot 0.8)/(2^{14} - 1) = 97.7\mu\text{V}$. The measured *INL* and *DNL* are within ±0.85 and $\pm 0.80\text{LSB}_{14\text{bit}}$, respectively. These low values are obtained thanks to the use of dual-quantization (with no need for correction/calibration of the DAC non-linearity) and a careful control of the distortion introduced by the front-end amplifier and switches.

4.7 Performance Summary

Table 4.8 summarizes the prototype features in the ADSL band ($30\text{kHz} - 1.1\text{MHz}$) and in the ADSL+ band ($30\text{kHz} - 2.2\text{MHz}$), at $32\times$ and $16\times$, respectively. The measured dynamic ranges are 85dB (13.8bit) and 78dB (12.7bit), respectively. Note that, although the prototype is sensitive to the on-chip switching activity, the measured resolution is degraded by only 3dB (0.5bit) from the expected value.

The total power dissipated in the ΣΔ modulator is 65.8mW, from which 12% (7.8mW) is consumed in the digital blocks, 4.5% (3mW) in the mixed-

TABLE 4.8 Summary of measured prototype performance.

		1.1-MHz band (ADSL)	2.2-MHz band (ADSL+)
Topology		2-1-1(3b)	
Oversampling ratio		32	16
Reference voltage		1.5V	
Clock frequency		70.4MHz	
Digital output rate		2.2MS/s	4.4MS/s
Dynamic range		85dB	78dB
ENOB		13.8bit	12.7bit
SNDR peak		80.0dB	72.5dB
SFDR		90dB	
THD		–87dB	
Power consumption		65.8mW	
	Analog	55.0mW (includes 10mW in auxiliary blocks)	
	Mixed	3.0mW	
	Digital	7.8mW	
FOM_1		2.10	2.25
FOM_2		16.98	7.39
Active area		2.78mm^2 (pads excluded)	
Supply voltage		2.5V	
Technology		0.25-μm MS CMOS (1P5M) [MiM]	

signal section, and 83.5% (55mW) in the analog section. The latter includes the consumption of the on-chip auxiliary blocks (reference voltage generator, master bias generator, and anti-aliasing filter), which accounts for 10mW (18%) of the total analog power.

The corresponding values of the figures-of-merit FOM_1 and FOM_2— defined in equations (1.70) and (1.71), respectively— are also included in Table 4.8. The value of FOM_1 is 2.10 in the ADSL band and 2.25 in the ADSL+ band. The respective values of FOM_2 are 16.98 and 7.39.

Note that this prototype in 2.5-V 0.25-μm CMOS considerably improves the performance of that in 3.3-V 0.35-μm CMOS presented in Chapter 3 (see Table 3.12). In spite of the supply voltage reduction, the achieved resolution is around 1-bit larger, whereas the power dissipated in the modulator core is reduced in approximately 30%. This results in a significant improvement of the values of FOM_1 (the better, the smaller) and FOM_2 (the better, the larger).

Besides this, the following issues are also worth mentioning when comparing both prototypes:

- None of the prototypes uses low-Vt transistors or on-chip voltages higher than the nominal supply.
- The ratio full-scale/supply is maintained in both ΣΔ modulators—in our case $2 \cdot 2/3.3 \cong 2 \cdot 1.5/2.5 \cong 1.2$. That is, the prototype in 0.25-μm CMOS achieves the same relative full scale as the 0.35-μm prototype, although the scaling of the transistors threshold voltages with the technology shrinking is smaller than that of the supply voltage.
- The prototype in 0.25-μm CMOS has been designed to be compliant with industrial applications, considering technological corners and temperatures in the range of $[-40°C, +110°C]$.
- The ΣΔ modulator in 0.25-μm CMOS has been integrated together with its decimation filter in the same die. The performance of the complete ΣΔ ADC has been also validated.

4.8 Performance Comparison with the State of the Art

Table 4.9 shows the state of the art in high-speed low-pass ΣΔMs implemented in CMOS processes. Considered prototypes achieve resolutions larger than 11 bits at rates above 1MS/s. Their features are summarized including information about the technology and the modulator architecture. The list of abbreviations used can be found in Section 1.7 on page 56. For comparison purposes, the prototype presented here is included at the bottom of the table.

It must be noted that the two prototypes here exhibit the largest full scale relative to the supply voltage, which is 1.2 in both cases. The closest reported values are 1.1 in [Jiang02] (1.8-V supply) and [Feld98] (3.3-V supply), and 1.0 in [Grilo02] (2.7-V supply) and [Vleu01] (2.5-V supply). From the viewpoint of the practical implementation of a ΣΔM, the larger the value of this ratio, the more demanding the requirements of output swing on amplifiers and of linearity on amplifiers and switches, what obviously impacts the power consumption needed to fulfill them. Although this ratio thus affects the global features of the designs, it is not taken into account in the following comparison with the state of the art—albeit it would benefit both prototypes—, in order to use the two figures-of-merit (FOM_1 and FOM_2) commonly established in literature.

Fig. 4.28*a* to Fig. 4.28*d* depict the main performance parameters of each high-speed ΣΔM against its digital output rate. The performance of the prototypes is summarized in terms of the achieved $ENOB$, the normalized power consumption, and the values of the figures-of-merit.

TABLE 4.9 State-of-the-art high-speed low-pass ΣΔ prototypes in CMOS technologies.

REFs	*ENOB* (bit)	*DOR* (S/s)	*OSR*	Architecture	Technology	V_{ref} (V)	V_{supply} (V)	Power (W)	FOM_1	FOM_2
[Bran91b]	12	2.1M	24	2-1(3b)	1μm STD	1.5	5	41m	4.77	2.14
[Yin94b]	15.82	1.5M	64	2-1-1	2μm BiCMOS	2.5	5	180m	2.07	69.61
[Broo97]	14.5	2.5M	8	2(5b)-0(12b) [5-b flash with DDS, 12-b pipeline]	0.6μm MS [2P]	1	5 & 3	500m	8.63	6.70
[Feld98]	12.5	1.4M	16	2-2-2(1.5b)	0.7μm MS [2P]	1.8	3.3	81m	9.99	1.45
[Marq98a]	14.8	2M	24	2-1-1	1μm MS [2P]	2	5	230m	4.03	17.66
[Geer99]	15	2.2M	24	2-1-1	0.5μm MS [2P]	0.9	3.3	200m	2.77	29.48
[Mede99b]	13	2.2M	16	2-1-1(3b)	0.7μm STD	2	5	55m	3.05	6.70
[Fuji00]	15	2.5M	8	2(4b)-1(4b)-1(4b) [Bi-DWA in all stages]	0.5μm MS [2P, low-Vt]	2	5	105m	1.28	63.81
[Geer00]	15.8	2.5M	24	3rd-or(4b) [DWA]	0.65μm MS [2P]	1	5	295m	2.07	68.85
	11.5	12.5M	8					380m	10.50	0.69
[Mori00]	13	2.2M	24	2-2-2	0.35μm MS [2P]	1.2	3.3	150m	8.32	2.46
	12			2-2(5b)		1		99m	10.99	0.93
[Lamp01]	13.5	1.56M	32	2-2(3b) [LR]	0.35μm –	–	2.5	50m	2.76	10.47
[Vleu01]	15.5	4M	16	2(5b)-2(3b)-1(3b) [2S, P-DWA]	0.5μm MS [2P]	1.25	2.5	150m	0.81	142.94
[Grilo02]	13	1M	32	2nd-or(4b) [1st-or DEM]	0.35μm BiCMOS	1.4	2.7	11.88m	1.45	14.10
[Gupta02]	14.6	2.2M	29	2-1-1(2b) [2S]	0.35μm STD	1.5	3.3	180m	3.29	18.81
[Jiang02]	13.8	4M	8	5th-or(4b) [hybrid FIR-IIR, DWA]	0.18μm STD	1	1.8	149m	2.61	13.63
[Kuo02]	13.7	1.25M	12	4th-or(4b) [FB-FF, I-DWA]	0.25μm MS [MiM]	0.675	2.5	100m	6.01	5.53
	13.0	2M						105m	6.41	3.19
[Reut02]	14	2.5M	32	5th-or(1.5b) [FFS, LR]	0.25μm STD	–	2.5	24m	0.59	69.79
[Veld02]	11.3	4M	40	4th-or(1.5b) [RC-active/GmC, FFS]	0.18μm STD	–	1.8	6.6m	0.65	9.62
[Lee03]	14.16	1M	64	2-2	0.35μm MS [2P]	0.9	1.8	150m	8.20	5.57
	12	2M	32						18.31	0.56
[Mill03]	12.8	1.25M	18	2nd-or(6b) [mDWA]	0.18μm MS [dual-gate, MiM]	1.2	2.7	30m	3.30	5.51
	11.7	3.84M	12					50m	4.00	2.04
Work in Chapter 3 **(Work A)**	12.9	2M	16	2-1-1(4b)	0.35μm STD	2	3.3	73.7m	4.82	3.96
	11.9	4M						78.3m	5.12	1.86
This Work	13.8	2.2M	32	2-1-1(3b)	0.25μm MS [MiM]	1.5	2.5	65.8m	2.10	16.98
	12.7	4.4M	16						2.25	7.39

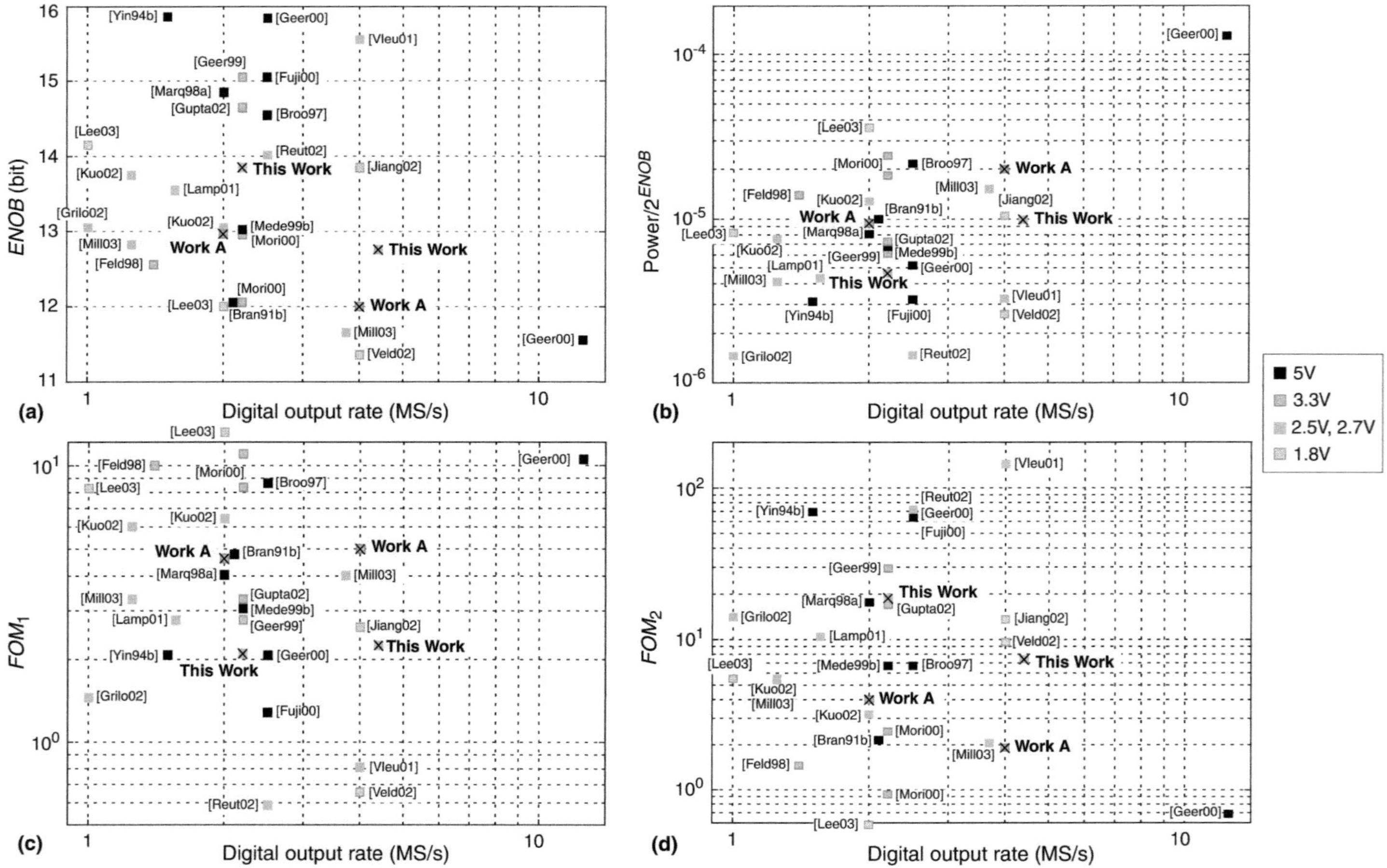

FIGURE 4.28 State of the art in high-speed ΣΔMs: (a) $ENOB$, (b) Normalized power consumption, (c) FOM_1, and (d) FOM_2 vs. digital output

Note from Fig. 4.28 that prototype in 2.5-V 0.25-μm CMOS presented in this chapter outperforms the 2.5-V ICs reported in [Grilo02], [Kuo02], and [Lamp01], and the 2.7-V design in [Mill03]. The ΣΔM here achieves larger bandwidth and resolution and better figures-of-merit, except for the value of FOM_1 in [Grilo02], which benefits from the higher efficiency of a BiCMOS process to achieve a low power consumption. Only four low-voltage ICs [Vleu01] [Jiang02] [Veld02] [Reut02] improve the global performance of this 0.25-μm prototype.

4.9 Summary

This chapter presents a high-speed ΣΔ modulator targeted to be incorporated in a CPE modem for ADSL and ADSL+ applications. The employed architecture is a 4th-order cascade (2-1-1 ΣΔM) with dual-quantization. Multi-bit quantization (3 bits) is used only in the last stage of the cascade, so that quantization error is reduced without jeopardizing the modulator linearity. The use of correction or calibration mechanisms is therefore avoided.

The prototype is implemented in a 0.25-μm CMOS process with MiM capacitors and operates from a single 2.5-V supply, with no need for higher on-chip voltages or low-Vt transistors. Standard transmission gates are used for the switches. The anti-aliasing filter and circuitry for the generation of the reference voltage are also integrated on-chip. The requirements of the main blocks have been derived using the models and equations developed in Chapter 2 for the effect of non-idealities in SC ΣΔMs. The design of the different cells has been discussed and presented at transistor level. Special attention has been paid to critical issues in low-voltage implementations, such as non-linearities associated to the amplifier DC gain and to the switch on-resistance.

The power consumption of the ΣΔM and its auxiliary blocks (clock phase and reference voltage generators, and I/O buffers) is limited to only 65.8mW and the total area occupation is 2.78mm^2. The prototype achieves an effective resolution of 13.8bit in the ADSL band and of 12.7bit in the ADSL+ band. High linearity is also obtained. The measured spurious-free dynamic range is 90dB, whereas integral and differential non-linearity measured with the code-histogram method are within ±0.85 and $\pm 0.80\text{LSB}_{14\text{bit}}$, respectively. The IC performance compares well with state-of-the-art high-speed ΣΔMs published in open literature.

This prototype proves the feasibility of implementing high-performance ΣΔ ADCs in low-voltage deep-submicron CMOS processes, with no need for correction techniques for improved device matching, non-standard low-Vt transistors, or higher on-chip voltages that can involve lifetime issues.

CHAPTER 5

A ΣΔ Modulator with Programmable Signal Gain for Automotive Sensor Interfaces

DURING THE LAST FEW YEARS there has been a significant increase in the use of electronic sensory systems for the automotive [Flem01]. This is partially fuelled by advances in micro-electro-mechanical systems (MEMS) technology and by the possibility to combine sensors together with DSPs in the same package and even in the same chip [Eddy98]. Linked to this trend, the need arises to design CMOS analog front-ends (AFEs), and specifically ADCs, capable of coping with the stringent requirements of the automotive industry.

Fig. 5.1 shows a typical conceptual block diagram for an automotive sensor chip. On the one hand, the converter must be compliant with a very large temperature range—typically [–40°C,+175°C]—and with many other hostile environmental issues [John99] [Malc00] [Wang01]. On the other hand, the AFE must cope with weak signals (ranging from microvolts to hundreds of millivolts) in the presence of large temperature and process-dependent offset. The changing input signal amplitude can be handled through programmable gain in the pre-amplifier + ADC system—similar to what happens in multi-purpose sensors [Wang01]. Given the stringent noise requirements that must be usually imposed on the pre-amplifier, the use of an ADC with selectable signal gain can considerably simplify the architecture and design of the pre-amplifier. Also, large accuracy and bandwidths up to 10kHz ~ 20kHz are needed at the ADC to increase the quality of the information delivered to the DSP, and hence the smartness of the system [Eddy98].

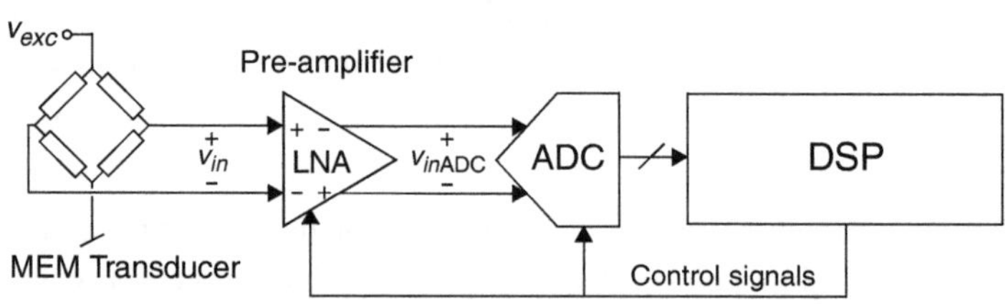

FIGURE 5.1 Conceptual block diagram of a 'smart' sensor chip.

In order to handle these challenges, it is advised to use ΣΔ-based ADCs rather than Nyquist ADCs due to several reasons:

- They are better suited than full Nyquist ADCs to achieve high resolution (16bit ~ 17bit) in the band of interest with moderate power consumption.
- The action of feedback renders ΣΔMs very linear, and high linearity is a must for automotive applications.
- The behavior of some sensing devices fits well to ΣΔ-based ADCs, thus easing partial or total integration of the sensor within the converter [Kulah00] [Maki02].
- ΣΔMs can incorporate programmable signal gain without significant performance degradation.

However, detailed modeling of circuit non-idealities and involved design plans are needed to take advantage of all these potentials while addressing the stringent environmental and robustness requirements. Actually, several ΣΔMs that digitize 20kHz ~ 25kHz signals with medium-to-high resolution have been reported [Rabii97] [Coban99] [Yang03] [Yao04]. To the best of our knowledge, the highest resolution (18.7bit) is featured by the modulator in [Yang03], whereas the lowest power consumption is reported in [Yao04], which targets 14bit within a 20-kHz signal bandwidth. In all these cases, the modulator signal gain is fixed to unity.

This chapter describes a chopper-stabilized SC 2-1 cascade ΣΔ modulator for automotive sensor interfaces in a 0.35-μm CMOS technology. To obtain a better fitting with the characteristics of the different sensor outputs, the circuit can be digitally programmed to yield four signal gain values—$\times 0.5$, $\times 1$, $\times 2$, and $\times 4$—and has been designed to operate within the stringent environmental conditions of automotive electronics—$[-40°C, +175°C]$. In order to relax the amplifiers dynamic requirements for the different modulator signal gains, switchable capacitor arrays are used for all the capacitors in the first integrator. The circuit is clocked at 5.12MHz and the overall power consumption is 14.7mW from a single 3.3-V supply. Experimental results show an overall dynamic range of 110.1dB within a 20-kHz signal bandwidth and 113.8dB for 10-kHz signals. These performance features place the reported circuit at the cutting edge of state-of-the-art high-resolution ΣΔ modulators. Also, the herein presented modulator IC is one of the few high-resolution circuits with embedded programmable signal gain reported to date.

The chapter is organized as follows. Section 5.1 describes general design considerations. Section 5.2 applies those considerations to the proper selection of the modulator architecture and obtains a set of optimized specifications for the constituent building blocks. These specifications are inputs for the electrical design of the circuit cells, which is covered in Section 5.3. Finally, experimental results are presented and the performance of the chip is compared with current state-of-the-art designs.

5.1 Basic Design Considerations

The in-band error power (IBE) of a ΣΔM can be approximately expressed as

$$IBE \cong P_Q + P_{CN} + P_{nl} + P_{st} \tag{5.1}$$

where P_Q, P_{CN}, P_{nl}, and P_{st} stand for the in-band error powers of quantization error, circuit noise, non-linearity errors, and defective settling error, respectively. ΣΔMs can be designed such that the quantization error dominates; i.e., such that $P_{CN} + P_{nl} + P_{st} \ll P_Q$. In our case, and given the high accuracy target, making the quantization error dominate would lead to huge power consumption and area occupation. Instead, it is more efficient to make the design such that $P_Q \cong P_{CN} + P_{nl} + P_{st}$. Two basic needs arise related to this criterion:

- To make a comparative balance among the several error contributions.
- To find expressions linking the dominant errors to circuit design parameters.

Regarding the balance among error contributions, and owing to the limited bandwidth of the application (20kHz), it can be assumed that $P_{st} \ll P_{CN}$ will be easily fulfilled. Besides, since the available supply voltage grants large enough room for signal excursion, it can be assumed that $P_{nl} \ll P_{CN}$. Under these assumptions, the basic design criterion reduces itself to equalize the in-band contributions of quantization error and circuit noise; i.e., $P_Q \cong P_{CN}$.

Let us now focus on the expression of the in-band error power associated to circuit noise as a function of the circuit design parameters. Consider for this purpose that the front-end integrator of the programmable-gain ΣΔM can be represented by the two-branch SC integrator in Fig. 2.41. In this conceptual schematic the input and feedback branches employ different capacitors in order to implement a modulator signal gain $G = g_1/g_1' = C_{S1}/C_{S2}$.

As discussed in Sections 2.4.2 and 2.4.3, the main sources of circuit noise are the thermal and flicker noise in the amplifier, and the thermal noise associated to the switches on-resistance. Also, noise coming from the reference voltage generation block must be accounted for as a combination of white noise and flicker noise. Flicker noise will be handled at architectural level by applying chopper techniques [Malc00] [Wang01], so that only white circuit noise remains to dictate the electrical design considerations. From eq(2.125), its input-referred in-band error power can be approximated to

$$P_{CN,\,in} \cong \frac{4kT}{C_{S1}OSR} \cdot \left(1 + \frac{C_{S2}}{C_{S1}}\right) + S^t_{op}\frac{2\pi GB_i}{2OSR} \cdot \left(1 + \frac{C_{S2}}{C_{S1}}\right)^2 + \\ + S^t_{ref}\frac{2\pi GB_{ref}}{2OSR} \cdot \left(\frac{C_{S2}}{C_{S1}}\right)^2 \tag{5.2}$$

where S^t_{op} and S^t_{ref} stand for the input-referred thermal noise PSD of the amplifier and the reference (in V^2/Hz), respectively, GB_i is the amplifier gain-bandwidth product during integration (in Hz), and GB_{ref} is the bandwidth of the reference buffer (in Hz). Note that the kT/C noise from the switches is multiplied by factor 2 if compared to eq(2.125) in order to account for the fully-differential implementation. Besides, the amplifier gain-bandwidth product during integration is given by $2\pi GB_i = g_m/C_{eq,\,i}$, with $C_{eq,\,i}$ being

$$C_{eq,\,i} = C_P + C_{S2}(1+G) + C_L\left[1 + \frac{C_P + C_{S2}(1+G)}{C_I}\right] \tag{5.3}$$

If referred to the ΣΔM output, the in-band error power due to white circuit noise can be obtained as $P_{CN,\,out} = P_{CN,\,in} \cdot G^2$, thus yielding

$$P_{CN,\,out} \cong \frac{4kT}{C_{S2}} \cdot \frac{1}{OSR} \cdot (1+G) + \frac{2\pi GB_i}{2OSR} \cdot S^t_{op} \cdot (1+G)^2 + \frac{2\pi GB_{ref}}{2OSR} \cdot S^t_{ref} \tag{5.4}$$

Note from eq(5.4) that the kT/C noise and the amplifier noise contributions increase with the modulator signal gain. If $G > 1$ this means that a larger C_{S2} and a smaller S^t_{op} than in the typical situation (with $G = 1$) are required in order to compensate for the increased noise. What's more, the amplifier equivalent load increases as well—see eq(5.3)—and, hence, the power budget to meet the integrators dynamic requirements will need to be considerably increased.

Seeking minimum power consumption dictates setting G as close as possible to unity. Actually, for $G = 1$ the two branches in Fig. 2.41 can be merged into one, further reducing P_{CN} and $C_{eq,\,i}$. This is the choice commonly found in literature. However, such a choice for a sensor A/D interface like that shown in Fig. 5.1 implies that the programmability must be incorporated into the pre-amplifier, what usually renders its design more complicated.

In the application under consideration, the automotive sensor interface must accommodate a huge range of signal amplitudes (130-dB below full scale) within a 20-kHz bandwidth. For this purpose, the pre-amplifier is designed with a fixed gain of $\times 10$, whereas the ΣΔ modulator can be digitally programmed to yield four signal gain values—$\times 0.5$, $\times 1$, $\times 2$, and $\times 4$. This solution simplifies the architecture and design of the pre-amplifier, which in this case consists of a resistor-feedback instrumentation topology—a priori a less demanding design than using a programmable-gain ($\times 5$, $\times 10$, $\times 20$, and $\times 40$) amplifier based on switchable elements (capacitors or resistors), given the stringent noise and linearity requirements. Table 5.1 summarizes the expected maximum signal levels at the pre-amplifier input (v_{in}), at the ΣΔM input ($v_{in\Sigma\Delta M}$), and at the ΣΔM output ($v_{out\Sigma\Delta M}$) for the different cases.

TABLE 5.1 Maximum signal levels at the sensor interface for the different ΣΔM signal gains.

ΣΔM signal gain, G	$v_{in,peak}$	$v_{in\Sigma\Delta M,peak}$	$v_{out\Sigma\Delta M,peak}$
0.5	0.283V	2.83V	1.41V
1	0.141V	1.41V	1.41V
2	0.071V	0.71V	1.41V
4	0.035V	0.35V	1.41V

5.2 Architecture Selection and High-Level Sizing

The ΣΔM architecture has been selected among cascade candidates pertaining to the expandible $2-1^{L-2}$ family described in Appendix A. An instance of this family is univocally described by three parameters: the oversampling ratio (OSR), the modulator order (L), and the internal quantizer resolution (B).

How are the values of these parameters chosen? A first observation is that, for every value of L and B, as OSR increases, there is a transition from one region where $P_Q \gg P_{CN}$ into another where $P_Q \ll P_{CN}$. This is illustrated in Fig. 5.2, which shows the effective resolution obtained by behavioral simulation for three different ΣΔMs as a function of OSR. The same front-end integrator was considered in the three modulators. Note from Fig. 5.2 that there is a transition for each curve from a slope of $(L+0.5)$bit/octave—IBE dominated by P_Q—to another of 0.5bit/octave—IBE dominated by P_{CN}. The intersection of these lines defines the breakpoint where $P_Q \cong P_{CN}$. The only way to shift these breakpoints up is by enlarging the sampling capacitor, since kT/C noise ultimately limits P_{CN}.

This observation provides rationale supporting the choice of $P_Q \cong P_{CN}$, namely:

- Within the region dominated by quantization error, the sampling capacitor, and hence the demands on building block dynamics, area, and power, are larger than needed.

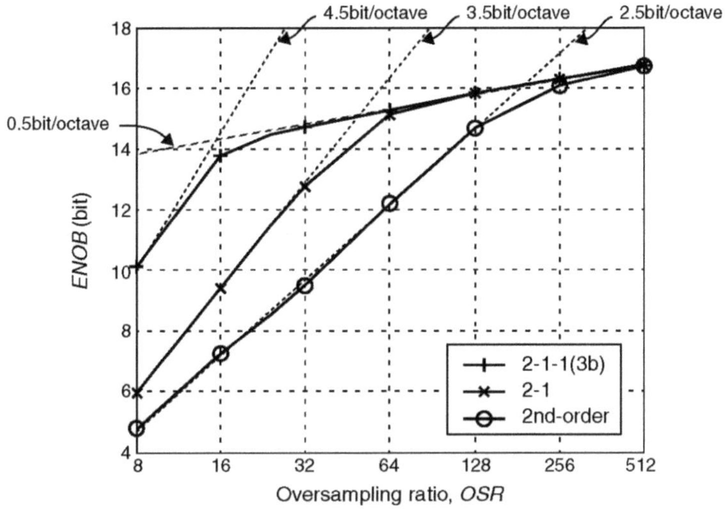

FIGURE 5.2 *ENOB* vs. *OSR* for three ΣΔMs: a 2-1-1(3b) cascade ($L = 4, B = 3$), a 2-1 1-bit cascade ($L = 3, B = 1$), and a 2nd-order 1-bit ΣΔM.

- Within the region dominated by white noise, oversampling becomes less efficient and either L or B, or both, may allow a reduction without significant impact on resolution.

Actually, most state-of-the-art circuits are designed to operate close to this breakpoint.

A second qualitative observation is related to the choice of OSR. On the one hand, the condition $P_Q \cong P_{CN}$ defines an univocal relationship between each $\{L, B\}$ pair and OSR. On the other hand, since the target bandwidth for the ΣΔM is medium/low, constraints on OSR are not strong. This means that there are fairly large degrees of freedom to set L, OSR, and B. As a rule-of-thumb, since the targeted resolution is only moderately high, the oversampling ratio does not need to be extremely high. This means that the breakpoint in Fig. 5.2 can be moved to the left, providing room for many possible combinations of B and L.

The actual selection of the ΣΔM architecture is driven by power minimization. An analytical procedure has been developed to estimate the power consumption of cascade ΣΔMs pertaining to the 2-1^{L-2} family. Appendix B details the underlying expressions, which contemplate both architectural and technological features. The procedure schematically consists of the following steps:

- For given values of L, OSR, B, V_{ref}, and G, we calculate P_Q and select C_{S2} so that $P_Q + P_{CN}$ is smaller than the maximum allowed in-band error power. At this step, the noise leakages due to capacitor mismatch, finite amplifier DC gain, or errors in multi-bit quantizers are also contemplated.

- $C_{eq,\,i}$ is estimated from eq(5.3). Then, a linear settling model with settling time constant $C_{eq,\,i}/g_m$ is used to estimate the required amplifier transconductance, taking into account that it takes a number $\ln(2^{ENOB})$ of time constants to settle within $ENOB$ resolution.
- The amplifier transconductance is related to its power dissipation, for which the candidate amplifier topology must be known a priori. Suitable selections are closely linked to the process technology: supply voltage, minimal device length, etc. Potential choices are folded-cascode amplifiers for supply voltages above 3V and two-stage amplifiers below 2.5V [Raza00].
- Once the first integrator power dissipation is estimated, that of the remaining integrators—which have, in practice, less demanding specifications— can be estimated as a fraction of it. The overall modulator power is then obtained by adding up all the contributions, together with the dynamic power in the SC stages.

5.2.1 Modulator architecture

Previous considerations have been applied to estimate the power consumption required by a large number of ΣΔ architectures in order to fulfill the target specifications of $DR \geq 110\text{dB}$ at 40kS/s. All ΣΔMs considered correspond to single-bit cascades of the $2\text{-}1^{L-2}$ family or 2nd-order single-bit loops.

Table 5.2 shows the three ΣΔMs with the lowest estimated power dissipation for each modulator signal gain. The comparison of the different architectural alternatives led us to choose the 2-1 cascade ΣΔM with $OSR \geq 128$. Note from Table 5.2 that the 2-1 ΣΔM (shown in Fig. 5.3) obtains the best results, except for $G = 4$. In this case, the lowest power consumption is obtained by a 2nd-order ΣΔM with $OSR = 512$, the third-order cascade being the second one in the ranking. However, the 2nd-order loop does not qualify for some values of G.

5.2.2 SC implementation

Fig. 5.4 shows the fully-differential SC schematic of the selected ΣΔM. Note that the SC branch connected to the input signal in the first integrator uses double sampling to achieve an extra signal gain of two, without increasing circuit noise [Enz96]. The other branch receives the DAC outputs. Making use

TABLE 5.2 Outcome of the ΣΔM architecture selection procedure.

ΣΔM signal gain, G	Order, L	Oversampling ratio, OSR	Estimated power consumption (mW)
0.5	3	128	6.77
	3	256	8.12
	4	128	8.27
1	3	128	8.48
	3	256	9.77
	4	128	10.32
2	3	128	12.69
	3	256	13.85
	2	512	13.88
4	2	512	23.24
	3	128	24.21
	3	256	25.02

of the spare connection of this branch, an external DC signal (V_{off}) can be applied during ϕ_1 to center the sensor signal in the modulator full-scale range. This solution renders unnecessary a third branch for offset compensation; it can hence be eliminated, and results in no further increase of thermal noise.

The integrator in the second stage of the modulator incorporates only two input branches for implementing weights g_3, g_3', and g_3'', since the values used (see Fig.5.3) allow to distribute g_3 between the two SC branches.

The modulator operation is mainly controlled by two non-overlapped clock phases: ϕ_1 (sampling) and ϕ_2 (integration). Delayed versions of these phases (ϕ_{1d}, ϕ_{2d}) are also provided in order to attenuate signal-dependent charge injection. As illustrated in Fig. 5.4, the delay is incorporated only to the falling edges of the signals (turn-off of the switches), while the rising edges are

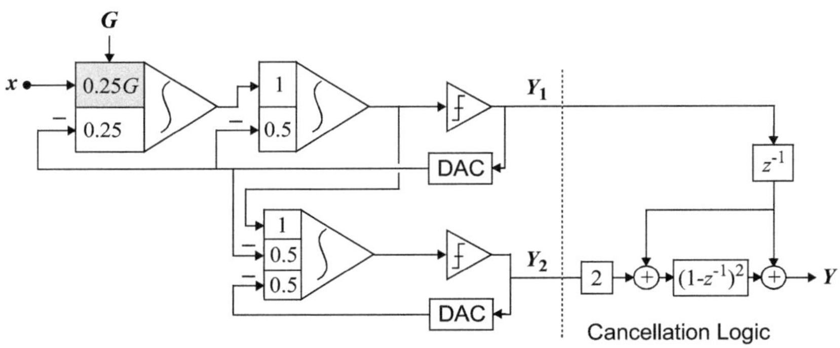

FIGURE 5.3 Block diagram of the programmable signal gain 2-1 cascade ΣΔM.

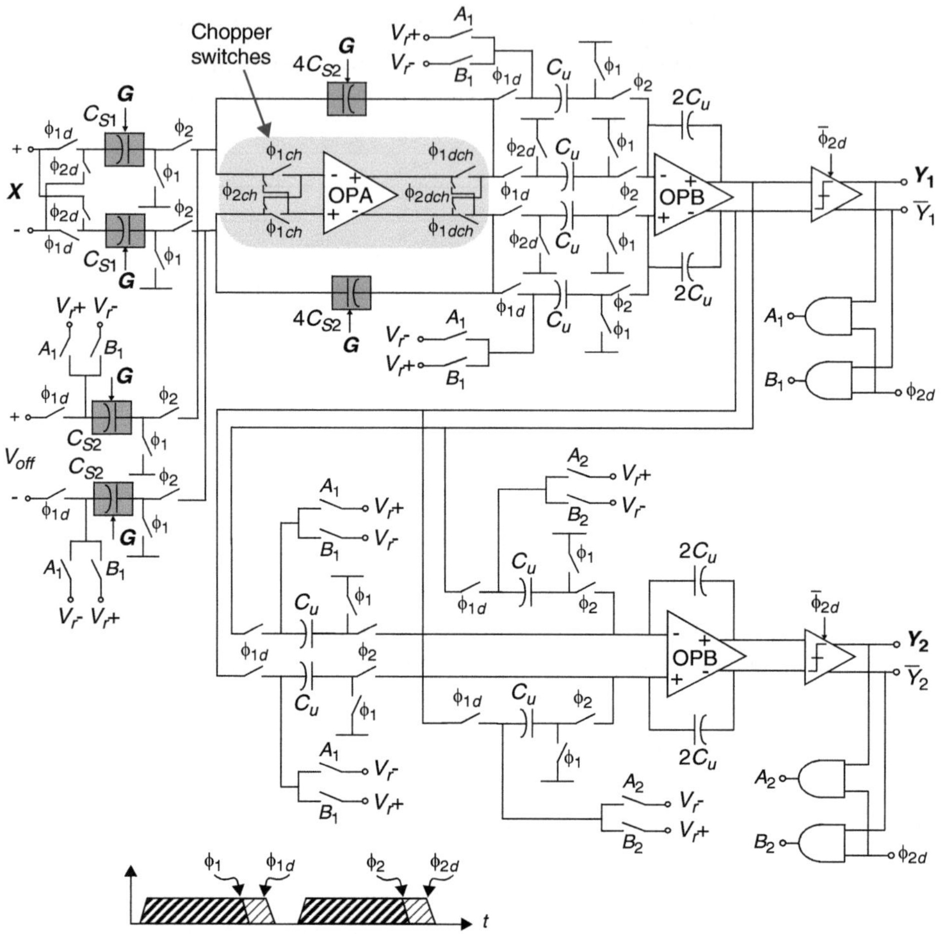

FIGURE 5.4 Fully-differential SC implementation of the programmable signal gain 2-1 ΣΔM.

synchronized for increasing the effective time slot for the modulator operations [Marq98a]. The comparators are activated at the end of phase ϕ_2—using $\bar{\phi}_{2d}$ as a strobe signal—to avoid any possible interference due to the transient response of the integrators outputs at the beginning of the sampling phase. This timing guarantees a single delay per clock cycle. Besides these clock phases, additional one are required to control the chopper switches that are used in the first integrator to attenuate flicker noise (see Fig. 5.4). As will be described in Sections 5.3 and 5.4, these chopper phases are controlled by a master clock with programmable frequency in order to have a better control of the flicker noise experimentally for the different signal conditions.

G	C_{S1}	C_{S2}	C_I	G_a	G_b	G_c	G_d
0.5	C_u	$4C_u$	$16C_u$	1	0	0	0
1	$2C_u$	$4C_u$	$16C_u$	0	1	0	0
2	$6C_u$	$6C_u$	$24C_u$	0	1	1	0
4	$16C_u$	$8C_u$	$32C_u$	0	1	1	1

(a)

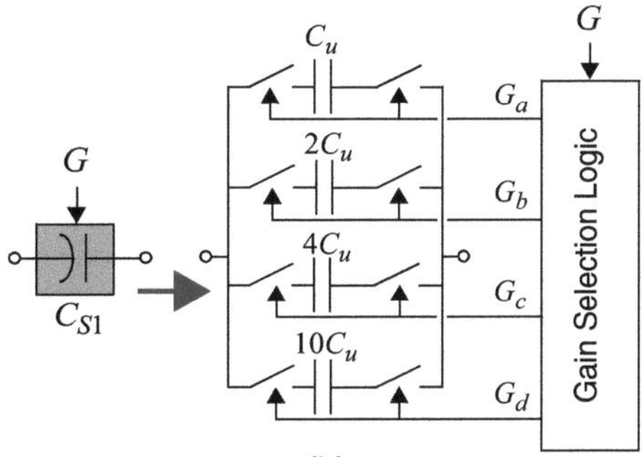

(b)

FIGURE 5.5 Programmable capacitors in the first integrator: (a) Capacitor arrays; (b) Implementation of C_{S1}.

The programmable modulator signal gain ($G = 0.5, 1, 2, 4$) has been mapped onto switchable capacitor arrays in the front-end integrator, each of them formed by a variable number of unit capacitors ($C_u = 1.5\text{pF}$), as shown in Fig. 5.5. Such numbers have been selected for minimum power dissipation, bearing in mind the circuit noise limitation and the maximum temperature target ($+175°\text{C}$). In order to keep the amplifier dynamic requirements as relaxed as possible for all cases of modulator signal gain, we propose to switch all unit capacitors instead of just those forming the input sampling capacitors. This is based on the following issues:

- For large gains ($G > 1$) C_{S1} is larger than C_{S2}. Hence, the output-referred thermal noise power is amplified with respect to the unity-gain case as it is multiplied by $(1 + C_{S1}/C_{S2})$—see eq(5.4).
- For low gain ($G < 1$) the situation is the contrary. Now the output-referred noise power is attenuated with respect to the unity-gain case, what allows us to decrease the capacitance values, mainly C_{S2}. This strategy also relaxes the required amplifier dynamics because its equivalent capacitive load will also be decreased.
- In order to simultaneously handle the modulator signal gain ($2C_{S1}/C_{S2}$ †[1]) and the first integrator feedback weight (C_{S2}/C_I always equal to 0.25), the value of all capacitances must be changed by re-arranging the number of unit capacitors forming them.

1. An extra signal gain of 2 is considered as a consequence of using double sampling in the first integrator.

5.2.3 High-level sizing and building-block specifications

The top-down methodology described in Section 3.1 has been followed for the design of the ΣΔM. In this flow [Mede99a], the modulator specifications are mapped onto building-block specifications using statistical optimization for design parameter selection and compiled equations (capturing non-ideal building-block behavior) for evaluation. This process is fine-tuned by behavioral simulation using ASIDES in order to cover non-idealities more accurately than when compiled equations are used. Worst cases for speed (largest capacitor values) and for thermal noise (highest temperature and lowest capacitor values) have been also contemplated.

The outcome of this sizing process is summarized in Table 5.3, where OPA denotes the amplifier used at the front-end integrator and OPB refers to the amplifier used at the second and third integrators (see Fig. 5.4). The specifications of the building blocks collected in Table 5.3 define the starting point for their design, which is described in Section 5.3.

The system-level performance of the modulator has been verified for the parameters in Table 5.3. Fig. 5.6*a* shows the *SNDR* curves obtained by behavioral simulation for the different modulator signal gains. Note that the $SNDR_{peak}$ is larger than 100dB in all cases. Fig. 5.6*b* shows the simulated in-band output spectra for a –20dBV@5kHz input sinewave. The cumulative error power within the signal band is also depicted and shows that the modulator dynamic range is clearly limited by white noise (10-dB/dec slope). The in-band error power is around -103dB in all cases, as required for the proposed application. Considering the modulator full-scale of ± 2V, the former value provides a *DR* of 106dB (17.3bit) for $G = 1$, which is boosted by the modulator signal gain to a *DR* of 118dB (19.3bit) for $G = 4$.

5.3 Design of the Building Blocks

This section details the architecture selection and sizing of the main modulator building blocks at transistor level, according to the requirements formerly obtained in Section 5.2.

5.3.1 Amplifiers

The key features for the design of the amplifiers are their open-loop DC gain, dynamic requirements, and output swing. Regarding the latter, the set of integrator weights used for the ΣΔM (see Fig. 5.3) allow us to relax the required

TABLE 5.3 High-level sizing of the programmable signal gain 2-1 ΣΔM.

MODULATOR	Topology		2-1
	Internal quantization		1bit
	Oversampling ratio		128
	Clock frequency		5.12MHz
	Differential reference voltage		2V
FRONT-END INTEGRATOR	Sampling capacitor	$G = 0.5$	1.5pF
		$G = 1$	3.0pF
		$G = 2$	9.0pF
		$G = 4$	24.0pF
	Unit capacitor		1.5pF
	Capacitor standard deviation (1.5-pF MiM cap)		0.1%
	Capacitor non-linearity		25ppm/V^2
	Bottom parasitic capacitor		5%
	Switch on-resistance		650Ω
AMPLIFIERS	OPA	Open-loop DC gain	2500 (68dB)
		DC-gain non-linearity	15%/V^2
		Gain-bandwidth product (44.2-pF load)	15MHz
		Slew rate (44.2-pF load)	17V/μs
		Differential output swing	±2.5V
	OPB	Open-loop DC gain	1400 (63dB)
		DC-gain non-linearity	15%/V^2
		Gain-bandwidth product (8.9-pF load)	15MHz
		Slew rate (8.9-pF load)	28V/μs
		Differential output swing	±2.5V
COMPARATORS	Hysteresis + Offset		30mV
	Resolution time		50ns

output swing to be only slightly larger than the reference voltage (±2V), which is feasible when operating with a 3.3-V supply in differential mode.

A single-stage folded-cascode architecture, shown in Fig. 5.7, was selected for the two amplifiers considered in the ΣΔM (OPA and OPB). N-channel input transistors are employed to take advantage of the twin-well technology feature in removing the body effect of nMOS transistors. The common-mode feedback (CMFB) nets have been implemented using SC circuits, which provide fast, linear operation with small power dissipation.

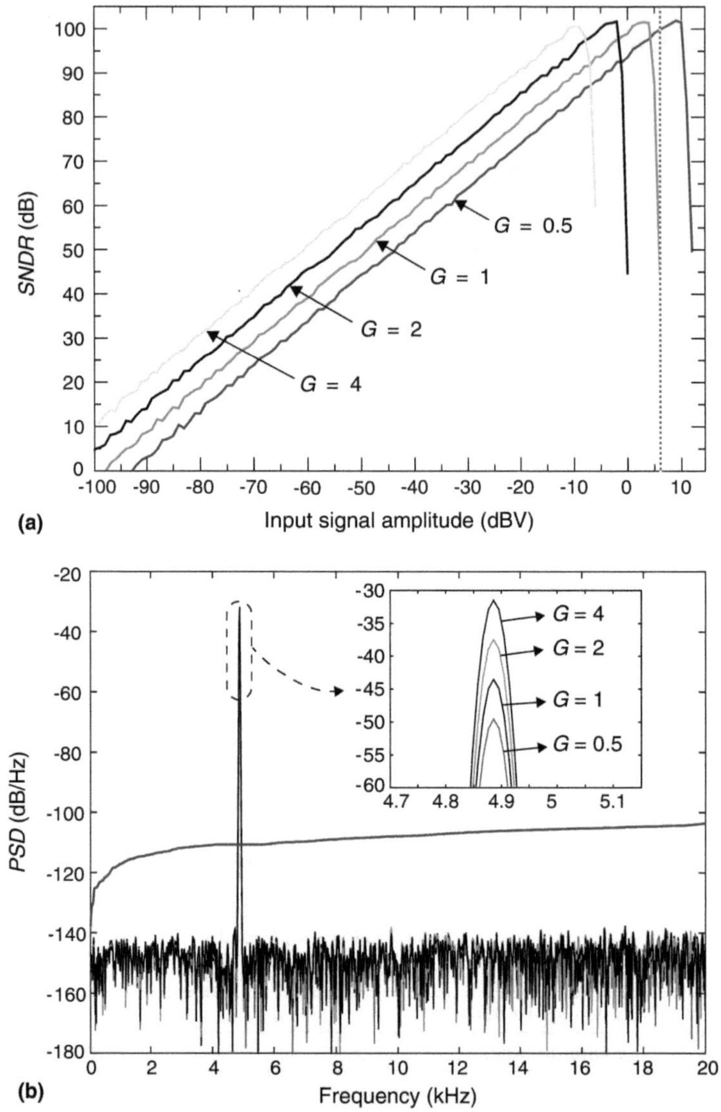

FIGURE 5.6 Behavioral simulation results for the different signal gains of the ΣΔM: (a) *SNDR* curves; (b) In-band output spectra.

Although both amplifiers share the same topology, two different sizings have been completed in order to optimize the power consumption, given that the requirements for OPA are more demanding than for OPB (see Table 5.3).

Table 5.4 and Table 5.5 summarize the obtained electrical performance for OPA and OPB after full sizing, regarding the target values imposed during the

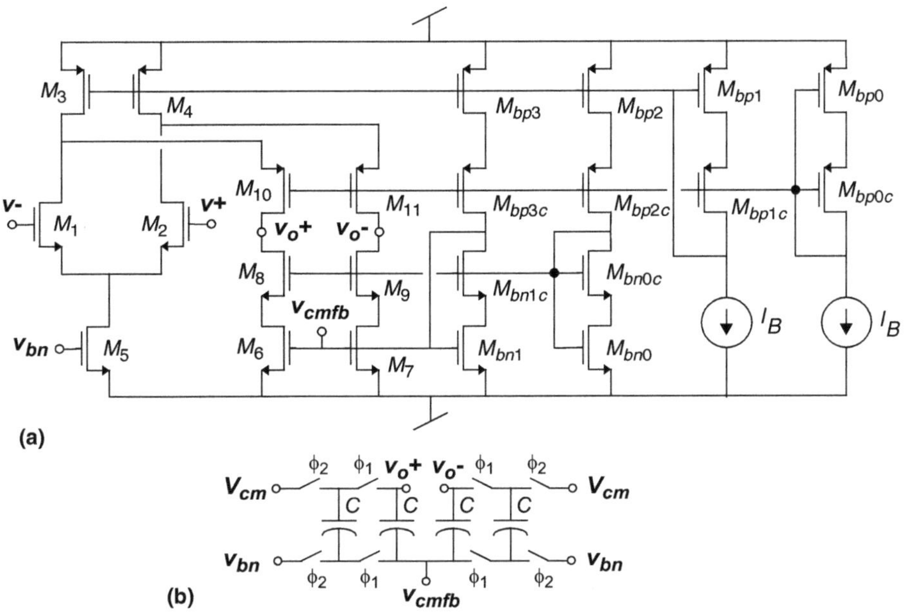

FIGURE 5.7 Fully-differential folded-cascode amplifier: (a) Amplifier core and bias stage; (b) SC CMFB net.

TABLE 5.4 Electrical simulation results for OPA (45-pF load).

	Target	Typical	Worst-case
DC gain	>68dB	74.0dB	71.1dB
Gain-bandwidth product	>15MHz	22.6MHz	15.8MHz
Phase margin	>60°	86.4°	85.5°
Slew rate	>17V/μs	22.1V/μs	21.1V/μs
Differential output swing	>±2.50V	±2.75V	±2.50V
Power consumption	minimum	7.1mW	7.2mW

TABLE 5.5 Electrical simulation results for OPB (9-pF load).

	Target	Typical	Worst-case
DC gain	>63dB	68.3dB	65.1dB
Gain-bandwidth product	>15MHz	34.4MHz	23.8MHz
Phase margin	>60°	83.5°	81.4°
Slew rate	>28V/μs	38.1V/μs	35.7V/μs
Differential output swing	>±2.50V	±2.75V	±2.50V
Power consumption	minimum	2.3mW	2.4mW

optimization procedure with FRIDGE [Mede99a]. The tables depict the value of the amplifier electrical parameters for typical operation conditions, as well as the worst-case value of each parameter in a corner analysis—considering fast and slow device models, ±10% variation in the 3.3-V supply, and temperatures in the range [–40°C, +175°C].

The amplifier non-linear features (mainly non-linear DC gain and dynamics) deserve special attention in a high-linear implementation. In order to accurately account for them in behavioral simulations, we have resorted to the table look-up procedure described in Section 4.4.1—which is based on curves obtained by electrical simulation—for validating the amplifier DC curves and the transient response of the front-end integrator.

5.3.2 Comparators

The comparators at the end of the modulator stages demand a voltage resolution smaller than 30mV and a resolution time around 50ns—a quarter of the clock period. Fig. 5.8 shows the selected comparator topology, based on a regenerative latch with a pre-amplifying stage [Yin92]. It consists of a nMOS input differential pair ($M_{1,2}$), a CMOS regenerative latch, and a SR latch. The regenerative latch is composed of a nMOS flip-flop ($M_{3,4}$) with a pair of nMOS switches ($M_{9,10}$) for strobing and a nMOS switch (M_{17}) for resetting, and a pMOS flip-flop ($M_{5,6}$) with a pair of pMOS pre-charge switches ($M_{7,8}$). Different voltage supplies have been used for the pre-amplifier and for the regenerative latch in order to reduce the comparator sensitivity to injected digital switching noise and supply bounce.

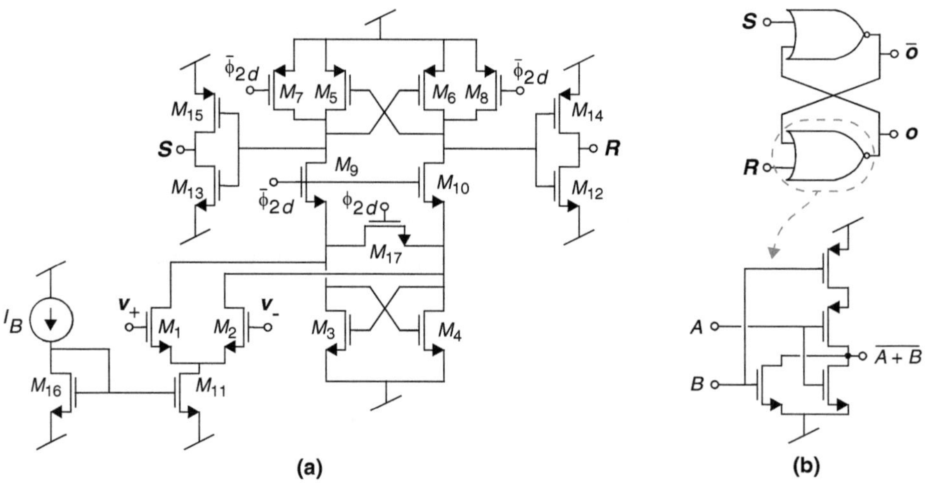

FIGURE 5.8 Comparator: (a) Pre-amplifier and regenerative latch; (b) SR latch.

TABLE 5.6 Electrical simulation results for the comparators.

	Typical	Worst-case
Offset	0.75mV	9.92mV
Hysteresis	30μV	120μV
Resolution time, LH	4.10ns	8.60ns
Resolution time, HL	3.90ns	6.15ns
Power consumption	0.43mW	

The positive feedback is activated at the end of the integration phase (when $\bar{\phi}_{2d}$ goes high) in order to make the latch react before the integrators outputs change at the beginning of ϕ_1. At the beginning of the next integration phase, the latch is forced to the state $(S, R) = (0, 0)$ for maintaining the comparator output until the next strobing.

After full sizing, the performance of the comparator was evaluated using Monte Carlo simulations and corner analysis. Table 5.6 summarizes the worst-case values of the electrical parameters. As expected, the resolution is limited by mismatching, with the offset being dominant with a worst-case value of 9.92mV—compliant with the specifications in Table 5.3.

5.3.3 Switches

The main design issue related to the analog switches is their finite on-resistance (R_{on}), which can be initially constricted by dynamic considerations. Incomplete settling caused by the switches is traditionally reduced by making $R_{on}C_{S1}f_s \ll 1$. In our circuit R_{on} values up to 650Ω can be tolerated with minor degradation of the modulator performance. These values can be obtained using CMOS switches with aspect ratios of 6.5/0.35 for the nMOS transistor and 23.5/0.35 for the pMOS, operating with the nominal 3.3-V supply.

Besides, high linearity is a must in the application considered. Thus, in addition to the approximate value of switch on-resistance, its voltage-dependence must be carefully addressed by analyzing the non-linear sampling process at the front-end integrator. This non-linearity causes a dynamic distortion that is larger, the larger the sampling capacitor and the signal frequency [Yu99]. In our case this distortion is particularly noticeable for the switch size given above if $G = 4$, where $C_{S1} = 24\text{pF}$. In order to push distortion below the specifications of the application, the value of R_{on} was decreased by resorting to transistors with larger aspect ratios. This possibility was preferred to including clock-boosting [Wu98] [Bult00], since the latter increases complexity and may compromise the robustness of the design.

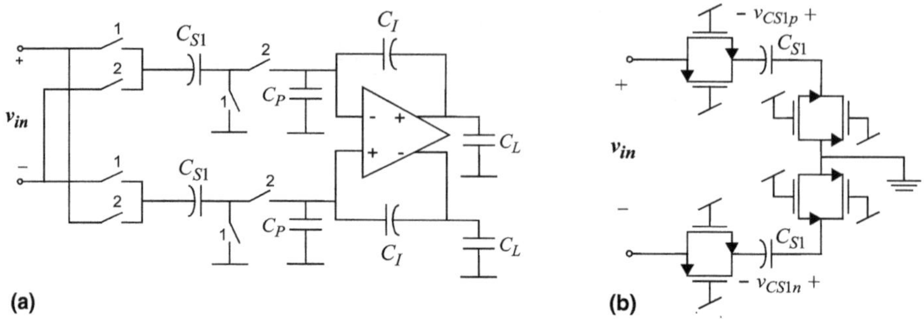

FIGURE 5.9 (a) Front-end SC integrator (only the input branch is depicted); (b) Circuit for evaluating distortion.

TABLE 5.7 *THD* caused by the analog switches for a $1V_{pd}$ @ 20kHz input tone.

MOST model	Temperature	*THD*
Slow	-40ºC	-100dB
	+25ºC	-101dB
	+175ºC	-102dB
Typical	-40ºC	-104dB
	+25ºC	-105dB
	+175ºC	-107dB
Fast	-40ºC	-109dB
	+25ºC	-110dB
	+175ºC	-112dB

Nevertheless, resorting to larger aspect ratios increases the switch parasitics and, thus, the overall power dissipation. For this reason, we have studied the sampling process at the front-end integrator (see Fig. 5.9*a*) in depth, seeking optimization. Electrical simulations of the circuit in Fig. 5.9*b* have been done using corner analysis for a 0dB@20kHz input sinewave. The differential voltage stored in capacitors C_{S1} was collected at a rate of 5.12MHz, and the Kaiser-windowed FFT was processed. Table 5.7 summarizes the *THD* values obtained for CMOS switches with aspect ratios of 29.1/0.35 for the nMOS transistor and 105.9/0.35 for the pMOS, which exhibit an $R_{on} \sim 60\Omega$. Note that the worst-case *THD* generated by the analog switches is -100dB—compliant with the required specifications.

Table 5.8 encloses the final sizing of the CMOS switches in the ΣΔM. These sizings ensure a dynamic distortion and a defective settling that are low enough for the application considered and, thus, make unnecessary the use of clock-boosting or similar techniques.

TABLE 5.8 Sizing of the CMOS switches in the ΣΔ modulator.

	nMOS (μm/μm)	pMOS (μm/μm)
1st Integ.	29.1/0.35	105.9/0.35
2nd, 3rd Integ.	6.5/0.35	23.5/0.35

5.3.4 Capacitor arrays

Capacitors have been implemented using metal-insulator-metal (MiM) structures available in the intended technology, which allows thin inter-metal oxide between metal-2 and metal-3. With this structure, the 1.5-pF unit capacitor used to implement the integrator weights is $31.6\mu m \times 31.6\mu m$ in size and exhibits an estimated mismatch of $\sigma(\Delta C/C) = 0.1\%$.

Given that the noise leakage caused by capacitor mismatch is a critical issue in cascade ΣΔMs, special attention has been paid to the layout of the capacitor ratios in common-centroid arrangements—a task that is complicated in our case by the programmable modulator signal gain. All the capacitors at the first integrator are made up of unit capacitors that are connected or disconnected from an array structure depending on the value of the signal gain—see Fig. 5.5. The active unit instances in the programmable common-centroid structure for the different modulator signal gains are symbolically depicted in Fig. 5.10.

5.3.5 Auxiliary blocks

Besides the programmable signal gain ΣΔM itself, some auxiliary blocks have been also included on chip; namely, the clock phase generator, the master-bias current generator, and the gain-selection logic.

Two clock phase generators have been incorporated to the prototype: one of them generates the master clock phases (Fig. 5.11*a*), whereas the other provides the chopper phases required by the first integrator (Fig. 5.11*b*). Both generate two non-overlapped clock phases (ϕ_1, ϕ_2—ϕ_{1ch}, ϕ_{2ch}) from a common external clock signal CLK. Versions of these phases with delayed falling edges (ϕ_{1d}, ϕ_{2d}— ϕ_{1dch}, ϕ_{2dch}) are also generated in order to attenuate signal-dependent charge injection [Lee85] while maximizing the effective time slot for the modulator operations [Marq98a]. Given that the ΣΔM employs CMOS switches, the complementary versions of all these phases are also provided. At the back-end of both generators a buffer tree is used to equalize the assorted capacitive loads among the signals. In order to relax the integrators dynamic requirements, especially for the case $G = 4$, the duty cycle of the external clock signal is not set to 0.5: around 77.5% of the clock

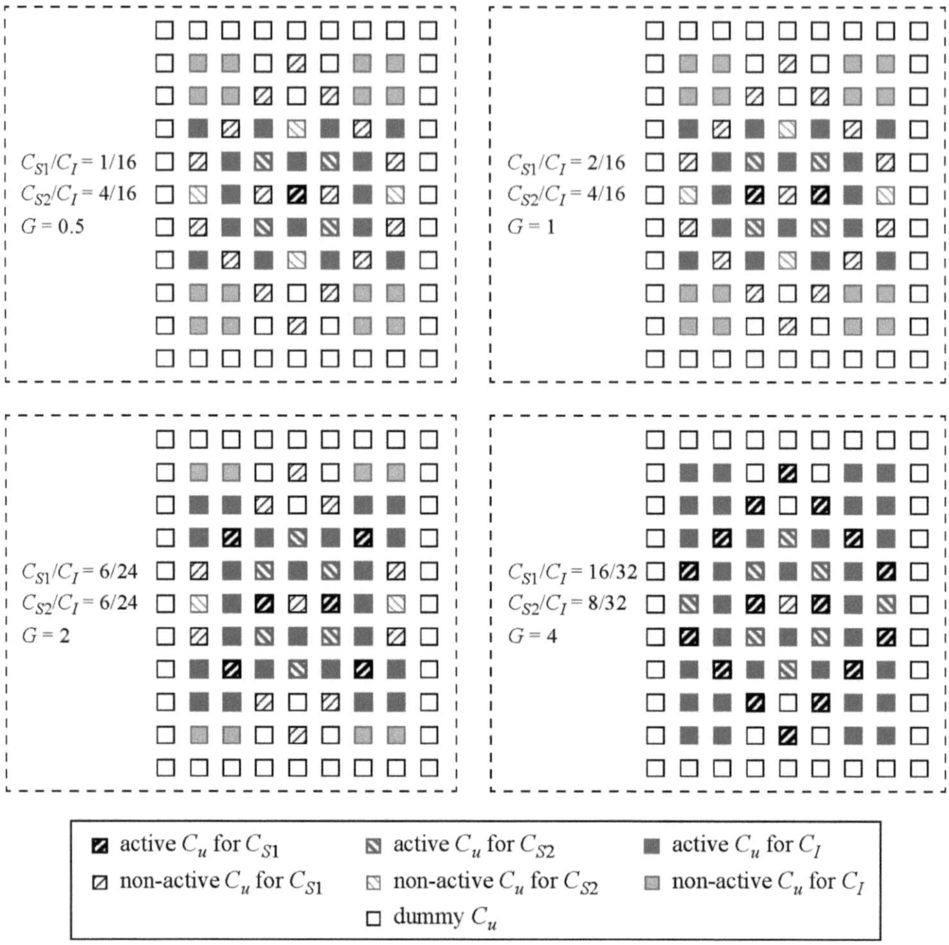

FIGURE 5.10 Conceptual layout of the programmable capacitor array at the front-end integrator.

period is devoted to integration and 22.5% to sampling. The complete clock scheme is conceptually shown in Fig. 5.11*c*. The non-overlapping and the delay times are approximately 0.2ns.

All the bias currents needed in the ΣΔ modulator are internally generated by the circuit shown in Fig. 5.12. An external 15.4-kΩ resistor is used to generate a single master current of 110μA, which is mirrored and properly scaled for biasing the amplifiers in the integrators and the pre-amplifying stages of the comparators.

As previously described, the programmable modulator signal gain is implemented by connecting or disconnecting unit capacitors from a capacitor

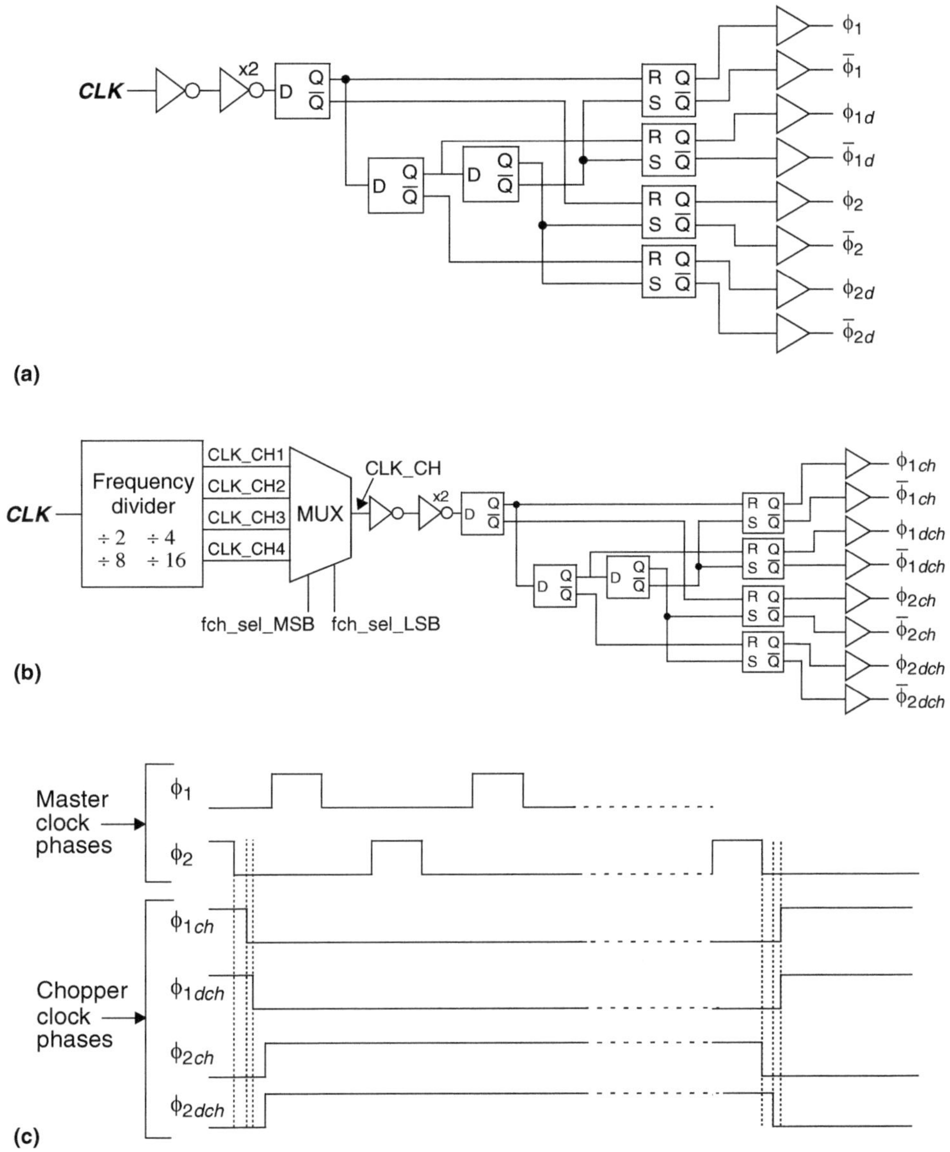

FIGURE 5.11 Clock phase generators: (a) Master clock; (b) Chopper clock; (c) Clock phase scheme.

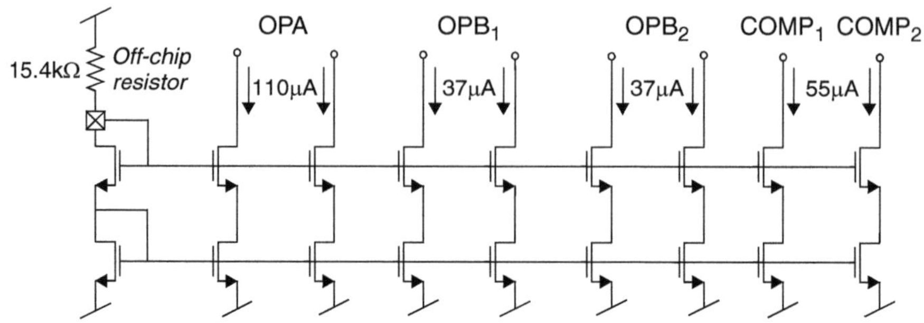

FIGURE 5.12 Master current generator.

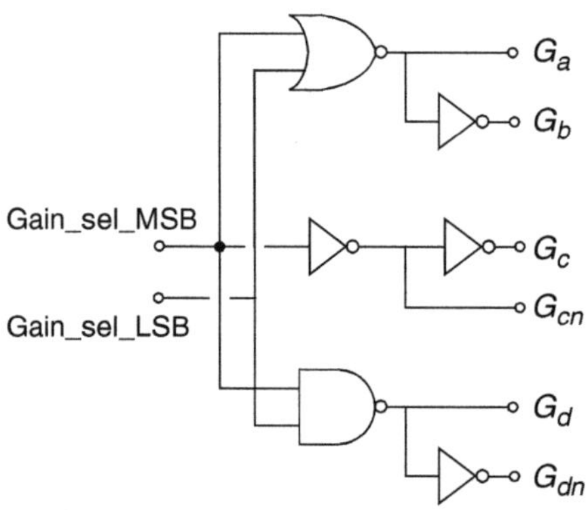

G	Gain_sel _MSB	Gain_sel _LSB	G_a	G_b	G_c	G_d
0.5	0	0	1	0	0	0
1	0	1	0	1	0	0
2	1	0	0	1	1	0
4	1	1	0	1	1	1

FIGURE 5.13 Gain-programming logic circuit.

array at the front-end integrator. The selection of the modulator signal gain is controlled by the logic circuit shown in Fig. 5.13, where Gain_sel_MSB and Gain_sel_LSB are the most significant bit (MSB) and the least significant bit (LSB) of an external control signal, respectively. In practical implementations of the complete 'smart' sensor, this signal is directly controlled by the DSP (see Fig. 5.1), thus allowing the automatic accommodation of the A/D interface to the different signal amplitudes provided by the sensor.

5.4 Layout and Prototyping

The modulator has been designed and fabricated in a single-poly, five-metal 0.35-μm CMOS technology. Fig. 5.14 shows the prototype layout and a microphotograph of its main parts. The layout has been carefully designed according to the following considerations:

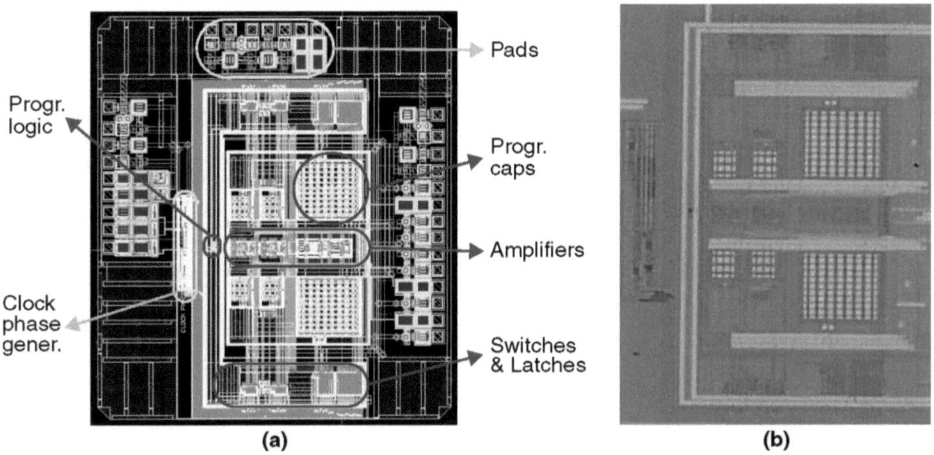

FIGURE 5.14 (a) Layout of the ΣΔM; (b) Microphotograph of the main parts of the chip.

- The front-end analog blocks and the digital section have been placed in opposite parts of the chip in order to maximize their relative distance and, thus, attenuate the impact of the switching activity.
- Separated analog, mixed, and digital supplies have been used. Analog power supplies are employed for the current biasing of the analog blocks (mainly, the amplifiers and the pre-amplifiers stages of the comparators), as well as the voltage biasing of the substrate and wells in the analog section of the chip. Mixed supplies are dedicated to the integrator switches, the SC CMFB nets of the amplifiers, and the comparators regenerative latches. Digital power supplies are used for the biasing of the clock phase generators and for the two buffers driving the output of the modulator stages out of the chip.
- The analog, mixed, and digital sections of the chip include guard-rings surrounding them, in order to maintain a low impedance return path and to avoid the spreading of digital switching noise to the sensitive parts of the modulator.
- The layout has been kept symmetrical and centroid layout techniques with unit transistors have been employed for matched transistors in the amplifiers, and in the pre-amplifying stages and regenerative latches of the comparators.

The complete modulator occupies an area of 5.7mm^2 (pads included) and has been encapsulated in 64-pin plastic quad flat package—see Fig. 5.15*a*. Multiple pads and pins have been used at the power supplies in order to reduce wire inductance and supply bounce.

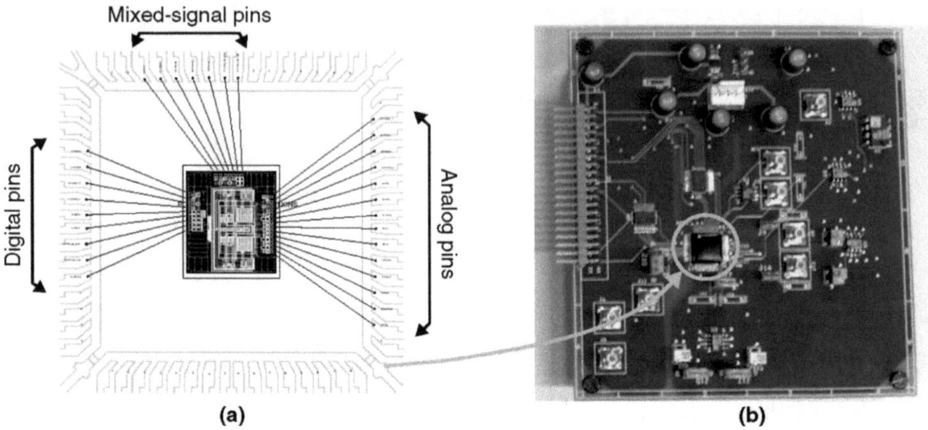

FIGURE 5.15 (a) Bonding diagram of the ΣΔM; (b) PCB used for testing purposes.

5.5 Experimental Results

The circuit has been tested using a printed circuit board (PCB), shown in Fig. 5.15*b*, that includes intensive filtering and decoupling strategies, as well as proper impedance termination to avoid signal reflections. The performance of the ΣΔM was evaluated using a high-resolution (–100dB THD) sinusoidal source to generate the input signal and a digital tester to generate the clock signal and to acquire the output bit-streams of the modulator stages. The same unit provides the supply and reference voltages. After the acquisition, which is automatically performed by controlling the test set-up through proprietary C-routines, data are transferred to a work-station to perform the digital post-processing using MATLAB. The digital filtering was performed with a *sinc* filter implemented by software.

Fig. 5.16*a* shows the effect of varying the chopper frequency on the modulator performance. The figure depicts the measured 64k-point Kaiser-windowed FFT of the modulator output, clocked at 5MHz and considering a –20dBV@5kHz input sinewave, a modulator signal gain of $G = 1$, and chopper frequencies equal to $f_{ch} = f_s/16, f_s/4, f_s/2$. Note that, the lower f_{ch}, the more flicker noise appears in the baseband, thus degrading the modulator performance. This is better illustrated in Fig. 5.16*b*, which shows the measured $SNDR$ versus the input signal amplitude for $G = 1$ and different cases of f_{ch}. It can be noted that the best performance is clearly achieved for $f_{ch} = f_s/2$. For this reason, all the measurement results that are henceforth discussed have been obtained for this value of f_{ch}.

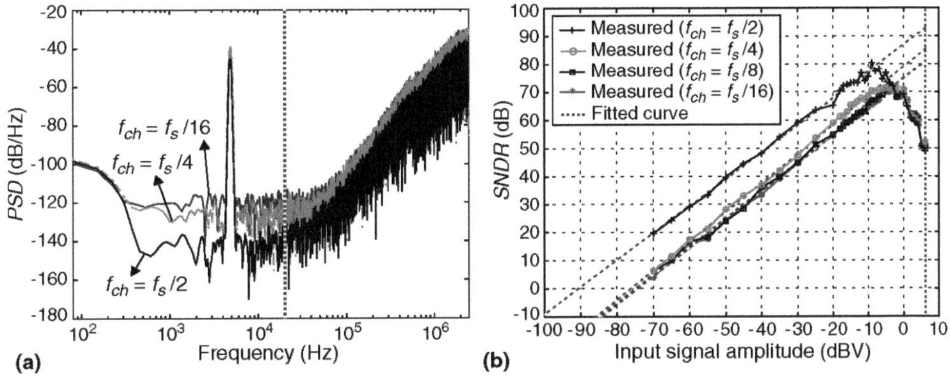

FIGURE 5.16 Effect of the chopper frequency on the measured results: (a) Output spectra; (b) *SNDR* vs. input signal amplitude for different values of f_{ch} $(G = 1)$.

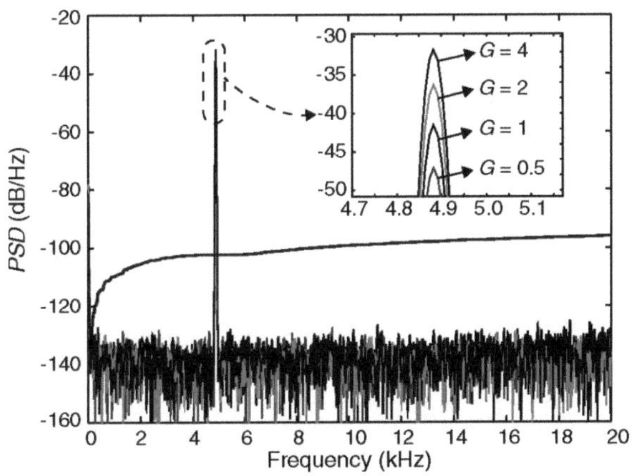

FIGURE 5.17 Measured in-band output spectra for the different modulator signal gains.

Fig. 5.17 shows the measured in-band output spectra for a –20dBV@5kHz input signal and the different modulator signal gains $(G = 0.5, 1, 2, 4)$. The measured in-band spectra are close to those obtained by behavioral simulation (see Fig. 5.6*b*), with a measured in-band error power around –96dB in all cases. Considering the modulator full-scale of $\pm 2\text{V}$, the former value provides a *DR* of 99dB (16.2bit) for $G = 1$. This resolution can be notably improved by the programmable modulator signal gain. This is illustrated in Fig. 5.18, where the measured *SNR* and *SNDR* are represented against the input signal amplitude. Note that, for $G = 4$, the input-referred *DR* is approximately 104dBV; i.e., 110-dB below the full-scale reference voltage.

However, it must be noted from Fig. 5.18 that, in most cases of signal gain, the *SNR* and *SNDR* peaks are reached at signal amplitudes smaller than

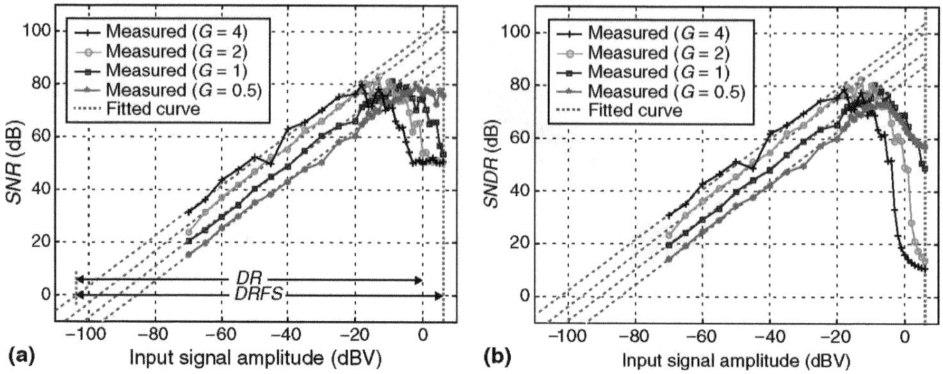

FIGURE 5.18 (a) *SNR* and (b) *SNDR* curves for the different modulator signal gains.

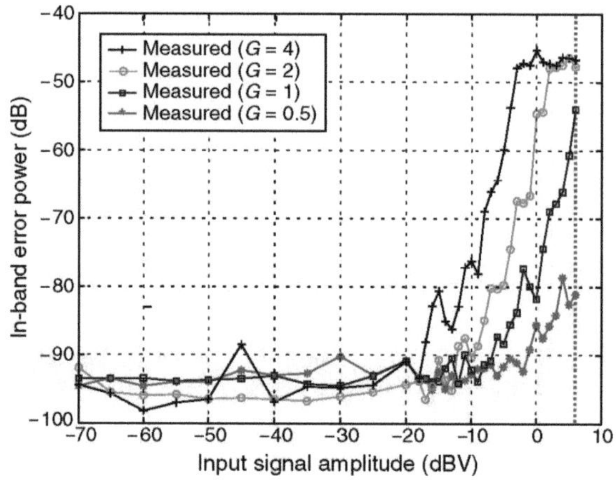

FIGURE 5.19 In-band error power as a function of the input signal amplitude.

expected. Indeed, the in-band error power clearly increases for large input amplitudes, as illustrated in Fig. 5.19. Behavioral simulations reveal that this non-linear phenomenon is due to an incorrect operation of the chopper circuitry, which seems to be caused by the dielectric relaxation of the MiM capacitors at the first integrator [Fatt90]. This effect, often not properly characterized in most technology processes, may lead to an underestimation of the in-band error power during the design phase, especially in high-resolution ADCs. Nevertheless, if the reference voltage is reduced from 2V (nominal) to 1V, the SNR_{peak} improves in approximately 5dB, as illustrated in Fig. 5.20.

The performance of the modulator has also been measured for a signal bandwidth of 10kHz. Fig. 5.21 plots the *SNR* versus the input signal amplitude for the different modulator signal gains and reference voltages of 1V and 2V. Note that, in addition to the resolution improvement that is obtained by

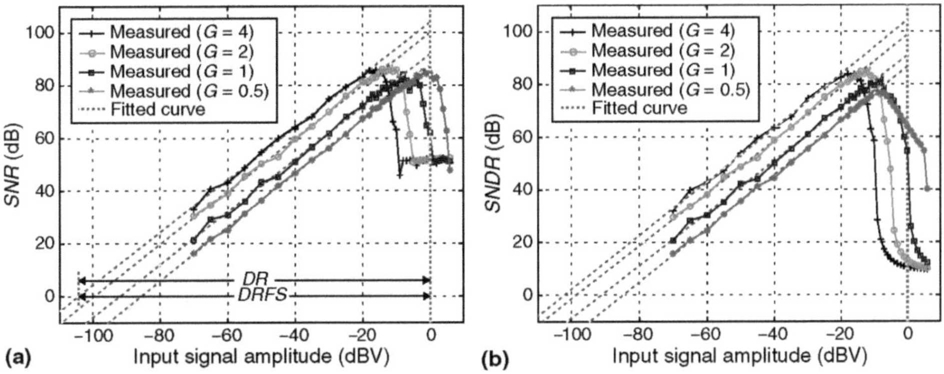

FIGURE 5.20 (a) *SNR* and (b) *SNDR* curves for $V_{ref} = 1\text{V}$.

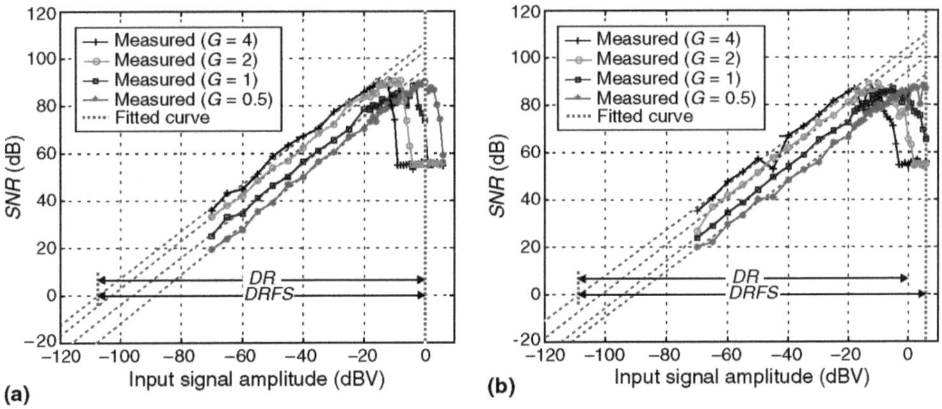

FIGURE 5.21 *SNR* vs. input amplitude in the 10-kHz bandwidth: (a) $V_{ref} = 1\text{V}$; (b) $V_{ref} = 2\text{V}$.

doubling the oversampling ratio, the modulator behavior near full scale is better than that in Fig. 5.18*a* and Fig. 5.20*a*.

Table 5.9 summarizes the measured modulator performance by displaying its most significant figures. As a matter of conclusion, these features are compared in Fig. 5.22 with current state-of-the-art ΣΔMs by using the figure-of-merit FOM_2—defined in eq(1.71)—to quantify the quality of ΣΔMs. Note that the better the ΣΔM, the larger the value of FOM_2.

Table 5.10 compares the performance of those ΣΔMs in Fig. 5.22 that achieve $FOM_2 > 200$. All ΣΔMs in Table 5.10 use SC techniques and achieve medium/high resolutions within signal bandwidths smaller than 25kHz. Note that, thanks to the combined use of modulator signal gain programmability and high-resolution†2, the prototype achieves the largest FOM_2 reported to date.

TABLE 5.9 Summary of measured prototype performance.

V_{ref} (V)	*BW* (kHz)	*G*	*SNR* / *SNDR* peak [a] (dB)	*DR* (dBV) / *DRFS* (dB)
2	20	0.5	80.2 / 73.5	85.1 / 91.6
		1	81.3 / 80.4	91.1 / 97.1
		2	82.9 / 82.5	96.9 / 102.9
		4	78.8 / 70.8	104.1 / **110.1**
	10	0.5	88.0 / –	90.3 / 96.3
		1	86.8 / –	94.3 / 100.3
		2	89.0 / –	100.1 / 106.1
		4	87.3 / –	107.8 / **113.8**
1	20	0.5	85.2 / 76.8	85.7 / 85.7
		1	84.0 / 81.8	91.9 / 91.9
		2	87.3 / 85.7	100.1 / 100.1
		4	85.8 / 83.7	104.1 / 104.1
	10	0.5	88.5 / –	90.1 / 90.1
		1	88.8 / –	100.0 / 100.0
		2	90.7 / –	103.0 / 103.0
		4	90.1 / –	108.0 / 108.0
Topology	2-1			
Sampling frequency	5.12MHz			
Power consumption	14.7mW			
Active area	5.7mm^2 (pads included)			
Supply voltage	3.3V			
Technology	0.35-μm MS CMOS (1P5M) [MiM]			

a. In the case of $BW = 10\text{kHz}$, measurements are given for an input signal frequency of 5kHz and hence, only *SNR* is computed.

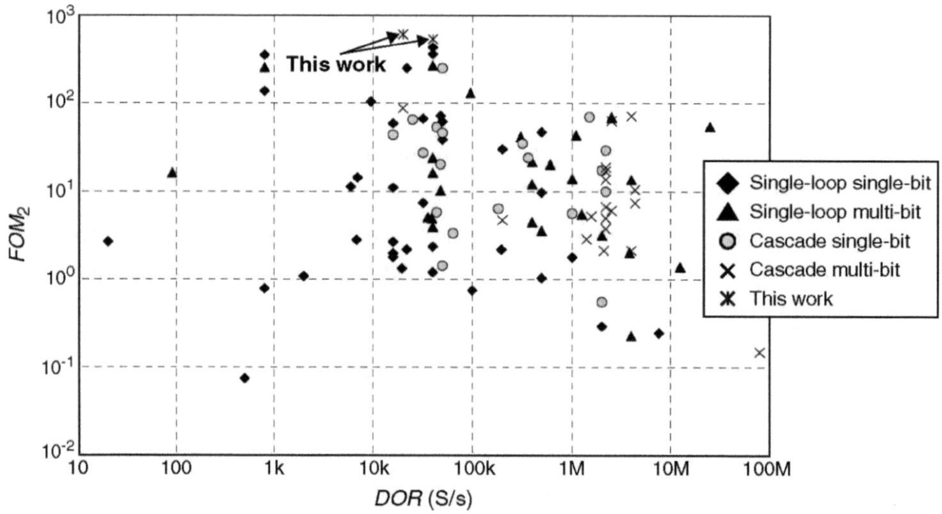

FIGURE 5.22 Comparison with state-of-the-art ΣΔMs.

TABLE 5.10 Performance comparison for reported ΣΔMs with $FOM_2 > 200$.

REFs	*ENOB* (bit)	*DOR* (kS/s)	Power (mW)	FOM_2
[Snoe01]	16.65	22	2.5	232.3
[Rabii97]	16.1	50	2.5	246.3
[Nys97]	19	0.8	2.175	252.4
[Yang03]	18.7	40	68	266.3
[Kerth94]	21	0.8	25	351.3
[Yao04]	14.4	40	0.13	358.9
[Coban99]	16	40	1	428.8
This work	18.1	40	14.7	536.1
	18.7	20		615.9

ACKNOWLEDGMENTS

The authors would like to thank Sara Escalera and Oscar Guerra, at IMSE, for their invaluable support with the design, layout, and characterization of the prototype. Also thanks to Edgard Laes, Luc Vandervoorde, and Ronny Vanhooren, at AMI Semiconductors, for their constructive comments on the 0.35-μm technology process.

5.6 Summary

This chapter describes the design of an SC ΣΔM to be included in an automotive sensor interface. The modulator targets a dynamic range larger than 110dB at 40kS/s and presents a programmable signal gain of $\times 0.5$, $\times 1$, $\times 2$, and $\times 4$. The main design considerations have been discussed and applied for selecting the most appropriate modulator architecture in terms of resolution, speed, and power consumption. The topology selected is a 3rd-order cascade (2-1 ΣΔM) with single-bit quantization.

The prototype has been implemented in a 0.35-μm CMOS process with MiM capacitors and operates at 3.3-V supply. The requirements of the building blocks have been derived and their design has been presented at transistor level.

2. It is important to note that the dynamic range of the ΣΔM presented here is enhanced by the action of the modulator signal gain programmability. This is one of the key features of the proposed modulator and it is very important for sensor applications in which there is a changing signal range. Note that if the unity-gain case is considered, the FOM_2 decreases, being 25.4 for 20-kHz signals.

Experimental results show an overall dynamic range of 110.1dB within a 20-kHz signal bandwidth and of 113.8dB for 10-kHz signals, with a power consumption of 14.7mW.

These performances place the reported circuit at the cutting edge of state-of-the-art ΣΔMs for similar applications, achieving the largest value of FOM_2 reported to date.

APPENDIX A

An Expandible Family of Cascade $\Sigma\Delta$ Modulators

Given the high signal bandwidths required in nowadays telecom applications, the oversampling ratio of $\Sigma\Delta$Ms must be restricted to low values in order to run the modulator at a feasible clock rate. Thus, high-order shaping and/or multi-bit quantization must be usually used in order to achieve the required modulator resolution. Among the variety of existing alternatives, the combination of both cascade $\Sigma\Delta$ architectures and dual-quantization techniques have proved to be a feasible and efficient approach to enhance the limited dynamic range attainable at low oversampling. These architectures circumvent the stability problems associated to high-order $\Sigma\Delta$ loops by cascading only 1st- and 2nd-order stages, whereas mechanisms for correcting the non-linearity of the multi-bit DAC can be avoided provided that multi-bit quantization is not used in the modulator front-end stage [Bran91b] [Mede99b] [Mori00] [Lamp01] [Rio01a] [Gupta02].

Further investigation of the potentialities of these architectures have led us to propose an easily expandible, modular family of high-order cascade $\Sigma\Delta$Ms. Thanks to a proper selection of the integrator coefficients, this family of cascades preserves a low systematic loss of resolution and a high overload level, regardless the overall modulator order.

A.1 Topology Description

Fig. A.1 shows the generic block diagram of the proposed family of high-order cascades, henceforth called 2-1^{L-2} $\Sigma\Delta$M. An Lth-order modulator is formed with a 2nd-order stage followed by $L-2$ identical 1st-order stages. As in all cascade $\Sigma\Delta$Ms, the outputs of the $L-1$ stages are combined and processed in the digital domain through simple operators to cancel out the quantization noise generated in each stage but the last one. Linear analysis shows that the output of the 2-1^{L-2} $\Sigma\Delta$M can be expressed in the z-domain as

$$Y(z) = z^{-L}X(z) + 2(1 - z^{-1})^L E_N(z) \quad \text{(A.1)}$$

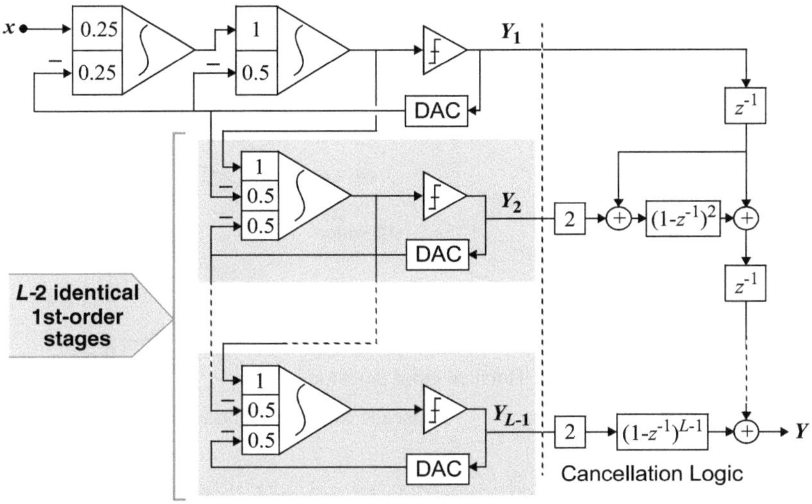

FIGURE A.1 Lth-order ΣΔ modulator using a $2\text{-}1^{L-2}$ cascade.

where $X(z)$ stands for the input signal, which is simply delayed, and $E_N(z)$ stands for the last-stage quantization error, which is Lth-order shaped. Note that the cascade response equals that of an ideal Lth-order ΣΔM, except for the scaling factor 2. This factor derives from the signal scaling required to avoid premature overload when transmitting the signal from one stage to the next.

By integrating the error term in eq(A.1) over the signal band, the in-band quantization error power is obtained as

$$P_Q = 4\sigma_Q^2 \cdot \frac{\pi^{2L}}{(2L+1)OSR^{2L+1}} \tag{A.2}$$

where $\sigma_Q^2 = \Delta^2/12$ is the power associated to the quantization error in the last-stage single-bit quantizer ($\Delta = 2V_{ref}$ stands for its full scale). Note that the factor 2 in eq(A.1) quadruples the in-band quantization error power and, thus, leads to a reduction of 6dB (1bit) in the dynamic range of the $2\text{-}1^{L-2}$ cascade in comparison with an ideal Lth-order ΣΔM. This systematic loss of resolution is one of the smallest possible and considerably smaller than that of other high-order cascades [Feld98] [Miao98] [Mori00]. More importantly, it is constant, regardless of the number of stages.

In fact, the most appealing feature of this architecture—with the set of coefficients proposed—is that it can be easily set to any order, just by changing the number of identical 1st-order stages. As shown in Fig. A.2, a correct operation is maintained with constant overload level, regardless the overall modulator order.

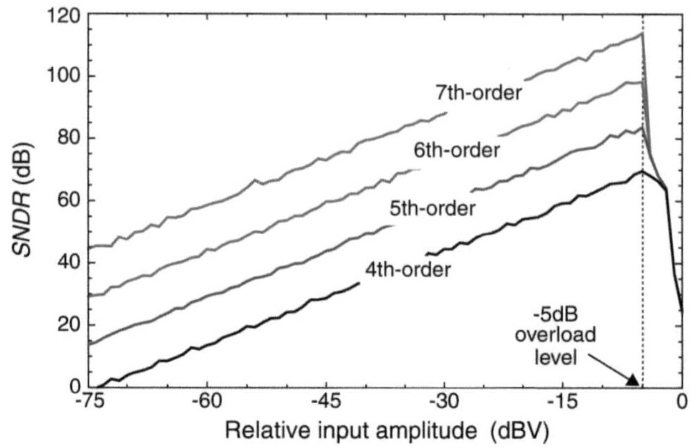

FIGURE A.2 *SNDR* curves of the $2\text{-}1^{L-2}$ cascade for several modulator orders ($OSR = 16$).

The set of integrator coefficients depicted in Fig. A.1 presents also the following interesting properties:

- The output swing required in all integrators does not exceed the quantizer full scale. Such an appealing feature for low-voltage implementations is illustrated in Fig. A.3 for the 5th-order cascade.
- The largest coefficient of each three-weight integrator can be obtained as the summation of the others, so that three-branch SC integrators are not required. By proper sharing of the SC input stages, all coefficients can be implemented with just two-branch integrators, which minimizes the total number of unit capacitors.
- All 1st-order stages contain the same coefficients, so that they can be electrically identical. This considerably simplifies the electrical and physical implementation of the modulator.

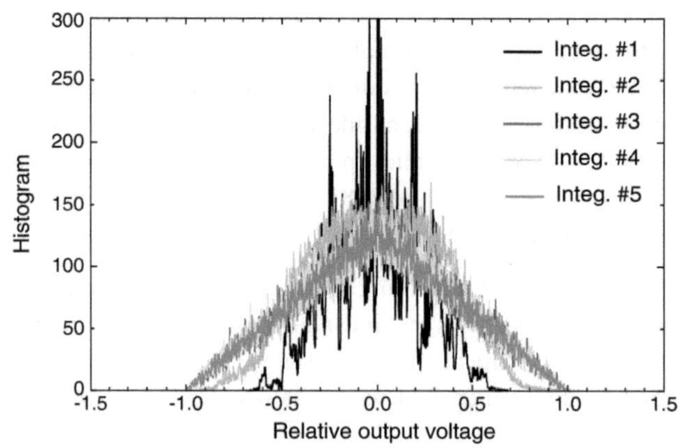

FIGURE A.3 Histogram of the integrator outputs relative to the reference voltage for $L = 5$.

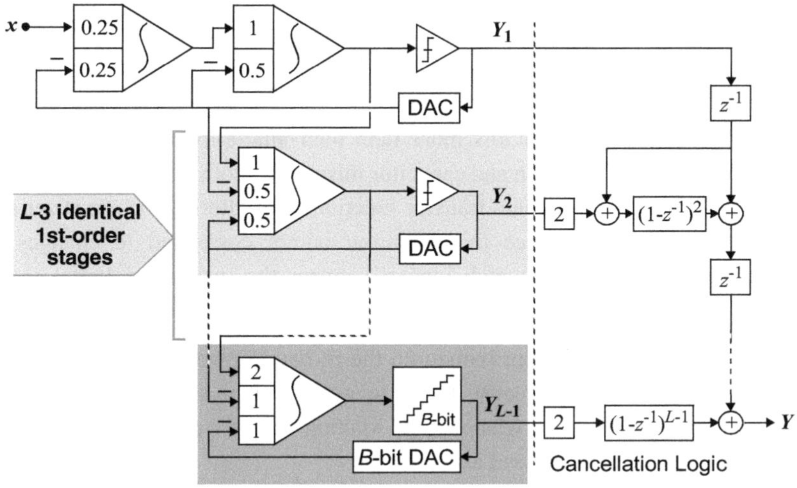

FIGURE A.4 Lth-order $2\text{-}1^{L-2}$ cascade ΣΔM employing dual quantization.

Also, as shown in Fig. A.4, a dual-quantization operation [Bran91b] [Dias93] [Tan93] can be easily achieved in the $2\text{-}1^{L-2}$ cascade ΣΔM by including multi-bit quantization only in the last stage, while the remaining are single-bit. This being the case, the coefficients in the last-stage integrator can be multiplied by a factor 2 in order to have a loop gain of 1 if the full scales of the multi-bit ADC and DAC coincide, what considerably simplifies their design.

If errors in the multi-bit DAC are considered in the linear analysis, the output of the dual-quantization $2\text{-}1^{L-2}$ ΣΔM can be obtained as

$$Y(z) = z^{-L}X(z) + 2(1-z^{-1})^L E_N(z) + 2(1-z^{-1})^{(L-1)} E_D(z) \tag{A.3}$$

where $E_D(z)$ stands for the error of the last-stage multi-bit DAC in the z-domain, which is $(L-1)$th-order shaped. Thus, provided that errors in the multi-bit DAC are considerably high-pass filtered, its non-linearity can be tolerated to some extent with no need for calibration/correction mechanisms.

The in-band quantization error power can be estimated as

$$P_Q = 4\sigma_Q^2 \cdot \frac{\pi^{2L}}{(2L+1)OSR^{2L+1}} + 4\sigma_D^2 \cdot \frac{\pi^{2(L-1)}}{(2L-1)OSR^{2L-1}}$$
$$\text{with} \quad \sigma_Q^2 = \frac{1}{12} \cdot \left(\frac{\Delta}{2^B - 1}\right)^2 \quad , \quad \sigma_D^2 \cong \frac{1}{2} \cdot \Delta^2 \cdot \left(\frac{INL}{100}\right)^2 \tag{A.4}$$

where σ_Q^2 is the power of the last-stage quantization error ($\Delta = 2V_{ref}$ stands for the multi-bit quantizer full scale and B for its resolution) and σ_D^2 is the power associated to the DAC errors, with INL being the DAC integral non-linearity expressed in %FS.

A.2 Non-Ideal Performance

As stated in Chapter 2, SC implementations of cascade modulators suffer from certain non-ideal behaviors more than their single-loop counterparts; namely, finite amplifier DC gain and capacitor mismatch. Both non-idealities modify the ideal integrator z-domain transfer function, thus altering the quantization error transfer function. Since this variation is not correlated to changes of the cancellation logic, mismatch appears between the analog and digital processing that precludes perfect cancellation of the low-order quantization errors.

Into first-order approximation, the in-band power of the error leakages is independent of L, because they are generated in the modulator first stage, which is the same for whatever L. Making use of equations (3.7), (3.12), and (3.13), it can be expressed as

$$\Delta P_Q(\mu, \varepsilon_g) = \frac{\Delta^2}{12} \cdot \left(\frac{25}{16A_{DC}^2} \cdot \frac{\pi^2}{3OSR^3} + 36\sigma_C^2 \frac{\pi^4}{5OSR^5}\right) \tag{A.5}$$

where A_{DC} stands for the 1st-stage amplifier DC gain and σ_C is the capacitor standard deviation. Comparing equations (A.2) [or (A.4)] and (A.5) for a given OSR, it is clear that for certain values of A_{DC}, σ_C, and L, leakages may dominate the in-band error power, thus imposing an upper bound to the practical values of L.

In order to estimate this limit under realistic circuit imperfections, Fig. A.5*a* shows the simulated half-scale $SNDR$ as a function of the amplifier DC gain for $OSR = 16$. Fig. A.5*b* shows the $SNDR$ histograms obtained from Monte Carlo behavioral simulation assuming 0.1% sigma in capacitor ratios—0.05% is currently featured by metal-insulator-metal (MiM) capacitors in CMOS processes. Under these conditions, mainly because of the matching sensitivity, the 7th-order architecture seems not worth implementing for $OSR = 16$. Nevertheless, the 6th-order modulator provides 90-dB worst-case $SNDR$ with a DC gain of 2500. Specially robust is the 5th-order cascade requiring a DC gain of 1000 to achieve 80-dB worst-case $SNDR$ for $OSR = 16$. It is important to remark that these gains are basically needed for the 1st-stage amplifiers. The DC-gain requirement for the integrators in the remaining $L-2$ stages of the cascade are much more relaxed. This is also applicable to other circuit imperfections such as electronic noise, finite dynamics, non-linearity, mismatch, etc. This practice allows us to use simpler circuit topologies and layouts for these stages, thus saving area and power consumption.

In the same way as the modulator order, in practice, the number of bits in the last-stage quantizer (B) cannot be arbitrarily large. As shown in Fig. A.6, for a given oversampling ratio, the evolution of the overall effective resolution with

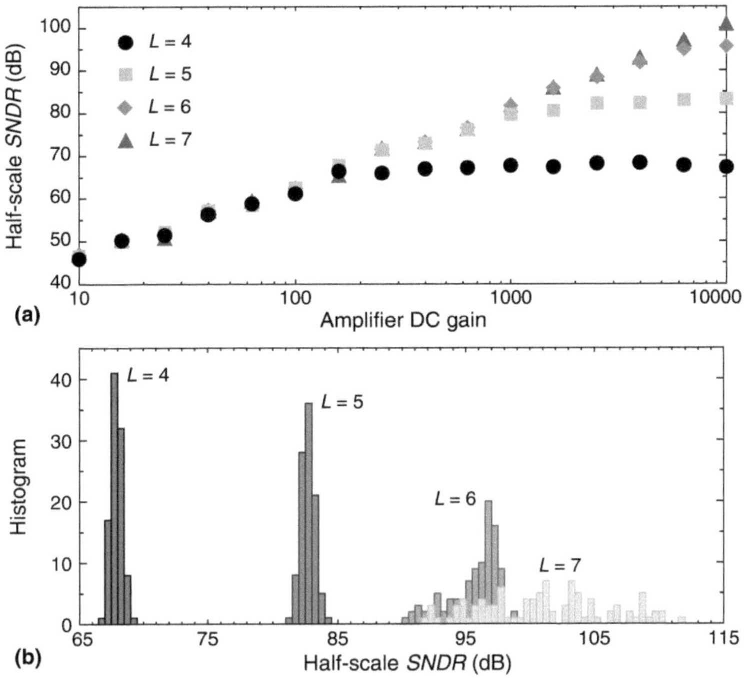

FIGURE A.5 Effect of (a) finite DC gain and (b) capacitor mismatch on the *SNDR* of single-bit 2-1^{L-2} ΣΔMs for $OSR = 16$.

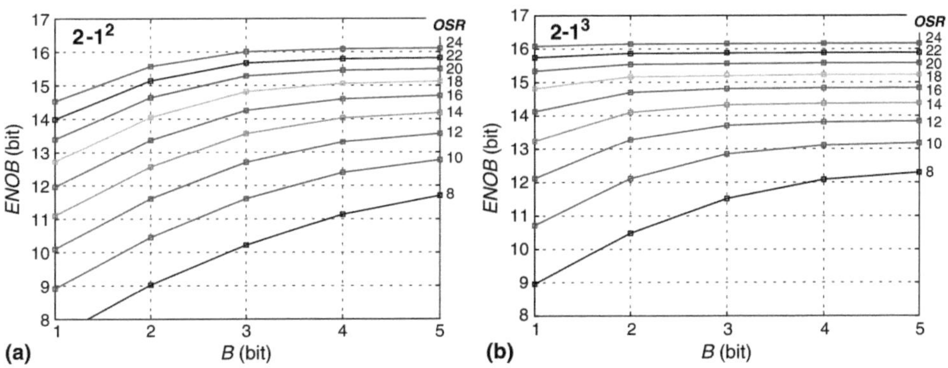

FIGURE A.6 *ENOB* of 2-1^{L-2} cascade ΣΔMs versus the resolution in the last-stage quantizer, for oversampling ratios from 8 to 24: (a) 2-1^2 ΣΔM, (b) 2-1^3 ΣΔM. ($A_{DC} = 2500$, $\sigma_C = 0.12\%$, and DAC $INL = 0.4\%\text{FS}$).

B tends to saturate due to the presence of leakages. In fact, when increasing the last quantizer resolution in 1bit generates less than 1-bit increase in the overall modulator resolution, the dual-quantization architecture starts losing efficiency and may become unsuitable for the specifications and technology considered.

Nevertheless, depending on the signal bandwidth, the reduction in oversampling ratio that can be achieved by resorting to multi-bit quantization may define the border between feasible and unfeasible implementations. For instance, let us consider the 4th-order cascade (2-1^2 ΣΔM) in order to obtain 14-bit effective resolution. According to Fig. A.6*a*, its single-bit version would require $OSR = 24$, whereas a 3-bit version with $OSR = 16$ is also feasible. For a signal bandwidth of 2.2MHz, the oversampling ratios mean 105.6-MHz and 70.4-MHz clock rate, respectively. Apart from an eased testing, certain power saving can be expected by using the multi-bit modulator.

APPENDIX B

Power Estimator for Cascade ΣΔ Modulators

In this Appendix we present an analytical procedure to estimate the power consumption of single-bit and multi-bit cascades ΣΔMs based on the expandible Lth-order $2\text{-}1^{L-2}$ topology proposed in Appendix A.

Both architecture and technological features are contemplated in the underlying expressions, together with simplifying assumptions inspired in practical design solutions.

B.1 Dominant Error Mechanisms

Let us start assuming that, in whatever practical design of a high-frequency ΣΔ modulator, the dominant sources of in-band error power are quantization error, white circuit noise, and incomplete settling error. The latter is specially important for telecom converters, in which a sampling frequency at the edge of the CMOS feasibility will have to be used.

Under these initial assumptions, the dynamic range (DR) of the ΣΔ modulator can be roughly expressed as follows

$$DR = 3 \cdot 2^{2ENOB-1} \cong \frac{V_{ref}^2/2}{P_Q + P_{CN} + P_{st}} \qquad \text{(B.1)}$$

where V_{ref} is the reference voltage that determines the modulator full scale ($\Delta = 2V_{ref}$) and P_Q, P_{CN}, and P_{st} are the in-band powers of quantization error, white circuit noise or thermal noise, and settling error, respectively.

The selection of the reference voltage impacts P_Q and, although more indirectly, also P_{CN} and P_{st}. Moreover, V_{ref} is obviously constrained by the supply voltage, because it imposes a given output swing requirement in integrators, which must be feasible in the intended technology. In conclusion, the selection of V_{ref} is closely related to the amplifier topology and its capability to trade DC gain, speed, and output swing [Raza00] [Malo01]. In practice, an upper bound for a feasible selection of V_{ref} is given by †1

$$V_{ref} = V_{\text{supply}} - n_{ob} V_{sat} \tag{B.2}$$

where V_{sat} is the saturation voltage of the amplifier output devices and n_{ob} is the number of transistors in the output branch, which again depends on the specific amplifier topology. If a single-stage amplifier is used, cascode devices will be required to achieve enough DC gain, so that $n_{ob} \geq 4$. This common choice is not adequate in low-voltage implementations, where an excessive value of V_{sat} will result in a very small value for V_{ref}.

Among the alternatives, two-stage amplifiers offer the possibility to still yield a large open-loop DC gain, while their output branches can contain only two transistors ($n_{ob} = 2$). This allows to increase the value of the reference voltage and to set the modulator full scale to an useful level.

Next, for the sake of simplicity, we will assume for the time being that settling error can be controlled by design so that $P_{st} \ll P_Q, P_{CN}$; i.e., eq(B.1) simplifies to

$$DR \cong \frac{V_{ref}^2/2}{P_Q + P_{CN}} \tag{B.3}$$

With respect to P_Q, it is formed by three main error mechanisms:

- Last-stage quantization error,
- Last-stage DAC non-linearity (only if multi-bit quantization is used), and
- Non-cancelled portion of the low-order quantization errors caused by integrator leakage and capacitor mismatch.

A close expression including the former non-idealities can be obtained for the expandible 2-1^{L-2} cascade ΣΔM by adding up equations (A.4) and (A.5), what results in

$$\begin{aligned} P_Q = {} & 4\sigma_Q^2 \cdot \frac{\pi^{2L}}{(2L+1)OSR^{2L+1}} + 4\sigma_D^2 \cdot \frac{\pi^{2(L-1)}}{(2L-1)OSR^{2L-1}} + \\ & + \frac{(2V_{ref})^2}{12} \cdot \left(\frac{25}{16A_{DC}^2} \cdot \frac{\pi^2}{3OSR^3} + 36\sigma_C^2 \frac{\pi^4}{5OSR^5} \right) \end{aligned} \tag{B.4}$$

with

$$\sigma_Q^2 = \frac{1}{12} \cdot \left(\frac{2V_{ref}}{2^B - 1} \right)^2 \qquad \sigma_D^2 \cong \frac{1}{2} \cdot (2V_{ref})^2 \cdot \left(\frac{INL}{100} \right)^2 \tag{B.5}$$

1. A fully-differential modulator employing symmetrical references $\pm V_{ref}$ is assumed.

where σ_Q^2 is the power associated to the last-stage quantization error (B stands for the resolution of the multi-bit quantizer), σ_D^2 is the error power associated to the DAC, with INL being the DAC integral non-linearity in %FS, A_{DC} stands for the 1st-stage amplifier DC gain, and σ_C is the capacitor standard deviation.

Concerning P_{CN}, it is usually dominated by the white noise injected by the switches and the front-end amplifier, whose PSD is folded-back over the baseband by undersampling. A conservative expression for the in-band power of white noise can be derived (see Section 2.4.2)

$$P_{CN} = P_{kT/C} + P_{op} \cong 2 \cdot \frac{2kT}{C_S} \cdot \frac{1}{OSR} + \frac{4kT}{3C_S} \cdot \frac{1}{OSR} = \frac{16kT}{3C_S} \cdot \frac{1}{OSR} \tag{B.6}$$

where C_S is the value of the sampling capacitor.

B.2 Estimation of Power Consumption

Equations (B.3) to (B.6) show that the dynamic range of a cascade ΣΔM can be roughly expressed as a function of the following design parameters: L, OSR, C_S, A_{DC}, and σ_C, to which we have to add B and INL if the last-stage quantizer is multi-bit. So, for given values of A_{DC}, σ_C, and INL, the minimum value of the capacitor C_S required to obtain a given DR can be obtained as a function of L, OSR, and B. Once C_S is known, the equivalent load for the amplifier in the integrator can be estimated as

$$C_{eq} \cong C_S + C_P + C_L\left(1 + \frac{C_S + C_P}{C_I}\right) \tag{B.7}$$

where C_I, the integrator feedback capacitance, is related to C_S through the integrator weight ($C_I = C_S/g_i$) whereas C_P and C_L stand for the integrator summing node and output node parasitics, respectively. Estimating the latter two capacitances is a difficult task because of their extreme dependence on the actual amplifier design.

Usually, the main contribution to C_P is the amplifier input parasitic. In a fully-differential topology, it is formed by the input transistor gate-to-source capacitance C_{gs} (both channel and overlap contributions) and its overlap gate-to-drain capacitance C_{gd}^{ov} amplified by Miller effect [Raza00]. Neglecting C_{gb},

$$\begin{aligned} C_P &\cong C_{gs}^{ch} + C_{gs}^{ov} + C_{gd}^{ov}(1 + A_{DC_1}) = \\ &= \frac{2}{3}C_{ox}'W_{in}L_{in} + C_{ox}'W_{in}\Delta L_{in}(A_{DC_1} + 2) \end{aligned} \tag{B.8}$$

where C_{ox}' is the gate oxide capacitance density and ΔL_{in} stands for the lateral diffusion of drain/source regions below the gate, both technology-dependent parameters. Apart from the input transistor dimensions (W_{in}, L_{in}), the other unknown variable in eq(B.8) is its input-to-output gain A_{DC_1}. This is equal to the complete amplifier gain for single-stage amplifiers or to the 1st-stage gain if multi-stage topologies are used. It can even be around unity if cascode devices are used, such as in telescopic or folded-cascode amplifiers [Raza00] [Malo01]. Now, making use of the well-known (as much as inadequate) square-low expression for the input transistor drain current

$$C_P = \frac{2L_{in}I_{D,\,in}}{\mu V_{OVD}^2}\left[\frac{2}{3}L_{in} + \Delta L_{in}(A_{DC_1} + 2)\right] \tag{B.9}$$

where $V_{OVD} \equiv V_{GS} - V_T$ is the input transistor overdrive voltage.

The other unknown capacitance in eq(B.7), C_L, has two main contributions: the first one is due to the bottom parasitic of the integration capacitor C_I and the second one is due to the amplifier itself. The former contribution can significantly vary depending on the type of capacitors. With modern MiM structures it turns out to be very small, ranging from less than 1% to 5% of C_I. Because of this, C_L tends to be dominated by the amplifier output parasitic load, which strongly depends on the actual output devices and, overall, on the amplifier topology. Even the supply voltage, via output swing and DC-gain requirements, makes an impact on the transistor sizes and hence on C_L. For a given amplifier schematic, the latter influence makes C_L slightly increase under technology scaling and shrinking supply voltages, because wider output devices are required to accommodate similar output swings. All things considered, a reliable estimation of this capacitance prior to the sizing of the amplifier is not possible. Based on previous design experiences, we will assume a constant value equal to 2.5pF.

Returning to the settling error power, P_{st}, an accurate estimation would involve the following calculations. For example, just for a single-pole amplifier model, complicate expressions are derived [Mede99a] if a non-linear (slew-rate limited) settling is considered. Further complexity arises from considering both sampling and integration incomplete charge-transference and the contribution of the non-zero switch on-resistance (see Section 2.3). Hence, the treatment will be simplified assuming that the slew rate of the amplifier is large enough and the switch on-resistance small enough to neglect their impact on the integrator transient response, so that the settling is linear with time constant equal to C_{eq}/g_m. This being the case, it takes a number $\ln(2^{ENOB})$ of time constants to settle within $ENOB$ resolution; i.e., the following relationship should be fulfilled

$$\ln[2^{(ENOB+1)}]C_{eq}/g_m \le T_s/2 \tag{B.10}$$

where T_s is the sampling period. Note that an extra bit has been added in order to make room for the inaccuracy of this simplified model. The above expression can be used to estimate the minimum value of the transconductance parameter

$$g_m = 2\ln[2^{(ENOB+1)}]C_{eq}f_s \tag{B.11}$$

where $f_s \equiv 1/T_s$ is the sampling frequency. This is the transconductance required for a single-stage amplifier with equivalent output load C_{eq}. For multi-stage amplifiers, the previous relationship must be carefully tackled because both parameters, total transconductance and equivalent output load, lose control of the amplifier dynamics. However, provided that the main pole of the amplifier is set by the input stage and an eventual inter-stage compensation capacitor, eq(B.11) can still be used to determine the input stage transconductance, that is related to the input transistor current as follows

$$g_m = \frac{2I_{D,\,in}}{V_{OVD}} \tag{B.12}$$

Equations (B.7), (B.9), (B.11), and (B.12) can be handled in an iterative way to determine the current required through the input transistors of the amplifier, whose actual topology sets the power consumption. Whenever possible, a single-stage amplifier should be used because of its better performance/power figure. However, as discussed previously, as technologies scale down and supply voltages shrink, two-stage amplifiers are gaining ground. Moreover, in practice two gain stages are not enough to achieve the overall gain requirement, so that the first one often includes cascode devices in a telescopic-cascode configuration. Let us consider this topology as an archetype in modern deep-submicron technologies. The current through the first stage has been already estimated as $2I_{D,\,in}$. Assuming, for the sake of simplicity, a fixed ratio η_{io} between the currents flowing through the input and output branches, the total current through the amplifier can be estimated as

$$I_B \cong 2I_{D,\,in} + 2\eta_{io}I_{D,\,in} + I_{D,\,in} = [2(1+\eta_{io})+1]I_{D,\,in} \tag{B.13}$$

where an extra $I_{D,\,in}$ has been added to account for the biasing stage of the amplifier.

The power dissipation of the first amplifier can be estimated with eq(B.13). That of the remaining amplifiers in the cascaded stages can be decreased, following the scaling rule commonly applied to amplifier requirements in ΣΔ modulators. This power reduction may come from either a relaxed set of specifications or the subsequent simplification of the amplifier topology. Sometimes, even when a two-stage amplifier may be required for the first integrator, it is possible to use a single-stage topology for the rest of integrators. So, we can write

$$I_{B,\,\mathrm{total}} = I_B\left(1 + \sum_{i=2}^{L} \chi_i\right) \tag{B.14}$$

with χ_i being the ratio of the current absorption of the i-th amplifier to the first one. From this, the static power dissipated in the amplifiers is:

$$P_{op,\,\mathrm{sta}} = I_B V_{\mathrm{supply}}\left(1 + \sum_{i=2}^{L} \chi_i\right) \tag{B.15}$$

Besides this static consumption, which usually accounts for around 80% of the total power, there are other contributing blocks, namely:

- $L-1$ latched comparators used as single-bit quantizers and those in the last-stage multi-bit quantizer, usually implemented by a flash ADC; i.e., (2^B-1) more latches. This consumption must include the static power dissipated in a convenient pre-amplifying stage.
- Last-stage multi-bit DAC (if $B>1$). The relaxed requirements for this block allows us to implemented it with a resistor ladder. Its main design considerations are resistor matching and linearity (both causing INL) and the current it must drive, which must be large enough to provide a good settling. The current requirement scales with the sampling frequency and the capacitive load involved. The latter can be considered almost constant, because the last-stage capacitors should be set to the minimum required to achieve certain level of matching (thermal noise playing a secondary role). So, we can empirically write

$$P_{DAC} \cong V_{\mathrm{supply}} I_{DAC}{}^* \cdot \frac{f_s}{f_s{}^*} \tag{B.16}$$

  where $I_{DAC}{}^*$ is the current through the DAC required for operating at a certain frequency of reference, $f_s{}^*$.
- Dynamic power in SC stages. The dynamic power dissipated to switch a capacitance C_u between the reference voltages at a frequency f_s can be estimated as $C_u f_s V_{ref}^2$, which tends to increase in high-speed high-resolution converters. Its actual value depends on the integrator weights used. In our case, the following expression provides a good estimate

$$P_{SC} = 2[5C_{u_1} + 4(L-1)C_{u_2}]f_s V_{ref}^2 \tag{B.17}$$

  where the factor 2 comes from the differential implementation; C_{u_1} is the unit capacitor used in the first integrator, whereas C_{u_2} is the one used in the rest of integrators, usually smaller than C_{u_1}.

- Small digital blocks: flip-flops, gates, cancellation logic, etc. Apart from being small, they do not make any difference for the architectures considered and will be neglected here. Of course, this does not apply to the decimation filter, whose power consumption is comparable to that of the ΣΔ modulator. Moreover, since the order of the digital filter must equal $L+1$, high-order ΣΔMs require more complex filters than low-order ones. However, an increase of the modulator order entails a decrease of the oversampling ratio and the filter can be operated at a lower frequency, dissipating less power. To our purpose, we can consider an essentially constant power consumption in the digital filter.

By adding up all the contributions, the power dissipation of the cascade ΣΔ modulator can be estimated as:

$$\text{Power} \cong P_{op,\,\text{sta}} + P_{DAC} + [(L-2) + (2^B - 1)]P_{comp} + P_{SC} \tag{B.18}$$

REFERENCES

[Adams86] R.W. Adams, "Design and Implementation of an Audio 18-bit Analog-to-Digital Converter Using Oversampling Techniques". *Journal of Audio Engineering Society*, vol. 34, pp. 153-166, March 1986.

[Adams91] R.W. Adams, P.F. Ferguson, A. Ganesan, S. Vincelette, A. Volpe, and R. Libert, "Theory and practical implementation of a fifth-order sigma-delta A/D converter". *Journal of Audio Engineering Society*, vol. 39, pp. 515-528, July 1991.

[Adams95] R.W. Adams and T.W. Kwan, "Data-directed scrambler for multi-bit noise-shaping D/A converters". U.S. Patent 5,404,142, 1995.

[Adams97a] R.W. Adams and R. Schreier, "Stability Theory ΣΔ Modulators", Chapter 4 in *Delta-Sigma Data Converters: Theory, Design and Simulation (S.R. Norsworthy, R. Schreier, and G.C. Temes, Editors)*. IEEE Press, 1997.

[Adams97b] R.W. Adams, "The Design of High-Order Single-Bit ΣΔ ADCs", Chapter 5 in *Delta-Sigma Data Converters: Theory, Design and Simulation (S.R. Norsworthy, R. Schreier, and G.C. Temes, Editors)*. IEEE Press, 1997.

[Adar87] S.R. Adarlan and J.J. Paulos, "An Analysis of Nonlinear Behaviour in Sigma-Delta Modulators". *IEEE Transactions on Circuits and Systems*, vol. 34, pp. 593-603, June 1987.

[Agra83] B.P. Agrawal and K. Shenoi, "Design methodology for Sigma-Delta modulators". *IEEE Transactions on Communications*. vol. 31, pp. 360-370, March 1983.

[Ahuja83] B.K. Ahuja, "An Improved Frequency Compensation Technique for CMOS Operational Amplifiers". *IEEE Journal of Solid-State Circuits,* vol. 18, pp. 629-633, December 1983.

[Apar02] R. Aparicio and A. Hajimiri, "Capacity Limits and Matching Properties of Integrated Capacitor". *IEEE Journal of Solid-State Circuits*, vol. 37, no. 3, pp. 384-393, March 2002.

[Arag99] X. Aragonès, J.L. González, and A. Rubio, *Analysis and Solutions for Switching Noise Coupling in Mixed-Signal ICs*. Kluwer Academic Publishers, 1999.

[Au97] S. Au and B.H. Leung, "A 1.95-V, 0.34-mW, 12-b Sigma-Delta Modulator Stabilized by Local Feedback Loops". *IEEE Journal of Solid-State Circuits*, vol. 32, no. 3, pp. 321-328, March 1997.

[Aziz93] P. Aziz, H.V. Sorensen, and J.V.D. Spiegel, "Multiband Sigma-delta Modulation". *Electronics Letters*, vol. 29, pp. 760-762, April 1993.

[Baher92] H. Baher and E. Afifi, "Novel fourth-order sigma-delta convertor". *Electronics Letters*, vol. 28, pp. 1437-1438, July 1992.

[Bajd02] O. Bajdechi and J.H. Huijsing, "A 1.8-V ΔΣ Modulator Interface for an Electret Microphone With On-Chip Reference". *IEEE Journal of Solid-State Circuits*, vol. 37, pp. 279-285, March 2002.

[Baird94] R.T. Baird and T.S. Fiez, "Stability Analysis of High-Order Delta-Sigma Modulation for ADC's". *IEEE Transactions on Circuits and Systems - II*, vol. 41, pp. 59-62, January 1994.

[Baird95] R.T. Baird and T. Fiez, "Linearity Enhancement of Multibit ΔΣ A/D and D/A Converters using Data Weighted Averaging". *IEEE Transactions on Circuits and Systems - II*, vol. 42, pp. 753-762, December 1995.

[Baird96] R.T. Baird and T.S. Fiez, "A Low Oversampling Ratio 14-b 500-kHz ΔΣ ADC with a Self-Calibrated Multibit DAC". *IEEE Journal of Solid-State Circuits*, vol. 31, pp. 312-320, March 1996.

[Balm00] P. Balmelli, Q. Huang, and F. Piazza, "A 50-mW 14-bit 2.5-MS/s Σ–Δ modulator in a 0.25 μm digital CMOS technology". *Proc. of the Symposium on VLSI Circuits*, pp. 142-143, 2000.

[Benn48] W.R. Bennett. "Spectra of quantized signals". *Bell System Technical Journal*, vol. 27, pp. 446-472, July 1948.

[Berr99] J. Berrie, "The Defensive Design of Printed-Circuit Boards". *IEEE Spectrum*, pp. 76-81, September 1999.

[Black80] B. Black and D. Hodges, "Time-interleaved A/D conversion". *IEEE Journal of Solid-State Circuits*, vol. SC-15, pp. 1022-1029, December 1980.

[Blan02] P.G. Blanken and S.E.J. Menten, "A 10μV-Offset 8kHz Bandwidth 4th-Order Chopped ΣΔ A/D Converter for Battery Management". *Proc. of the IEEE International Solid-State Circuits Conference*, vol. 1, pp. 338-339, 2002.

[Boser88] B.E. Boser and B.A. Wooley, "The Design of Sigma-Delta Modulation Analog-to-Digital Converters". *IEEE Journal of Solid-State Circuits*, vol. 23. pp. 1298-1308, December 1988.

[Bran91a] B. Brandt, D.W. Wingard, and B.A. Wooley, "Second-Order Sigma-Delta Modulation for Digital-Audio Signal Acquisition". *IEEE Journal of Solid-State Circuits*, vol. 23. pp. 618-627, April 1991.

[Bran91b] B.P. Brandt and B.A. Wooley, "A 50-MHz Multibit Sigma-Delta Modulator for 12-b 2-MHz A/D Conversion". *IEEE Journal of Solid-State Circuits*, vol. 26, pp. 1746-1756, December 1991.

[Bran97a] B. Brandt, "High-Speed Cascaded ΣΔ ADCs", Chapter 7 in *Delta-Sigma Data Converters: Theory, Design and Simulation (S.R. Norsworthy, R. Schreier, and G.C. Temes, Editors)*. IEEE Press, 1997.

[Bran97b] B. Brandt, P.F. Ferguson, and M. Rebeschini, "Analog Circuit Design of ΣΔ ADCs", Chapter 11 in *Delta-Sigma Data Converters: Theory, Design and Simulation (S.R. Norsworthy, R. Schreier, and G.C. Temes, Editors)*. IEEE Press, 1997.

[Brig02] S. Brigati, F. Francesconi, P. Malcovati, and F. Maloberti, "A fourth-order single-bit switched-capacitor sigma-delta modulator for distributed sensor applications". *Proc. of the IEEE Instrumentation and Measurement Technology Conference*, vol.1, pp. 253-256, 2002.

[Broo97] T.L. Brooks, D.H. Robertson, D.F. Kelly, A. Del Muro, and S. W. Harston, "A Cascaded Sigma-Delta Pipeline A/D Converter with 1.25 MHz Signal Bandwidth and 89 dB SNR". *IEEE Journal of Solid-State Circuits*, vol. 32, pp. 1896-1906, December 1997.

[Broo02] T. Brooks, "Architecture Considerations for Multi-Bit ΣΔ ADCs", in *Analog Circuit Design — Structured Mixed-Mode Design, Multi-Bit Sigma-Delta Converters, Short Range RF Circuits (M. Steyaert, A.H.M. van Roermund, and J.H. Huijsing, Editors)*. Kluwer Academic Publishers, 2002.

[Bult90] K. Bult and G.J.G.M. Geelen, "A Fast-Settling CMOS Op Amp for SC Circuits with 90-dB DC Gain". *IEEE Journal of Solid-State Circuits*, vol. 25, pp. 1379-1384, December 1990.

[Bult00] K. Bult, "Analog Design in Deep Sub-Micron CMOS". *Proc. of the IEEE European Solid-State Circuits Conference*, pp. 11-17, 2000.

[Burm96] T.V. Burmas, K.C. Dyer, P.J. Hurst, and S.H. Lewis, "A second-order double-sampled delta-sigma modulator using additive-error switching". *IEEE Journal of Solid-State Circuits*, vol. 31, pp. 284-293, March 1996.

[Candy81] J.C. Candy and O.J. Benjamin, "The structure of quantization noise from sigma-delta modulation". *IEEE Transactions on Communications*. vol. 29, pp. 1316-1323, September 1981.

[Candy85] J.C. Candy, "A Use of Double Integration in Sigma-Delta Modulation". *IEEE Transactions on Communications*. vol. 33, pp. 249-258, March 1985.

[Candy97] J.C. Candy, "An Overview of Basic Concepts", Chapter 1 in *Delta-Sigma Data Converters: Theory, Design and Simulation (S.R. Norsworthy, R. Schreier, and G.C. Temes, Editors)*. IEEE Press, 1997.

[Carl88] L.R. Carley and J. Kenney, "A 16-bit 4'th order noise-shaping D/A converter". *Proc. of the IEEE Custom Integrated Circuits Conference*, pp. 21.7.1-4, 1988.

[Carl97] L.R. Carley, R. Schreier, and G.C. Temes, "Delta-Sigma ADCs with Multibit Internal Converters", Chapter 8 in *Delta-Sigma Data Converters: Theory, Design and Simulation (S.R. Norsworthy, R. Schreier, and G.C. Temes, Editors)*. IEEE Press, 1997.

[Casi01] H.J. Casier, "Requirements for Embedded Data Converters in an ADSL Communication System". *Proc. of the IEEE International Conference on Electronics, Circuits and Systems*, vol. I, pp. 489-492, September 2001.

[Cata89] T. Cataltepe, A.R. Kramer, L.E. Larson, G.C. Temes, and R.H. Walden, "Digitally corrected multi-bit ΣΔ data converters". *Proc. of the IEEE International Symposium Circuits and Systems*, pp. 647-650, 1989.

[Chao90] K.C.-H. Chao, S. Nadeem, W.L. Lee, and C. Sodini, "A Higher Order Topology for Interpolative Modulators for Oversampling A/D Converters". *IEEE Transactions on Communications*, vol. 37, pp. 309-318, March 1990.

[Char01] E. Charbon, R. Gharpurey, P. Miliozzi, R.G. Meyer, and A. Sangiovanni-Vincentelli, *Substrate Noise: Analysis and Optimization for IC Design*. Kluwer Academic Publishers, 2001.

[Chen95] F. Chen and B.H. Leung, "A High Resolution Multibit Sigma-Delta Modulator with Individual Level Averaging". *IEEE Journal of Solid-State Circuits*, vol. 30, pp. 453-460, April 1995.

[Chou89] W. Chou, P. Wong, and R. Gray, "Multi-Stage Sigma-Delta Modulation". *IEEE Transaction on Information Theory*, vol. 35, pp. 784-796, July 1989.

[Coban99] A.L. Coban and P.E. Allen, "A 1.5V 1.0mW Audio Delta Sigma Modulator with 98dB Dynamic Range". *Proc. of the IEEE International Solid-State Circuits Conference*, pp. 50-51, 1999.

[Corm94] R.F. Cormier Jr., T.L. Sculley, and R.H. Bamberger, "Combining Sub-band Decomposition and Sigma-delta Modulation for Wide-band A/D Conversion". *Proc. of the IEEE International Symposium Circuits and Systems*, pp. 357-360, 1994.

[Croc83] R.E. Crociere and L.R. Rabiner, *Multirate Digital Signal Processing*. Prentice-Hall, 1983.

[Cutl60] C.C. Cutler, *Transmission System Employing Quantization*. U.S. Patent No. 2,927,962, 1960.

[Dedic94] I. Dedic, "A Sixth-Order Triple-Loop ΣΔ CMOS ADC with 90dB SNR and 100kHz Bandwidth". *Proc. of the IEEE International Solid-State Circuits Conference*, pp. 188-189, 1994.

[Dias92] V.F. Dias, G. Palmisano, P. O'Leary, and F. Maloberti, "Fundamental Limitations of Switched-Capacitor Sigma-Delta Modulators". *IEE Proceedings-G*, vol. 139, pp. 27-32, February 1992.

[Dias93] V.F. Dias and V. Liberali, "Cascade Pseudomultibit Noise Shaping Modulators". *IEE Proceedings-G*, vol. 140, no. 4, pp. 237-246, August 1993.

[Doer84] J. Doernberg, H. S. Lee, and D.A. Hodges, "Full-Speed Testing of A/D Converters". *IEEE Journal of Solid-State Circuits,* vol. 19, no. 6, pp. 820-827, December 1984.

[Dorr03] L. Dorrer, A. Di Giandomenico, and A. Wiesbauer, "A 10-Bit, 4 mW Continuous-Time Sigma-Delta ADC for UMTS in a 0.12 μm CMOS process". *Proc. of the IEEE International Symposium Circuits and Systems*, pp. 1057-1060, 2003.

[Eddy98] D.S. Eddy and D.R. Sparks, "Application of MEMs Technology in Automotive Sensors and Actuators". *Proc. of the IEEE*, pp. 1747-1755, August 1998.

[Enge99] J.V. Engelen and R. van de Plassche, *BandPass Sigma-Delta Modulators: Stability Analysis, Performance and Design Aspects*. Kluwer Academic Publishers, 1999.

[Enz96] C.C. Enz and G.C. Temes, "Circuit Techniques for Reducing the Effects of Op-Amp Imperfections: Autozeroing, Correlated Double Sampling, and Chopper Stabilization". *Proceedings of the IEEE*, vol. 84, no. 11, pp. 1584-1614, November 1996.

[Eshr96] A. Eshraghi and T. Fiez, "A Comparison of Three Parallel ΣΔ A/D Converters". *Proc. of the IEEE International Symposium Circuits and Systems*, pp. 517-520, 1996.

[Eshr03] A. Eshraghi and T. Fiez, "A Time-Interleaved Parallel ΣΔ A/D Converter". *IEEE Transactions on Circuits and Systems - II*, vol. 50, pp. 118-129, March 2003.

[Fatt90] J.W. Fattaruso, M. De Wit, G. Warwar, K.S. Tan, R.K. Hester, "The Effect of Dielectric Relaxation on Charge-Redistribution A/D Converters". *IEEE Journal of Solid-State Circuits*, pp. 1550-1561, December 1990.

[Feld98] A.R. Feldman, B.E. Boser, and P.R. Gray, "A 13-Bit, 1.4-MS/s Sigma-Delta Modulator for RF Baseband Channel Applications". *IEEE Journal of Solid-State Circuits*, vol. 33, pp. 1462-1469, October 1998.

[Felde99] M. Felder and J. Ganger, "Analysis of Ground-Bounce Induced Substrate Noise Coupling in a Low Resistive Bulk Epitaxial Process: Design Strategies to Minimize Noise Effects on a Mixed-Signal Chip". *IEEE Transactions on Circuits and Systems - II*, vol. 46, pp. 1427-1436, November 1999.

[Fern03] F.V. Fernández, R. del Río, R. Castro-López, O. Guerra, F. Medeiro, and B. Pérez-Verdú, "Design Methodologies for Sigma-Delta Converters". Chapter 15 at *CMOS Telecom Data Converters* (edited by A. Rodríguez-Vázquez, F. Medeiro, and E. Janssens). Kluwer Academic Publishers, 2003.

[Fisc82] J.H. Fischer, "Noise Sources and Calculation Techniques for Switched Capacitor Filters". *IEEE Journal of Solid-State Circuits,* vol. 17, no. 4, pp. 742-752, August 1982.

[Flem01] W. J. Fleming, "Overview of Automotive Sensors". *IEEE Sensors Journal*, pp. 296-308, December 2001.

[Fogl00] E. Fogleman, I. Galton, W. Huff, and H. Jensen, "A 3.3-V Single-Poly CMOS Audio ADC Delta-Sigma Modulator with 98-dB Peak SINAD and 105-dB Peak SFDR". *IEEE Journal of Solid-State Circuits*, vol. 35, pp. 297-307, March 2000.

[Fogl01] E. Fogleman, J. Welz, and I. Galton, "An audio ADC Delta-Sigma modulator with 100-dB peak SINAD and 102-dB DR using a second-order mismatch-shaping DAC". *IEEE Journal of Solid-State Circuits*, vol. 36, pp. 339-348, March 2001.

[Fuji97] I. Fujimori, K. Koyama, D. Trager, F. Tam, and L. Longo, "A 5-V Single-Chip Delta-Sigma Audio A/D Converter with 111 dB Dynamic Range". *IEEE Journal of Solid-State Circuits*, vol. 32, pp. 329-336, March 1997.

[Fuji00] I. Fujimori, L. Longo, A. Hairapetian, K. Seiyama, S. Kosic, J. Cao, and S.-L. Chan, "A 90-dB SNR 2.5-MHz Output-Rate ADC Using Cascaded Multibit Delta-Sigma modulation at 8x Oversampling Ratio". *IEEE Journal of Solid-State Circuits*, vol. 35, pp. 1820-1828, December 2000.

[Gaggl03] R. Gaggl, A. Wiesbauer, G. Fritz, C. Schranz, and P. Pessl, "A 85-dB Dynamic Range Multibit Delta-Sigma ADC for ADSL-CO Applications in 0.18-μm CMOS". *IEEE Journal of Solid-State Circuits*, vol. 38, pp. 1105-1114, July 2003.

[Galt95] I. Galton and H.T. Jensen, "Delta-Sigma Based A/D Conversion Without Oversampling". *IEEE Transactions on Circuits and Systems - II*, vol. 42, pp. 773-784, December 1995.

[Galt96] I. Galton and H.T. Jensen, "Oversampling Parallel Delta-Sigma Modulator A/D Conversion". *IEEE Transactions on Circuits and Systems - II*, vol. 43, pp. 801-810, December 1996.

[Geer99] Y. Geerts, A. Marques, M. Steyaert, and W. Sansen, "A 3.3-V, 15-bit, Delta-Sigma ADC with a Signal Bandwidth of 1.1 MHz for ADSL Applications". *IEEE Journal of Solid-State Circuits*, vol. 34, pp. 927-936, July 1999.

[Geer00] Y. Geerts, M. Steyaert, and W. Sansen, "A High-Performance Multibit ΔΣ CMOS ADC". *IEEE Journal of Solid-State Circuits*, vol. 35, pp. 1829-1840, December 2000.

[Geer02] Y. Geerts, M. Steyaert, and W. Sansen, *Design of Multi-Bit Delta-Sigma A/D Converters*. Kluwer Academic Publishers, 2002.

[Gerf03] F. Gerfers, M. Ortmanns, and Y. Manoli, "A 1.5-V 12-bit Power-Efficient Continuous-Time Third-Order ΣΔ Modulator". *IEEE Journal of Solid-State Circuits*, vol. 38, pp. 1343-1352, August 2003.

[Gero03] A. Gerosa and A. Neviani, "A Very Low-Power 8-bit ΣΔ Converter in a 0.8μm CMOS Technology for the Sensing Chain of a Cardiac Pacemaker, Operating down to 1.8V". *Proc. of the IEEE International Symposium on Circuits and Systems*, vol. 5, pp. 49-52, 2003.

[Gobet81] C.-A. Gobet and A. Knob, "Noise Analysis of Switched Capacitor Networks". *Proc. of the IEEE International Symposium on Circuits and Systems*, pp. 856-859, 1981.

[Gobet83] C.-A. Gobet and A. Knob, "Noise Analysis of Switched Capacitor Networks". *IEEE Transactions on Circuits and Systems,* vol. CAS-30, no. 1, pp. 37-43, January 1983.

[Gomez00] G.J. Gomez, "A 102-dB Spurious-Free DR ΣΔ ADC Using a Dynamic Dither Scheme". *IEEE Transactions on Circuits and Systems - II*, pp. 531-535, June 2000.

[Gomez02] G. Gomez and B. Haroun, "A 1.5V 2.4/2.9mW 79/50dB DR ΣΔ Modulator for GSM/WCDMA in a 0.13μm Digital Process". *Proc. of the IEEE International Solid-State Circuits Conference*, vol. 1, pp. 306-307, 2002.

[Good95] M. Goodson, B. Zhang, and R. Schreier, "Proving Stability of Delta-Sigma Modulators Using Invariant Sets". *Proc. of the IEEE International Symposium on Circuits and Systems,* vol. 2, pp. 633-636, 1995.

[Good96] F. Goodenough, "Analog Techniques of all Varieties Dominate ISSCC". *Electronic Design*, vol. 44, pp. 96-111, February 1996.

[Gray90] R.M. Gray, "Quantization Noise Spectra". *IEEE Transactions on Information Theory*, vol. 36, pp. 1220-1244, November 1990.

[Gray97] R.M. Gray, "Quantization Noise in ΣΔ A/D Converters", Chapter 2 in *Delta-Sigma Data Converters: Theory, Design and Simulation (S.R. Norsworthy, R. Schreier, and G.C. Temes, Editors)*. IEEE Press, 1997.

[Greg86] R. Gregorian and G.C. Temes, *Analog MOS Integrated Circuits for Signal Processing*. John Wiley & Sons, 1986.

[Grilo96] J. Grilo, E. Mac Robbie, R. Halim, and G. Temes, "A 1.8V 94dB Dynamic Range ΣΔ Modulator for Voice Applications". *Proc. of the IEEE International Solid-State Circuits Conference*, pp. 230-231, 1996.

[Grilo02] J. Grilo, I. Galton, K. Wang, and R.G. Montemayor, "A 12-mW ADC Delta-Sigma Modulator With 80 dB of Dynamic Range Integrated in a Single-Chip Bluetooth Transceiver". *IEEE Journal of Solid-State Circuits*, vol. 37, pp. 271-278, March 2002.

[Guil01] J. Guilherme, P. Figueredo, P. Azevedo, G. Minderico, A. Leal, J. Vital, and J. Franca, "A Pipeline 15-b 10Msample/s Analog-to-Digital Converter for ADSL Applications". *Proc. of the IEEE International Symposium on Circuits and Systems*, vol. 1, pp. 396-399, May 2001.

[Gupta02] S.K. Gupta and V. Fong, "A 64-MHz Clock-Rate ΣΔ ADC with 88-dB SNDR and -105-dB IM3 Distortion at a 1.5-MHz Signal Frequency". *IEEE Journal of Solid-State Circuits*, vol. 37, pp. 1653-1661, December 2002.

[Gust00] M. Gustavsson, J.J. Wikner, and N.N. Tan, *CMOS Data Converters for Communications*. Kluwer Academic Publishers, 2000.

[Haigh83] D. Haigh and B. Singh, "A Switching Scheme for Switched Capacitor Filters which Reduces the Effect of Parasitic Capacitances Associated with Switch Control Terminals". *Proc. of the IEEE International Symposium on Circuits and Systems,* pp. 586-589, 1983.

[Hair91] A. Hairapetian, G.C. Temes, and Z.X. Zhang, "Multibit sigma-delta modulator with reduced sensitivity to DAC nonlinearity". *Electronics Letters*, vol. 27, pp. 990-991, May 1991.

[Hair94] A. Hairapetian and G.C. Temes, "A Dual-Quantization Multi-Bit Sigma Delta Analog/Digital Converter". *Proc. of the IEEE International Symposium on Circuits and Systems*, vol. 5, pp. 437-440, 1994.

[Hall92] B. Hallgren, "Design of a Second-Order CMOS Sigma-Delta A/D Converter with 150MHz Clock Rate". *Proc. of the European Solid-State Circuits Conference*, pp. 103-106, 1992.

[Hall93] B. Hallgren, "Continuous-Time Implementation of Sigma-Delta Converters for High Frequency Operation". *Proc. of the European Conference on Circuit Theory and Design*, pp. 1601-1606, 1993.

[IEEE01] *IEEE Standard for Terminology and Test Methods for Analog-to-Digital Converters*. IEEE Std 1241-2000, 2001.

[Inge97] M. Ingels and M.S.J. Steyaert, "Design Strategies and Decoupling Techniques for Reducing the Effects of Electrical Interference in Mixed-Mode IC's". *IEEE Journal of Solid-State Circuits,* vol. 32, pp. 1136-1141, July 1997.

[Inose62] H. Inose, Y. Yasuda, and J. Murakami, "A Telemetering System by Code Modulation — Δ-Σ Modulation". *IRE Transactions on Space Electronics and Telemetry*, vol. 8, pp. 204-209, September 1962.

[Jiang02] R. Jiang, and T.S. Fiez, "A 1.8V 14b ΔΣ A/D Converter with 4MSamples/s Conversion". *Proc. of the IEEE International Solid-State Circuits Conference,* pp. 220-221, 2002.

[Johns97] D.A. Johns and K. Martin, *Analog Integrated Circuit Design*. John Wiley&Sons, 1997.

[John99] J.E. Johnston, "A 24-bit delta-sigma ADC with an ultra-low noise chopper-stabilized programmable gain instrumentation amplifier". *Proc. of the International Conference on Advanced A/D and D/A Conversion Techniques and Their Applications*, pp. 179-182, 1999.

[Kare90] T. Karema, T. Ritoniemi, and H. Tenhunen, "An oversampled sigma-delta A/D converter circuit using two-stage fourth order modulator". *Proc. of the IEEE International Symposium on Circuits and Systems*, vol. 4, pp. 3279-3282, 1990.

[Kasha99] D.B. Kasha, W.L. Lee, and A. Thomsen, "A 16-mW, 120-dB linear switched-capacitor delta-sigma modulator with dynamic biasing". *IEEE Journal of Solid-State Circuits*, vol. 34, pp. 921-926, July 1999.

[Kerth94] D.A. Kerth, D.B. Kasha, T.G. Mellissinos, D.S. Piasecki, and E.J. Swanson, "A 120dB Linear Switched-Capacitor Delta-Sigma Modulator". *Proc. of the IEEE International Solid-State Circuits Conference,* pp. 196-197, 1994.

[Kesk02] M. Keskin, U.-K. Moon, and G.C. Temes, "A 1-V 10-MHz Clock-Rate 13-Bit CMOS ΔΣ Modulator Using Unity-Gain-Reset Opamps". *IEEE Journal of Solid-State Circuits*, vol. 37, pp. 817-824, July 2002.

[Khoi93] R. Khoini-Poorfard and D.A. Johns, "Time-interleaved oversampling converters". *Electronics Letters*, vol. 29, pp. 1673-1674, September 1993.

[Khoi97] R. Khoini-Poorfard, L.B. Lim, and D.A. Johns, "Time-interleaved Oversampling A/D Converters: Theory and Practice". *IEEE Transactions on Circuits and Systems - II*, vol. 44, pp. 634-645, August 1997.

[King94] E. King, F. Aram, T. Fiez, and I. Galton, "Parallel Delta-Sigma A/D Conversion". *Proc. of the IEEE Custom Integrated Circuits Conference*, pp. 23.3.1-4, 1994.

[King98] E.T. King, A. Eshragi, I. Galton, and T.S. Fiez, "A Nyquist-Rate Delta-Sigma A/D Converter". *IEEE Journal of Solid-State Circuits*, vol. 33, pp. 45-52, January 1998.

[Kiss99] P. Kiss, *Adaptive Digital Compensation of Analog Circuit Imperfections for Cascaded Delta-Sigma Analog-to-Digital Converters*. Ph.D. Thesis, Politehnica University of Timisoara, 1999.

[Kiss00] P. Kiss, J. Silva, A. Wiesbauer, T. Sun, U.-K. Moon, J.T. Stonick, and G.C. Temes, "Adaptive Digital Correction of Analog Errors in MASH ADCs - Part II: Correction Using Test-Signal Injection". *IEEE Transactions on Circuits and Systems - II*, vol. 47, pp. 629-638, July 2000.

[Klaa75] K.B. Klaasen, "Digitally controlled absolute voltage division". *IEEE Transactions on Instrumentation and Measurement*, vol. 24, no. 3, pp. 106-112, June 1975.

[Kozak00] M. Kozak and I. Kale, "Novel Topologies for Time-Interleaved Delta-Sigma Modulators". *IEEE Transactions on Circuits and Systems - II*, vol. 47, pp. 639-654, July 2000.

[Kulah00] H. Kulah, N. Yazdi, and K. Najafi, "A CMOS Switched-Capacitor Interface Circuit for an Integrated Accelerometer". *Proc. of the IEEE Midwest Symposium on Circuits and Systems*, pp. 244-247, August 2000.

[Kung88] K.K. Kung, P.K. Ko, C. Hu, and Y.C. Cheng, "Flicker Noise Characteristics of Advanced MOS Technologies". *Proc. of the International Electron Device Meeting 1988, Technical Digest*, pp. 34-37, December 1988.

[Kuo99] T.-H. Kuo, K.-D. Chen, and J.-R. Chen, "Automatic coefficients design for high-order sigma-delta modulators". *IEEE Transactions on Circuits and Systems - II*, vol. 46, pp. 6-15, January 1999.

[Kuo02] T.-H. Kuo, K.-D. Chen, and H.-R. Yeng, "A Wideband CMOS Sigma-Delta Modulator With Incremental Data Weighted Averaging". *IEEE Journal of Solid-State Circuits*, vol. 37, pp. 11-17, January 2002.

[Lamp01] H. Lampinen and O. Vainio, "Low-voltage fourth-order CMOS sigma-delta modulator implementation". *Electronics Letters*, vol. 37, pp. 734-735, June 2001.

[Lars88] L.E. Larson, T. Cataltepe, and G.C. Temes, "Multi-bit oversampled Σ–Δ A/D converter with digital error correction". *Electronics Letters*, vol. 24, pp. 1051-1052, August 1988.

[Lee85] K.-L. Lee and R.G. Meyer, "Low-Distortion Switched-Capacitor Filter Design Techniques". *IEEE Journal of Solid-State Circuits,* vol. 20, no. 6, pp. 1103-113, December 1985.

[Lee87a] W.L. Lee, *A novel higher order interpolative modulator topology for high resolution oversampling A/D converter*s. Ph.D. Thesis, Massachusetts Institute of Technology, 1987.

[Lee87b] W.L. Lee and C.G. Sodini, "A Topology for Higher Order Interpolative Coders". *Proc. of the IEEE International Symposium on Circuits and Systems*, pp. 459-462, 1987.

[Lee01] S. Lee and K. Yang, "Design a Low-Jitter Clock for High-Speed A/D Converters". *Sensors*, vol. 18, no. 10, October 2001.

[Lee03] K.-S. Lee and F. Maloberti, "A 1.8V, 1MS/s 85dB SNR 2+2 Mash ΣΔ Modulator with ±0.9V Reference Voltage". *Proc. of the Symposium on VLSI Circuits*, pp. 71-74, June 2003.

[Lei98] T.F. Lei, J.Y. Cheng, S.Y. Shiau, T.S. Chao, and C.S. Lai, "Characterization of Polysilicon Oxides Thermally Grown and Deposited on the Polished Polysilicon Films". *IEEE Transactions on Electron Devices*, vol. 45, pp. 912-917, April 1998.

[Lesl90] T.C. Leslie and B. Singh, "An Improved Sigma-Delta Modular Architecture". *Proc. of the IEEE International Symposium on Circuits and Systems,* pp. 372-375, 1990.

[Leung92] B. Leung and S. Sutarja, "Multibit Σ–Δ A/D Converter Incorporating a Novel Class of Dynamic Element Matching Techniques". *IEEE Transactions on Circuits and Systems - II*, vol. 39, pp. 35-51, January 1992.

[Leung97] K.Y. Leung, E.J. Swanson, K. Leung, S.S. Zhu, "A 5V, 118dB ΔΣ Analog-to-Digital Converter for Wideband Digital Audio". *Proc. of IEEE International Solid-State Circuit Conference*, pp. 218-219, 1997.

[Lewis87] S.H. Lewis and P.R. Gray, "A Pipelined 5-Msample/s 9-bit Analog-to-Digital Converter". *IEEE Journal of Solid-State Circuits*, vol. 22, no. 6, pp. 954-961, December 1987.

[Lewis92] S.H. Lewis, H.S. Fetterman, G.F. Gross, R. Ramachandran, and T.R. Viswanathan, "A 10-b 20-Msample/s analog-to-digital converter". *IEEE Journal of Solid-State Circuits*, vol. 27, no. 3, pp. 351-358, March 1992.

[Libe00] V. Liberali, R. Rossi, and G. Torelli, "Crosstalk effects in mixed-signal ICs in deep submicron digital CMOS technology". *Microelectronics Journal,* vol. 31, pp. 893-904, December 2000.

[Lin99] C.-H. Lin, C. Shi, M. Ismail, and M. Gyu, "A 5MHz Nyquist Rate Continuous-Time Sigma-Delta Modulator for Wideband Wireless Communication". *Proc. of the IEEE International Symposium on Circuits and Systems,* vol. 2, pp. 368-371, 1999.

[Lin00] J.C.H. Lin, *TSMC 0.25μm Mixed-Signal 1P5M+ MIM Salicide 2.5V/5.0V Design Guideline*. Taiwan Semiconductors Manufacturing Co., LTD, March 2000.

[Lips92] S.P. Lipshitz, R.A. Wannamaker, and J. Vanderkoy, "Quantization and Dither: A Theoretical Survey". *Journal of the Audio Engineering Society,* vol. 40, pp. 355-375, May 1992.

[Longo88] L. Longo and M. Copeland, "A 13 bit ISDN-band oversampled ADC using two-stage third-order noise shaping". *Proc. of the IEEE Custom Integrated Circuit Conference*, pp. 21.2.1-4, 1988.

[Luh98a] L. Luh, J. Choma Jr., and J. Draper, "A 50-MHz Continuous-Time Switched-Current Sigma-Delta Modulator". *Proc. of the IEEE International Symposium on Circuits and Systems*, vol. 1, pp. 579-582, 1998.

[Luh98b] L. Luh and J. Choma, "A Continuous-Time Switched-Current Sigma-Delta Modulator with Reduced Loop Delay". *Proc. of the 8th Great Lakes Symposium on VLSI*, pp. 286-291, 1998.

[Luh00] L. Luh, J. Choma, and J. Draper, "A 400-MHz 5th-Order Continuous-Time Switched-Current ΣΔ Modulator". *Proc. of the European Conference on Solid-State Circuits,* pp. 72-75, 2000.

[Maki02] K.A.A. Makinwa and J.H. Huijsing, "A Smart CMOS Wind Sensor". *Proc. of the IEEE International Solid-State Circuits Conference*, pp. 432-479, 2002.

[Malc00] P. Malcovati and F. Maloberti, "An Integrated Microsystem for 3-D Magnetic Field Measurements". *IEEE Trans. on Instrumentation and Measurement,* pp. 341-345, April 2000.

[Malo01] F. Maloberti, *Analog Design for CMOS VLSI Systems*. Kluwer Academic Publishers, 2001.

[Marq98a] A.M. Marques, V. Peluso, M.S.J. Steyaert, and W. Sansen, "A 15-b Resolution 2-MHz Nyquist Rate ΔΣ ADC in a 1-μm CMOS Technology". *IEEE Journal of Solid-State Circuits*, vol. 33, pp. 1065-1075, July 1998.

[Marq98b] A.M. Marques, V. Peluso, M.S.J. Steyaert, and W. Sansen, "Optimal Parameters for ΔΣ Modulator Topologies". *IEEE Transactions on Circuits and Systems - II*, vol. 45, pp. 1232-1241, September1998.

[Mats87] Y. Matsuya, K. Uchimura, A. Iwata, T. Kobayashi, M. Ishikawa, and T. Yoshitome, "A 16-bit oversampling A-to-D conversion technology using triple-integration noise shaping". *IEEE Journal of Solid-State Circuits*, vol. 22, pp. 921-929, December 1987.

[Maul00] P.C. Maulik, M.S. Chadha, W.L. Lee, and P.J. Crawley, "A 16-Bit 250-kHz Delta-Sigma Modulator and Decimation Filter". *IEEE Journal of Solid-State Circuits*, vol. 35, pp. 458-467, April 2000.

[Mede95] F. Medeiro, B. Pérez-Verdú, A. Rodríguez-Vázquez, and J.L. Huertas, "A Vertically-Integrated Tool For Automated Design of ΣΔ Modulators". *IEEE Journal of Solid-State Circuits*, vol. 30, pp. 762-772, August 1995.

[Mede97] F. Medeiro, B. Pérez-Verdú, J.M. de la Rosa, and A. Rodríguez-Vázquez, "Using CAD Tools for Shortening the Design Cycle of High-Performance ΣΔM: A 16.4bit 1.71mW ΣΔM in CMOS 0.7μm Technology". *International Journal of Circuit Theory and Applications*, vol. 25, pp. 319-334, March-April 1997.

[Mede98a] F. Medeiro, B. Pérez-Verdú, J.M. de la Rosa, and A. Rodríguez-Vázquez, "Multi-bit cascade ΣΔ modulator for high-speed A/D conversion with reduced sensitivity to DAC errors". *Electronics Letters*, vol. 34, pp. 422-424, March 1998.

[Mede98b] F. Medeiro, B. Pérez-Verdú, J.M. de la Rosa, and A. Rodríguez-Vázquez, "Fourth-Order Cascade ΣΔ Modulators: A Comparative Study". *IEEE Transactions on Circuits and Systems - I*, vol. 45, pp. 1041-1051, October 1998.

[Mede99a] F. Medeiro, B. Pérez-Verdú, and A. Rodríguez-Vázquez, *Top-Down Design of High-Performance Modulators*. Kluwer Academic Publishers, 1999.

[Mede99b] F. Medeiro, B. Perez-Verdú, and A. Rodríguez-Vázquez, "A 13-bit, 2.2-MS/s, 55-mW Multibit Cascade ΣΔ Modulator in CMOS 0.7-μm Single-Poly Technology". *IEEE Journal of Solid-State Circuits*, vol. 34, no. 6, pp. 748-760, June 1999.

[Mede03] F. Medeiro, R. del Río, J.M. de la Rosa, B. Pérez-Verdú, and A. Rodríguez-Vázquez, "High-Order Cascade Multi-bit ΣΔ Modulators". Chapter 9 at at *CMOS Telecom Data Converters* (edited by A. Rodríguez-Vázquez, F. Medeiro, and E. Janssens). Kluwer Academic Publishers, 2003.

[Miao98] G. Miao, H.C. Yang, and T. Pushan, "An Oversampled A/D Converter Using Cascaded Fourth Order Sigma-Delta Modulation and Current Steering Logic". *Proc. of the IEEE International Symposium on Circuits and Systems*, vol. 1, pp. 412-415, 1998.

[Mill03] M.R. Miller and C.S.A. Petrie, "A Multibit Sigma-Delta ADC for Multimode Receivers". *IEEE Journal of Solid-State Circuits*, vol. 38, no. 3, pp. 475-482, March 2003.

[Mori00] J.C. Morizio, M. Hoke, T. Kocak, C. Geddie, C. Hughes, J. Perry, S. Madhavapeddi, M.H. Hood, G. Lynch, H. Kondoh, T. Kumamoto, T. Okuda, H. Noda, M. Ishiwaki, T. Miki, and M. Nakaya, "14-bit 2.2-MS/s Sigma-Delta ADC's". *IEEE Journal of Solid-State Circuits*, vol. 35, pp. 968-976, July 2000.

[Motc93] C.D. Motchenbacher and J.A. Conelly, *Low-Noise Electronic System Design*. John Wiley & Sons, 1993.

[Mous94] S.M. Moussavi and B.H. Leung, "High-Order Single-Stage Single-Bit Oversampling A/D Converter Stabilized with Local Feedback Loops". *IEEE Transactions on Circuits and Systems - I*, vol. 41, pp. 19-25, January 1994.

[Nade94] S. Nadeem, C.G. Sodini, and L. Hae-Seung, "16-Channel Oversampled Analog-to-Digital Converter". *IEEE Journal of Solid-State Circuits*, vol. 29, pp. 1077-1085, September 1994.

[Nors97a] S.R. Norsworthy, R. Schreier, and G.C. Temes (Editors), *Delta-Sigma Data Converters: Theory, Design and Simulation*. IEEE Press, 1997.

[Nors97b] S.R. Norsworthy and R.E. Crochiere, "Decimation and Interpolation for ΣΔ Conversion", Chapter 13 in *Delta-Sigma Data Converters: Theory, Design and Simulation (S.R. Norsworthy, R. Schreier, and G.C. Temes, Editors)*. IEEE Press, 1997.

[Nys97] O. Nys and R. Henderson, "A 19-bit Low-Power Multi-Bit Sigma-Delta ADC Based on Data Weighted Averaging". *IEEE Journal of Solid-State Circuits*, vol. 32, pp. 933-942, July 1997.

[Olia00] O. Oliaei, "Thermal Noise Analysis of Multi-Input SC-Integrators for Sigma-Delta Modulator Design". *Proc. of the IEEE International Symposium on Circuits and Systems*, vol. 4, pp. 425-428, 2000.

[Olia02] O. Oliaei, P. Clement, and P. Gorisse, "A 5-mW Sigma-Delta Modulator With 84-dB Dynamic Range for GSM/EDGE". *IEEE Journal of Solid-State Circuits*, vol. 37, pp. 2-10, January 2002.

[OptE90] F. Op't Eynde, *High-Performance Analog Interfaces for Digital Signal Processors*. Ph.D. Thesis, Katholieke Universiteit Leuven, 1990.

[OptE91] F. Op't Eynde, G.M. Yin, and W. Sansen, "A CMOS Fourth-order 14b 500k-sample/s Sigma-delta ADC Converter". *Proc. of IEEE International Solid-State Circuit Conference*, pp. 62-63, 1991.

[OptE93] F. Op't Eynde and W. Sansen, *Analog Interfaces for Digital Signal Processing Systems*. Kluwer Academic Publishers, 1993.

[Paul87] J.J. Paulos, G.T. Brauns, M.B. Steer, and S.H. Ardalan, "Improved signal-to-noise ratio using tri-level delta-sigma modulation". *Proc. of the IEEE International Symposium on Circuits and Systems*, pp. 463-466, 1987.

[Pelu97] V. Peluso, M. Steyaert, and W. Sansen, "A 1.5-V 100μ–W ΣΔ Modulator with 12-bit Dynamic Range Using the Switched-Opamp Technique". *IEEE Journal of Solid-State Circuits*, vol. 32, pp. 943-952, July 1997.

[Pelu98] V. Peluso, P. Vancorenland, A.M. Marques, M.S.J. Steyaert, and W. Sansen, "A 900-mV Low-Power ΔΣ A/D Converter with 77-dB Dynamic Range". *IEEE Journal of Solid-State Circuits*, vol. 33, pp.1887-1897, December 1998.

[Pies02] T. Piessens, M. Steyaert, and E. Bach, "A Difference Voltage Buffer for ΔΣ Converters". *Analog Integrated Circuits and Signal Processing*, vol. 31, no. 1, pp. 31-37, April 2002.

[Plas94] R. van de Plassche, *Integrated Analog-to-Digital and Digital-to-Analog Converters*. Kluwer Academic Publishers, 1994.

[Rabii97] S. Rabii and B.A. Wooley, "A 1.8V Digital-Audio Sigma-Delta Modulator in 0.8μm CMOS". *IEEE Journal of Solid-State Circuits,* vol. 32, pp. 783-796, June 1997.

[Raza95] B. Razavi, *Principles of Data Converter System Design*. IEEE Press, 1995.

[Raza00] B. Razavi, *Design of Analog CMOS Integrated Circuits*. McGraw-Hill, 2000.

[Rebe90] M. Rebeschini, N.R. van Bavel, P. Rakers, R. Greene, J. Caldwell, J.R. Haug, "A 16-b 160 kHz CMOS A/D Converter using Sigma-Delta Modulation". *IEEE Journal of Solid-State Circuits,* vol. 25, pp. 431-440, April 1990.

[Rebe97] M. Rebeschini, "The Design of Cascaded ΣΔ ADCs", Chapter 6 in *Delta-Sigma Data Converters: Theory, Design and Simulation (S.R. Norsworthy, R. Schreier, and G.C. Temes, Editors)*. IEEE Press, 1997.

[Redm94] W. Redman-White and A.M. Durham, "Integrated fourth-order ΣΔ convertor with stable self-tuning continuous-time noise shaper". *IEE Proceedings — Circuits, Devices and Systems*, vol. 141, pp. 145-150, June 1994.

[Reut02] R. Reutemann, P. Balmelli, and Q. Huang, "A 33mW 14b 2.5MSample/s ΣΔ A/D converter in 0.25μm digital CMOS". *Proc. of IEEE International Solid-State Circuits Conference*, vol. 1, p. 316, 2002.

[Reve03] R. Le Reverend, I. Kale, G. Delight, D. Morling, and S. Morris, "An Ultra-Low Power Double-Sampled A/D MASH ΣΔ Modulator". *Proc. of the IEEE International Symposium on Circuits and Systems*, vol. 1, pp. 1001-1004, 2003.

[Ribn91a] D.B. Ribner, "A Comparison of Modulator Networks for High-Order Oversampled ΣΔ Analog-to-Digital Converters". *IEEE Transactions on Circuits and Systems*, vol. 38, pp. 145-159, February 1991.

[Ribn91b] D.B. Ribner, R.D. Baertsch, S.L. Garverick, D.T. McGrath, J.E. Krisciunas, and T. Fuji, "A Third-Order Multistage Sigma-Delta Modulator with Reduced Sensitivity to Nonidealities". *IEEE Journal of Solid-State Circuits*, vol. 26, pp. 1764-1774, December 1991.

[Rio98] R. del Río, F. Medeiro, B. Pérez-Verdú, and A. Rodríguez-Vázquez, "High-Order Cascade Multi-bit ΣΔ Modulators for High-Speed A/D Conversion". *Proc. of the Design of Circuits and Integrated Systems Conference*, pp. 76-81, 1998.

[Rio99] R. del Río, F. Medeiro, J.M. de la Rosa, B. Pérez-Verdú, and A. Rodríguez-Vázquez, "Reliable Analysis of Settling Errors in SC Integrators — Application to High-Speed Low-Power Sigma-Delta Modulators". *Proc. of the Design of Circuits and Integrated Systems Conference*, pp. 727-732, 1999.

[Rio00] R. del Río, F. Medeiro, B. Pérez-Verdú, and A. Rodríguez-Vázquez, "High-order cascade multibit ΣΔ modulators for xDSL applications". *Proc. of the IEEE International Symposium on Circuits and Systems*, vol. 2, pp. 37-40, 2000.

[Rio01a] R. del Río, J.M. de la Rosa, F. Medeiro, B. Pérez-Verdú, and A. Rodríguez-Vázquez, "Top-down Design of a xDSL 14-bit 4-MS/s Sigma-Delta Modulator in Digital CMOS Technology". *Proc. of the Design, Automation and Test in Europe Conference*, pp. 348-351, 2001.

[Rio01b] R. del Río, J.M. de la Rosa, F. Medeiro, B. Pérez-Verdú, and A. Rodríguez-Vázquez, "High-performance sigma-delta ADC for ADSL applications in 0.35μm CMOS digital technology". *Proc. of the IEEE International Conference on Electronics, Circuits and Systems*, vol. 1, pp. 501-504, 2001.

[Rio02a] R. del Río, F. Medeiro, J.M. de la Rosa, B. Pérez-Verdú, and A. Rodríguez-Vázquez, "Correction-Free Multi-Bit Sigma-Delta Modulators for ADSL", in *Analog Circuit Design—Structured Mixed-Mode Design, Multi-Bit Sigma-Delta Converters, Short Range RF Circuits (M. Steyaert, A.H.M. van Roermund, and J.H. Huijsing, Editors)*. Kluwer Academic Publishers, 2002.

[Rio04] R. del Río, J.M. de la Rosa, B. Pérez-Verdú, M. Delgado-Restituto, R. Domínguez-Castro, F. Medeiro, and A. Rodríguez-Vázquez, "Highly Linear 2.5-V CMOS ΣΔ Modulator for ADSL+". *IEEE Transactions on Circuits and Systems - I*, vol. 51, pp. 47-62, January 2004.

[Ritc77] G.R. Ritchie, *High Order Interpolation Analog-to-Digital Converters*. Ph.D. Thesis, University of Pennsylvania, 1977.

[Rito94] T. Ritoniemi, E. Pajarre, S. Ingalsuo, T. Husu, V. Eerola, and T. Saramiki, "A Stereo Audio Sigma-Delta A/D-Converter". *IEEE Journal of Solid-State Circuits*, vol. 29, pp. 1514-1523, December 1994.

[Robe91] A. Robertini, *Linear and Nonlinear Distortion in SC Circuits due to Nonideal Amplifiers and Switches*. Hartung-Gorre Verlag, 1991.

[Rodr99] A. Rodríguez-Vázquez, M. Delgado-Restituto, and R. Domínguez-Castro, "Comparator Circuits", in *Wiley Encyclopedia of Electrical and Electronics Engineering (J.G. Webster, Editor)*. John Wiley & Sons, 1999.

[Romb01] P. Rombouts, W. de Wilde, and L. Weyten, "A 13.5-b 1.2-V Micropower Extended Counting A/D Converter". *IEEE Journal of Solid-State Circuits*, vol. 36, pp. 176-183, February 2001.

[Sans87] W.M.C. Sansen, H. Qiuting, and K.A.I. Halonen, "Transient Analysis of Charge Transfer in SC Filters — Gain Error and Distortion". *IEEE Journal of Solid-State Circuits,* vol. 22, pp. 268-276, April 1987.

[Sarh93] M. Sarhang-Nejad and G.C. Temes, "A High-Resolution ΣΔ ADC with Digital Correction and Relaxed Amplifiers Requirements". *IEEE Journal of Solid-State Circuits*, vol. 28, pp. 648-660, June 1993.

[Saue02] J. Sauerbrey, T. Tille, D. Schmitt-Landsiedel, and R. Thewes, "A 0.7-V MOSFET-Only Switched-Opamp ΣΔ Modulator in Standard Digital CMOS Technology". *IEEE Journal of Solid-State Circuits*, vol. 37, pp. 1662-1669, December 2002.

[Saue03] J. Sauerbrey, M. Wittig, D. Schmitt-Landsiedel, and R. Thewes, "0.65V Sigma-Delta Modulators". *Proc. of the IEEE International Symposium on Circuits and Systems*, vol. 1, pp. 1021-1024, 2003.

[Schn94] M.C. Schneider, C. Galup-Montoro, and J.C.M. Bermudez, "Explicit formula for harmonic distortion in SC filters with weakly nonlinear capacitors". *IEE Proceedings — Circuits, Devices and Systems,* vol. 141, no. 6, pp. 505-509, December 1994.

[Schr93] R. Schreier, "An Empirical Study of Higher Order Single Bit Sigma Delta Modulators". *IEEE Transactions on Circuits and Systems - II*, vol. 40, pp. 461-466, August 1993.

[Schr95] R. Schreier and B. Zhang, "Noise-shaped multibit D/A converter employing unit elements". *Electronics Letters*, vol. 31, pp. 1712-1713, September 1995.

[Schu64] L. Schuchman, "Dither signals and their effect on quantization noise". *IEEE Transactions on Communications*. vol. 12, pp. 162-165, December 1964.

[Send97] D. Senderowicz, G. Nicollini, S. Pernici, A. Nagari, P. Confalonieri, and C. Dallavalle, "Low-Voltage Double-Sampled ΣΔ Converters". *IEEE Journal of Solid-State Circuits*, vol. 32, pp. 1907-1919, December 1997.

[Stik88] E.F. Stikvoort, "Some remarks on the stability and performance of the noise shaper or Sigma-Delta modulator". *IEEE Transactions on Communications*. vol. 36, pp. 1157-1162, October 1988.

[Snoe01] M.F. Snoeij, O. Bajdechi, and J.H. Huijsing, "A 4th-order switched-capacitor sigma-delta A/D converter using a high-ripple Chebyshev loop filter". *Proc. of the IEEE International Symposium on Circuits and Systems*, vol. 1, pp. 615-618, 2001.

[Srip77] A.B. Sripad and D.L. Snyder, "A necessary and sufficient condition for quantization errors to be uniform and white". *IEEE Trans. Acoust., Speech, Signal Processing*, vol. ASSP-25, pp. 442-448, October 1977.

[Tan93] N. Tan and S. Eriksson, "Fourth-order two-stage delta-sigma modulator using both 1 bit and multibit quantisers". *Electronics Letters*, vol. 29, pp. 937-938, May 1993.

[Thanh97] C.K. Thanh, S.H. Lewis, and P.J. Hurst, "A second-order double-sampled delta-sigma modulator using individual-level averaging". *IEEE Journal of Solid-State Circuits*, vol. 32, pp. 1269-1273, August 1997.

[Tille01] T. Tille, J. Sauerbrey, and D. Schmitt-Landsiedel, "A 1.8-V MOSFET-Only Sigma-Delta Modulator Using Substrate Biased Depletion-Mode MOS Capacitors in Series Compensation". *IEEE Journal of Solid-State Circuits*, vol. 36, pp. 1041-1047, July 2001.

[Vaid90] P.P. Vaidyanathan, "Multirate digital filters, filter banks, polyphase networks, and applications: a tutorial". *Proceeding IEEE*, vol. 78, January 1990.

[Veld02] R.H.M. van Veldhoven, B.J. Minnis, H.A. Hegt, and A.H.M. van Roermund, "A 3.3-mW ΣΔ Modulator for UMTS in 0.18-μm CMOS With 70-dB Dynamic Range in 2-MHz Bandwidth". *IEEE Journal of Solid-State Circuits*, vol. 37, pp. 1645-1652, December 2002.

[Vleu01] K. Vleugels, S. Rabii, and B. Wooley, "A 2.5-V Sigma-Delta Modulator for Broadband Communications Applications". *IEEE Journal of Solid-State Circuits*, vol. 36, pp. 1887-1899, December 2001.

[Wald90] R.H. Walden, T. Cataltepe, and G.C. Temes, "Architectures for higher-order multi-bit ΣΔ modulators". *Proc. of the IEEE International Symposium Circuits and Systems*, vol. 5, pp. 895-898, 1990.

[Wang97] F. Wang and R. Harjani, "Nonlinear Settling Behavior in Oversampled Converters". *Proc. of the IEEE Custom Integrated Circuits Conference*, pp. 23.6.1-4, 1997.

[Wang00] X. Wang, W. Quin, and X. Ling, "Cascaded Parallel Oversampling Sigma-Delta Modulators". *IEEE Transactions on Circuits and Systems - II*, vol. 47, pp. 156-161, February 2000.

[Wang01] C.B. Wang, "A 20-bit 25-kHz delta-sigma A/D converter utilizing a frequency-shaped chopper stabilization scheme". *IEEE Journal of Solid-State Circuits*, vol. 36, pp. 566-569, March 2001.

[Widr60] B. Widrow, "Statistical analysis of amplitude-quantized sample-data systems". *Trans. AIEE - Part II: Applications and Industry*, vol. 79, pp. 555-568, January 1960.

[Wies96] A. Wiesbauer and G.C. Temes, "Adaptive compensation of analog circuit imperfections for cascaded ΣΔ modulators". *Proc. of the Asilomar Conference on Signals, Systems and Computers*, vol. 2, pp. 1073-1077, 1996.

[Will91] L.A. Williams III and B.A. Wooley, "Third-Order Cascaded Sigma-Delta Modulators". *IEEE Transactions on Circuits and Systems*, vol. 38, pp. 489-498, May 1991.

[Will94] L.A. Williams and B.A. Wooley, "A Third-Order Sigma-Delta Modulator with Extended Dynamic Range". *IEEE Journal of Solid-State Circuits,* vol. 29, pp. 193-202, March 1994.

[Wolff97] C. Wolff, J.G. Kenney, and L.R. Carley, "CAD for the Analysis and Design of ΔΣ Converters", Chapter 14 in *Delta-Sigma Data Converters: Theory, Design and Simulation (S.R. Norsworthy, R. Schreier, and G.C. Temes, Editors)*. IEEE Press, 1997.

[Wu98] J.-T. Wu and K.-L. Chang, "MOS Charge Pumps for Low-Voltage Operation". *IEEE Journal of Solid-State Circuits,* vol. 33, no. 4, pp. 592-597, April 1998.

[Yama94] K. Yamamura, A. Nogi, and A. Barlow, "A low power 20 bit instrumentation delta-sigma ADC". *Proc. of the IEEE Custom Integrated Circuits Conference*, pp. 519-522, 1994.

[Yang03] Y. Yang, A. Chokhawala, M. Alexander, J. Melanon, and D. Hester, "A 114-dB 68-mW Chopper-Stabilized Stereo Multibit Audio ADC in 5.62mm^2". *IEEE Journal of Solid-State Circuits*, pp. 2061-2068, December 2003.

[Yao04] L. Yao, M.S.J. Steyaert, and W. Sansen, "A 1-V 140-μW 88-dB Audio Sigma-Delta Modulator in 90-nm CMOS". *IEEE Journal of Solid-State Circuits*, pp. 1809-1818, November 2004.

[Yin92] G.M. Yin, F. Op't Eynde, and W. Sansen, "A High-Speed CMOS Comparator with 8-b Resolution". *IEEE Journal of Solid-State Circuits,* vol. 27, pp. 208-211, February 1992.

[Yin93] G. Yin, F. Stubbe, and W. Sansen, "A 16-b 320-kHz CMOS A/D Converter Using Two-Stage Third-Order ΣΔ Noise Shaping". *IEEE Journal of Solid-State Circuits,* vol. 28, pp. 640-647, June 1993.

[Yin94a] G. Yin, *High-Performance Analog-to-Digital Converters Using Cascaded ΣΔ Modulators*. Ph.D. Thesis, Katholieke Universiteit Leuven, 1994.

[Yin94b] G. Yin and W. Sansen, "A High-Frequency and High-Resolution Fourth-Order ΣΔ A/D Converter in BiCMOS Technology". *IEEE Journal of Solid-State Circuits*, vol. 29, pp. 857-865, August 1994.

[Yoon98] S.H. Yoon and H.S.M. Rezaul, "A sixth-order CMOS sigma-delta modulator". *Proc. of the IEEE International ASIC Conference*, pp. 63-66, 1998.

[Yu99] W. Yu, S. Sen, and B.H. Leung, "Distortion Analysis of MOS Track-and-Hold Sampling Mixers Using Time-Varying Volterra Series". *IEEE Transactions on Circuits and Systems - II*, vol. 46, pp. 101-113, February 1999.

[Yuka85] A. Yukawa, "A CMOS 8-bit High-Speed Converter IC". *IEEE Journal of Solid-State Circuits,* vol. 20, pp. 775-779, June 1985.

[Zhou01] J. Zhou, M. Cheng, and L. Forbes, "SPICE Models for Flicker Noise in p-MOSFETs in the Saturation Region". *IEEE Transactions on Computer-Aided Design of Integrated Circuits*, vol. 20, pp. 763-767, June 2001.

[Zwan96] E.J. van der Zwan and E.C. Dijkmans, "A 0.2-mW CMOS ΣΔ Modulator for Speech Coding with 80 dB Dynamic Range". *IEEE Journal of Solid-State Circuits*, pp. 1873-1880, December 1996.

[Zwan99] E.J. van der Zwan, R.H.M. van Veldhoven, P.A.C.M. Nuijten, E.C. Dijkmans, and S.D. Swift, "A 13mW 500kHz Data Acquisition IC with 4.5 Digit DC and 0.02% Accurate True-RMS Extraction". *Proc. of the IEEE International Solid-State Circuits Conference*, pp. 398-399, 1999.

Index

D

P

Q

R

S

T

MIX
Papier aus verantwortungsvollen Quellen
Paper from responsible sources
FSC® C105338

If you have any concerns about our products,
you can contact us on
ProductSafety@springernature.com

In case Publisher is established outside the EU,
the EU authorized representative is:
Springer Nature Customer Service Center GmbH
Europaplatz 3, 69115 Heidelberg, Germany

Printed by Libri Plureos GmbH
in Hamburg, Germany